Critical Masses

American and Comparative Environmental Policy
Sheldon Kamieniecki and Michael E. Kraft, editors

Critical Masses: Citizens, Nuclear Weapons Production, and Environmental Destruction in the United States and Russia
Russell J. Dalton, Paula Garb, Nicholas P. Lovrich, John C. Pierce, and John M. Whiteley

Critical Masses

Citizens, Nuclear Weapons Production, and Environmental Destruction in the United States and Russia

Russell J. Dalton, Paula Garb, Nicholas P. Lovrich,
John C. Pierce, and John M. Whiteley

The MIT Press
Cambridge, Massachusetts
London, England

This book was set in Sabon by Achorn Graphic Services, Inc.

Printed and bound in the United States of America.

Library of Congress Cataloging-in-Publication Data

Critical masses : citizens, nuclear weapons production, and environmental destruction in the United States and Russia / Russell J. Dalton . . . [et al.].
p. cm — (American and comparative environmental policy)
Includes bibliographical references and index.
ISBN 0-262-04175-8 (alk. paper). — ISBN 0-262-54103-3 (pbk. : alk. paper)
1. Nuclear weapons plants—Environmental aspects—Washington (State)—Richland. 2. Nuclear weapons plants—Environmental aspects—Russia (Federation)—Ural Mountains Region. 3. Environmental policy—Washington (State)—Richland—Citizen participation. 4. Environmental policy—Russia (Federation)—Ural Mountains Region—Citizen participation. I. Dalton, Russell J. II. Series.
TD195.N85C75 1999
363.17′998′0973—dc21 99-28632
CIP

Contents

Series Foreword

This is the first volume published in The MIT Press book series American and Comparative Environmental Policy. The book is the result of a major interdisciplinary and collaborative research project, and it analyzes the central issues surrounding cleanup of two former nuclear weapons production facilities: Mayak in Russia and Hanford in the United States. As readers will see, the nuclear weapons race produced some of the most harmful environmental hot spots in the world. In addition to damaging the natural environment, these two facilities have significantly affected public health. How and why this damage was allowed to occur and what citizens, environmentalists, and policymakers have done to try to limit adverse impacts on the environment and public health is the main subject of the book. The book raises fundamental questions about the nature of contemporary democracy and society's capacity to respond appropriately to such environmental and public health risks. Given the scope of the problem and the great expense required to clean up such sites, the study's findings should be of keen interest to citizens, environmental activists, public health officials, policymakers, and students of environmental policy.

This book illustrates well the kind of works we will include in the series. We intend to publish books that examine a broad range of environmental policy issues. We are particularly interested in books that focus on interdisciplinary research as well as on the links between policy and environmental problems, issues, and controversies in either American or cross-national settings. Future volumes will analyze the policy dimensions of relationships between humans and the environment from either an empirical or theoretical perspective. The series will include works that assess environmental policy successes and failures, evaluate new institutional arrangements and policy approaches, and help to clarify new directions for environmental policy. We plan to publish high-quality scholarly studies that are written for a wide audience that includes academics, policymakers, environmental scientists and professionals, business and labor leaders, environmentalists, and students concerned with environmental issues. We hope that these books contribute to people's understanding of the most important environmental problems, issues, and policies that society now faces and with which it must deal well into the twenty-first century.

Sheldon Kamieniecki, University of Southern California
Michael E. Kraft, University of Wisconsin-Green Bay

Figures

Tables

Preface

This is a book born of revolutionary experiences. In the late 1980s, part of our research team was working to assist environmentalist and other autonomous citizen groups in the Soviet Union. This activity came after a period of Cold War confrontation over the stationing of intermediate-range nuclear missiles in Europe, the prospect of Star Wars and the militarization of outer space, tensions over Afghanistan and other international hot spots, and a continuation of confrontational Cold War rhetoric on both sides of the East/West divide.

As the decade came to an end, we joined the rest of the world to watch the spectacle of a tidal wave of democratization sweep across Eastern Europe, eventually overwhelming the Soviet empire. The collapse of authoritarian control systems in Russia presented us with a unique research opportunity: to study the citizen environmental groups that were protesting the newly discovered environmental consequences of nuclear weapons production at the formerly secret facility, Mayak. Earlier work with Russian environmental groups, and with some reformers who entered the new Yeltsin administration, gave us access to these formerly closed regions at a time when the new freedoms of post-Soviet society gave rise to citizen mobilization.

At nearly the same time, a group of scholars at Washington State University had been studying environmental issues surrounding the Hanford facility in eastern Washington State. With the thawing of the Cold War, and after the release of formerly classified documents about the facility under the Freedom of Information Act, there was a parallel citizen mobilization related to Hanford's environmental legacy. Citizen groups sought to establish public involvement in environmental and public health issues relating to Hanford.

This project results from the serendipity of fate, the coincidence of shared intellectual interests, and the generous support of our research sponsors to examine these historically unprecedented processes of democratization and environmental change as they began. In part this is a history of the development of environmental controversies surrounding these two facilities, and the response of the facilities and their respective governments. This is also a broader theoretical study of how democratic societies and their citizens attempt to function in such dynamic environments.

We want to acknowledge those organizations and individuals who provided the resources for this study. The W. Alton Jones Foundation was an early and committed supporter of the democratic citizens movement in Russia, and it provided the funds for surveying the residents in the Chelyabinsk region and the fieldwork conducted there by several members of our research team. The National Council for Soviet and East European Research (now the National Council for Eurasian and East European Research) furnished essential support to continue this fieldwork, and support for an international conference involving American and Russian participants in this project. The Institute for Global Conflict and Cooperation at the University of California, San Diego, provided valuable support for the analyses of our results, and partial support for the American data collection. In addition, the UC Irvine participants in the project would like to acknowledge the support of the Center for the Study of Democracy, and the Global Peace and Conflict Studies Program at UCI. These two research units created the intellectual environments in which such collaborative research could develop and flourish, and supported this research project in multiple and essential ways.

For the work conducted at Hanford, we wish to acknowledge the financial support provided by the College of Liberal Arts at Washington State University, and the survey research services provided by the Division of Governmental Studies and Services at WSU. Support for the field research for the qualitative work on the Hanford Nuclear Reservation, the local tribal governments, and interest groups was provided by the Consortium for Environmental Risk Evaluation (CERE), with funding by the

U.S. Department of Energy. The CERE project, headquartered at Tulane University and Xavier University, contracted with researchers from Washington State University to prepare a site study of Hanford, featuring an assessment of the governmental response to the advent of opposition to the facility from area residents, disproportionately affected disadvantaged minority groups, and the Native American tribal governments in the vicinity of Hanford. Without the assistance of the tribal governments of the Yakama Indian Nation, the Confederated Tribes of the Umatilla Indian Reservation, and Nez Perce Tribe in Washington, Oregon, and Idaho, this work could not have been completed.

A number of colleagues provided useful criticism during this project, and we want to thank them for their advice. In September 1993, we held an international conference of the project participants; Mark Chao, Thomas Cochran, John Hopkins, Sheldon Kamieniecki, William Potter, and Edward Walsh served as discussants and provided valuable commentary on the project and initial research findings. Riley Dunlap was extremely generous in sharing results from the Gallup Health of the Planet Survey, and current research on Russian and American environmentalism. Ken Niles provided valuable advice on the manuscript. We want to thank Andrew Drummond for preparing the index. The anonymous MIT Press reviewers gave us extensive and highly expert advice on the project; this helped us tremendously.

We also want to speak to the human dimension of our project. We want to thank several friends in Russia who became active in the democratization movement when glasnost and perestroika were still largely unknown concepts in Soviet politics. These friends and other contacts openly or quietly provided support for our research and democracy building efforts in the USSR and Russia, at times when this may have involved serious risks to themselves. We do not mention their names, but we want to acknowledge their efforts nonetheless.

Finally, as we note in this book, Mayak has the unfortunate distinction of being called the most polluted place on earth. Our study documents the depths of this problem and compares it with the parallel experience at Hanford, using a wide array of statistics. Statistics aside, there is a powerful human element to this story. Residents in the village of Muslyumovo, for example, live on the polluted Techa River and watch their children play in a severely contaminated environment. The average male in Muslyumovo lives only forty-five years. Similarly, part of our research team visited hospitals in the Chelyabinsk region and provided medical supplies from American donors. One of the most touching experiences occurred at the Chelyabinsk Children's Hospital, where there was a complete ward of pediatric leukemia cases—nearly all the residents of that ward died before this book reached publication. There is a human dimension to this study that we must never forget, and so we dedicate this book to the residents of Chelyabinsk and eastern Washington State, who unknowingly were the silent victims of the Cold War between our two nations.

I

Introduction

1

Introduction

Russell J. Dalton, Paula Garb, Nicholas P. Lovrich, John C. Pierce, and John M. Whiteley

For forty years American children lived with nightmare fear of Soviet missiles that never flew. Now it seems certain that in the process of building our own nuclear arsenal, we have killed and wounded many of our fellow citizens. . . . Ultimately it was our own fears—and the nuclear arms complex we built to soothe them—that did us harm, not the Soviet Union's missiles.
—Stewart Udall

This is the tale of two special places—Hanford in the United States and Mayak in Russia—that uniquely reflect the legacy of the Cold War and the uncertain prospects of a new future for these societies. These two sites are the homes to the major nuclear weapons complex in their respective nations. The Hanford facilities in eastern Washington State produced the plutonium for the first nuclear explosion at the Trinity test at Alamogordo, New Mexico, in July of 1945, and the nuclear warhead material for the atomic bomb that fell on Nagasaki in August of that year. Over the next four decades, Hanford produced the bulk of the plutonium for the American nuclear arsenal. The products of research and manufacturing at Hanford provided the foundation for America's nuclear deterrent.

Mayak played a similar role in the former Soviet Union's defense effort. The USSR built a nuclear processing facility in a secret city in the Southern Urals about 40 miles north of the industrial city of Chelyabinsk. The "Mayak Chemical Combine" was identified through its post office box number, Chelyabinsk-65, and these secret facilities produced the plutonium that armed the USSR's nuclear weapons.[1] Construction on the site began in 1945, and the first reactor went into operation in mid-1948. The plutonium produced at the Mayak complex fueled the Soviets' first nuclear test in 1949. The research centers at the complex constituted one of the primary sites for the Soviets' efforts to develop the science and weapons associated with the nuclear age. For four decades, in direct parallel with Hanford, the Mayak facilities produced plutonium and conducted nuclear research and engineering that provided the Eastern bloc's protection under the USSR's nuclear umbrella.

The scientists, engineers, and technicians at Hanford and Mayak were at the front lines of the Cold War. In their own way, they saw themselves as contributing to the preservation of peace enjoyed by the present postwar generation. In both countries, the nuclear arms race provided a powerful incentive for government to invest in the development of university-based research and scientific training (Leslie 1993). The leaders of both nations believed that the deterrent value of nuclear weapons dampened conflict between the superpowers and that these weapons prevented a third world war. The managers of the two facilities and their employees took great pride in their contribution to the defense of their homelands.

Both the Hanford and Mayak facilities stand among the monuments of scientific and engineering achievements of their respective societies. The construction of the first nuclear reactor at Hanford, for example, was accomplished in a phenomenally short time under extremely adverse conditions; the American nuclear weapons program heavily relied upon the activities at Hanford.[2] Similarly, the Soviet Union's nuclear science was dependent on the production facilities of Mayak and the research facilities of Chelyabinsk-70. Igor Kurchatov led the nuclear physicists at Chelyabinsk in developing the first nuclear weapons of the Soviet Union, and then was one of the principals in giving direction to subsequent nuclear research in the USSR.

The leaders of both nations undoubtedly feel that these facilities were essential to their having survived the Cold War. In the West, the United States assumed a new global role and participated in a process of détente that eventually transformed the international order. From the contemporary American perspective, the Cold War ended with an unequivocal American victory; the Soviet viewpoint, in contrast, is much more ambivalent. Nuclear weapons and rocket technology provided the former Soviet Union with a sense of national security and a claim of superpower parity with the United States. The relatively stable peace enjoyed over the course of more than three decades also was a factor in the evolution of the Soviet system beyond the authoritarian brutality of the Stalinist era. Détente and growing international interdependence also exposed the Soviet Union to strong forces of modernization and liberalization. As Mikhail Gorbachev once claimed, the Soviet Union finally reached the point where it could end the Cold War by depriving the United States of its enemy.

But the Cold War competition came at an enormous cost. This research examines the environmental legacy of this competition, and more specifically how citizens in both areas became politically active in response to these legacies. Nuclear weapons production at both sites released radioactive wastes into the environment that threatened downwind populations, and both kept this information secret from the citizenry (Makhijani, Hu, and Yih 1995). Today both areas are burdened by massive continuing environmental problems that may exceed technological solutions. This book examines the process of political mobilization against the environmental legacy of the two facilities as this information became known, and the governmental response to these actions. We believe this study provides unique insights into the nature of environmentalism, the functioning of democracy, and the processes that enable citizens to empower themselves. Our findings are based on the experiences of these two sites, and of their downwind neighbors.

The Development of Citizen Activism

The thawing of the Cold War, and the transformation of Soviet society through the principles of perestroika and glasnost, enabled skeptics to raise new questions about the societal costs associated with both nuclear security programs. Patriotism no longer required an unquestioning demonstration of support for the nuclear facilities, if it ever did. In the mid-1980s, some residents in eastern Washington State began to discuss the apparent increase in health problems suffered by people and animals living downwind from the Hanford facilities. Although the initial response to such discussions was often a harsh rebuke from their neighbors, and sharp denials by Hanford officials, the questioning expanded in scope and continued to nag Hanford officials (D'Antonio 1993). In 1984, a small number of eastern Washington residents formed a grassroots citizen group focused on Hanford, called HEAL (Hanford Education Action League). This group questioned plans for the construction of a national long-term nuclear waste repository at Hanford and simultaneously raised concerns about Hanford's past honesty in discussing on-site nuclear accidents and off-site emission problems. A series of articles in the Spokane *Spokesman-Review* by Karen Dorn Steele in 1985 brought broad public attention to many hitherto unanswered questions. Other journalists, scholars, and environmental groups became increasingly involved in the expanding debate about Hanford. At each step, the U.S. Department of Energy and Hanford officials (with Westinghouse Corporation as the lead contractor) responded with assurances that all was well, and that the claims of environmentalists were unjustified exaggerations.

In February 1986, Michael J. Lawrence, the manager of the Hanford facilities for the U.S. Department of Energy, released 19,000 pages of documents on the forty-year history of Hanford in response to a Freedom of Information Act request filed by Karen Dorn Steele. Hanford and Department of Energy officials considered this to be a crucial step in restoring their public credibility. The primary outcome of this bold action, however, was that the documents validated many of the environmentalists' claims. Among the disclosures causing the greatest alarm was the admission that for nearly twenty-five years Hanford had used "single-loop" cooling in their reactors, a practice that released thousands of curies of radiation into the Columbia River on a daily basis. Early methods of waste disposal by burial at the facility had created serious long-term problems, and the environmentalists' fears about groundwater contamination gained immediate credence from the government's documents.

Perhaps the most damaging evidence of official deception came from a report on the 1949 "Green Run." As part of a scientific experiment, Hanford officials had intentionally released several thousand curies of Iodine-131 into the atmosphere in 1949. The purpose of the test was to release airborne emissions similar to those produced by the Mayak facility. Government officials hoped that with proper calibration of instruments and better knowledge of dispersed particle behavior, the United States might be able to monitor the plutonium production occurring at

Mayak. To the misfortune of Hanford officials and local area residents, poor weather and precipitation destroyed much of the scientific value of the experiment, and radioactive residue was deposited on local crops and downwind communities—including the area's major city of Spokane, located 125 miles from the point of release. However, neither local government officials nor the public were told about this potentially harmful exposure either before or during this planned event, and no steps were taken to mitigate its effects afterward. To appreciate the scale of these total emissions, it is useful to recall that Three Mile Island became a political firestorm in 1978 when 14 curies were released into the environment. The Hanford documents admitted to releases totaling several million curies!

Of all of the nuclear weapons facilities of the United States, Hanford will be the costliest, most complex, and most technologically challenging to clean up. Cleanup costs have run over $1 billion a year since 1989. Originally intended to be accomplished in thirty years, the cleanup is now projected to extend late into the twenty-first century. Estimates of the total cost have floundered on the lack of currently available technology to solve some problems, such as contaminated groundwater, as well as questions about "how clean is clean" and the future land use at Hanford. Hanford receives approximately 21 percent of the environmental management funds expended by the Department of Energy. Excluding the cost of problems where no known technological remedy exists, current estimates of cleanup costs at Hanford exceed $70 billion, a figure that is probably conservative (see the review in chapters 2 and 10). Hanford is on the National Priority List under SARA (Superfund Amendments and Reauthorization Act), and has already received nearly $10 billion for cleanup activities since 1989.

The history of governmental action at Mayak reveals a number of remarkable similarities despite the fundamental social and political differences between the United States and the former Soviet Union. Mayak released massive amounts of radiation into the surrounding environment during the early years of plutonium production. By the facility's own accounts, about 2.75 million curies of liquid radioactive waste were released into the open Techa River from 1948–1951. These releases necessitated the resettlement of downriver populations when the environmental consequences of these actions became evident. In addition, the processing of material for nuclear weapons generated other airborne and water pollutants. Remember that the radiation release in Hanford's 1949 Green Run experiment was an attempt to simulate the operating procedures of Mayak.

The Mayak facilities also experienced a series of catastrophic environmental events (Cochran, Norris, and Bukharin 1995; Donnay et al. 1995). Most noteworthy, in 1957 there was a large explosion of a nuclear storage facility that released 20 million curies of radiation into the atmosphere. Over 15,000 square kilometers had radiation levels of at least 0.1 Ci/km^2, and over 10,000 people were eventually relocated. Mayak also used adjacent lakes as storage areas for wastes; for example, over 100 million curies of radiation from waste byproducts rest on the bottom of Lake Karachay alone. In 1967, a drought exposed a portion of the dried bottom of Karachay, and the radioactive contamination from the lake was blown over a large downwind area. Unsafe practices for the storage of radioactive liquid wastes and other hazardous by-products have continued; many storage vessels for high-level wastes are unstable and continue to pose a threat to public safety and the environment. The accumulated problems of plutonium production have earned the entire Chelyabinsk area the unenviable distinction of being called the most polluted place on earth.

Government efforts to build a new nuclear power station at Mayak stimulated an open public discussion in 1989 about Mayak's environmental legacy. Under the auspices of Gorbachev's policy of glasnost, residents of the city of Chelyabinsk, which lies directly south of Mayak, began to discuss the continuing problems of nuclear waste storage and accidental radiation releases from the facility. The public also was concerned with the impact of known and suspected radiation exposure on the area's population. The Chernobyl reactor core meltdown disaster in 1986 had severely eroded public confidence in the infallibility of Soviet science and technology, just as Three Mile Island had led Americans to question the safety of nuclear energy and the government's ability to safeguard the public. Moreover, Gorbachev's policies of perestroika and glasnost enabled Soviet citizens to question the actions of their government. At almost the same time that grassroots groups were forming around

Hanford, the residents of the Chelyabinsk region were creating their own environmental organizations.

An important early event was a public meeting held in September 1989 to discuss halting the nuclear power plant that was already under construction at Mayak.[3] By the following year, with the first open elections, a set of new environmental groups had formed in the Chelyabinsk area (see chapter 6). The Nuclear Safety Movement became an active critic of both the plans for the power station and the accumulated environmental problems of Mayak. Other groups, such as the Democratic Green Party and the Kyshtym-57 Foundation, also focused on nuclear issues related to Mayak. These "green" groups prompted government agencies to respond to their queries. The resulting public debate on nuclear power policy was among the first in modern Russian history. Chelyabinsk environmentalists succeeded in placing an initiative on the March 1991 ballot to tap popular sentiments toward the planned power station. Nearly three-quarters of the electorate in the city of Chelyabinsk voted against the power station, and five-sixths voted against the reprocessing of nuclear waste (spent fuel rods) at Mayak. A citizens' movement against Mayak had developed as the Soviet Union declared its transition toward democracy.

Dealing with the legacy of nuclear weapons production raises fundamental questions for both the United States and Russia. There is, for example, a continuing debate about the extent of environmental damage and radiation release problems at Hanford and Mayak. Virtually everyone agrees, however, that the accumulated waste problems of Hanford and Mayak (and other nuclear weapons facilities in both nations) are staggering (see chapters 2 and 3). Nonetheless, the environmental damage and continuing problems at Mayak dwarf the American situation at Hanford. In 1990, a Russian governmental commission declared the Chelyabinsk region a national environmental disaster area (similar to the designation for the Chernobyl region).

The Lessons for Democracy

For social scientists, the experiences recorded at these two major governmental facilities raise fundamental questions about the nature of contemporary democracy, especially when the direct concerns of some citizens come into conflict with national defense interests.[4] For an established democracy, such as the United States, we must ask how the public can maintain its democratic control over a sensitive area of national defense, such as nuclear weapons production, when there is a curtain of national security covering the government's actions. Hanford and the other primary nuclear weapons production, facilities of the Department of Energy (e.g., Rocky Flats, Savannah River, Fernald, Oak Ridge, and the Idaho National Engineering Laboratory) were exempt from the environmental regulations applied to commercial nuclear power stations by the Nuclear Regulatory Commission. Citizens who questioned the claims of government officials operating these facilities often had their own patriotism questioned.

Of course, such questions of democratic control are even more relevant for an emerging democratic polity in Russia, where the suppression of dissent and the cloak of secrecy were heavily felt for decades (Eckstein et al. 1998). The ability of citizens to control the government's actions at Mayak provides a timely test of the fledgling Russian democracy. It is also important to know how the American democratic system performed on this test.

This test of governance of both nations is particularly important because of the potential long-term impact of weapons development activity on the public. Many have died as the result of nuclear weapons production at Mayak, although we will never know the exact counts because of the intentional suppression of health statistics. Many Americans were also placed at unreasonable risk, without their knowledge, by their own government, through the activities at Hanford. Now, more than forty years later, the governments in both nations are beginning to compile the necessary evidence to assess past exposure rates and estimate the likely health effects of the facilities.

In addition, there was potential abuse of minority populations in both the Mayak and Hanford cases. Three Native American tribes—the Confederated Tribes and Bands of the Yakama Indian Nation, the Confederated Tribes of the Umatilla Indian Reservation (CTUIR), and the Nez Perce Tribe—include Hanford within their native homelands and still retain rights to fish, hunt, and gather in the region under a 1855 treaty (see Liebow, Younger, and Broyles 1987). These minority populations—

along with many Mexican-American farm laborers residing in the region—were placed at special risk because of their proximity to the complex and because of their heavy reliance on the fish, flora, and fauna of the area. Similarly, the Tatar and Bashkir minorities in the Chelyabinsk region may have suffered disproportionately from the activities at Mayak. There is a real human dimension to our study, which we take particular care to observe.

Citizen environmental action around Hanford and Mayak also offers a unique opportunity to compare the development of environmentalism in these two nations. The environmental movement has become a major force for social and political change in Western democracies, offering a new philosophy on how economic activity should be organized, lifestyles altered, and social relations changed. It is unclear whether environmentalism entails the same values and political orientations in the formerly Communist nations of Eastern Europe. The ideology of Marxism-Leninism in Russia is difficult to relate to the value contrasts underlying environmentalism in the West. By concentrating on two geographic areas where a common environmental problem has focused public attention, we have a rare opportunity to compare how selected groups of Americans and Russians think about critical environmental issues. We can also determine the common elements of environmentalism and "green thought" that transcend even the wide ideological and political divide that has separated the United States and Russia.

Another set of questions involves the process of environmental mobilization in both nations. Citizen activists in Washington State faced challenges in creating an organization, developing the expertise to evaluate government reports, and communicating their views to the broader public and policy makers. This was made even more challenging by the high-security nature of the nuclear issue and the mask of secrecy covering the nuclear weapons complex. Yet there are established models and methods of citizen action that American groups could draw upon, as well as other well-established political allies available to assist them. Environmental groups in Chelyabinsk faced a much greater challenge than did their American counterparts—namely, to develop a grassroots democratic movement in a system largely lacking the norms and infrastructure of democratic politics. The similarities and differences in how both movements met these challenges provide insights into the sources and broader significance of citizen-based political action in both democratic nations.

Finally, there are important lessons about the political process to be learned from these two experiences. Two extremely powerful forces shaped social thought and public policy at both facilities: national security concerns born of the Cold War and a largely unfettered belief in the power and veracity of modern science and technology. Numerous Russian and American commentators claim that these forces gave rise to a corporate culture within the government agencies managing both facilities that seriously impeded conventional oversight activities. The pressures to produce large quantities of weapons-grade plutonium in order to defend society, and an unassailable belief that science could solve any technical problem, meant that government agencies and officials at the weapons facilities acted largely without the normal safeguards and oversight typical of the private-sector nuclear power facilities in the United States. Decisions about substantial risks or social costs were sometimes made without sufficient debate and deliberation, sometimes even without informing the next higher level of the administrative structure. Even in the relatively open system of American democracy, a corporate culture was created at the Hanford facility in which a broader public accountability was systematically de-emphasized from the top down.[5]

A subtheme of our study is how this abrogation of authority developed. Equally important is the question of whether government agencies steeped in secrecy and deception can really change their fundamental character. Beginning in the late 1980s, both the American and Russian governments claimed that they were modifying this culture, thereby indicating that they deserved renewed public confidence in the agencies managing nuclear weapons plants, and in the political process more generally. If such change has indeed occurred, and if public confidence has been restored to a significant degree, this would be an extremely positive sign for the vitality of the political processes of both nations.

This book marshals evidence to address each of these questions. It reflects the culmination of a multiyear joint research effort by American and Russian scholars representing several academic disciplines. We tell the tale of these two important localities, as well as describe how this aspect of the Cold War affected and tested our two nations, and what insights may be derived about what lies ahead for our respective societies.

The Design of the Project

The Hanford/Mayak project has followed a long, complex (and sometimes arduous) route in developing a comparative study of citizen responses to the environmental consequences of nuclear weapons production in America and Russia. This study involved the development of two sizeable research teams to conduct data collection in two sensitive areas of public policy during a period of great political change and uncertainty in both countries. The breadth of our interests also required collaboration among interdisciplinary research teams, including political scientists, anthropologists, psychologists, and economists. Indeed, the ability to develop a collaborative American-Russian project on such a sensitive topic, and to see it through to completion, indicates how much both nations have changed in recent years.

The project began with a 1991 University of California, Irvine conference organized by John Whiteley to facilitate technical exchange between American, Russian, and Chinese officials responsible for environmental cleanup of their respective weapons facilities (Whiteley 1991). This initial meeting led to contacts between UC Irvine faculty and environmental activists in Chelyabinsk. The Center for the Study of Democracy and the Global Peace and Conflict Studies program at UC Irvine supported further discussions about the theoretical and political significance of the Chelyabinsk and Hanford experiences. This discussion expanded the project to include Russell Dalton and Paula Garb. Whiteley and Garb began to assemble a team of Russian coinvestigators who were interested in studying Chelyabinsk.

After the Russian research project was organized, the UC Irvine team contacted Nicholas Lovrich and John Pierce at Washington State University (WSU) to collaborate on a parallel study of Hanford. This collaboration grew through a series of visits, meetings, and conferences in 1992 and 1993 involving all of the participants from both universities and key Russian colleagues.

These efforts spawned five data collections that provide the basis for the analyses presented in subsequent chapters. First, members of the UC Irvine and Washington State University research teams conducted in-depth interviews with environmentalists in both nations, focusing on the political controversies related to the environmental damage produced at the two facilities. Several dozen environmentalists were interviewed in Russia, including almost all of the leading activists in the Mayak movement and several environmental leaders in Moscow. Similarly, virtually all of the leaders of the groups active in the Hanford area were interviewed using protocols developed from those used in the Mayak study.

Second, we collected parallel public opinion surveys of residents in both regions. With the assistance of the Kaluga Institute of Sociology, we conducted a survey in Chelyabinsk during the winter of 1992–93 (see appendix A for details of the methodology of the survey). The survey focused on the public's environmental attitudes, their perceptions of the nuclear facilities, and their views of the environmental movement. We surveyed 1,187 randomly sampled respondents in the region, conducting personal interviews lasting approximately one hour. Interviews were conducted in the closed city of Chelyabinsk-70, the city of Chelyabinsk, two villages near the nuclear facilities (Muslyumovo and Kyshtym), and Chebarkul, a town in a distant part of the oblast (to act as a control site).

During winter 1993–94 the Division of Governmental Studies and Services at Washington State University surveyed residents in eastern Washington State. Using mail and telephone questionnaires, we interviewed residents of Richland (which lies adjacent to the Hanford facilities), the city of Spokane, three small towns on the Yakama Indian Reservation (White Swan, Wapato, and Toppenish), and Wenatchee, a town in a distant part of the region to serve as a control (see appendix A). These parallel Russian and American surveys provide a valuable resource to examine public attitudes toward the facilities, to assess the degree of popular support for a "New Environmental Paradigm" (NEP), and to describe citizen opinions toward green movements in these sensitive areas.

Third, we conducted additional in-depth interviews with members of the general population and representatives of the important ethnic and political groupings in both areas. In Russia, a team of anthropologists interviewed members of the Tatar, Bashkir, and Russian communities. In America, members of the WSU research team held two conference sessions with official representatives of the Yakama, Umatilla, and Nez Perce tribes. The WSU research team also held focus group sessions with two groups of Mexican-American residents of the area, a minority group that predominates in the agricultural hinterland of the Tri-Cities area (Richland, Kennewick, and Pasco). These interviews, conference sessions, and focus group meetings concentrated on environmental attitudes, perceptions of the facilities, and beliefs about how ethnic groups have been affected by the facilities.

Fourth, members of the research teams interviewed government officials and plant officials in both locales. Interviews with government officials in Chelyabinsk and Moscow inquired about past policies and their views on citizen activities protesting the facilities. The interviews with plant officials also assessed orientations regarding a role-change from weapons producers to postproduction environmental cleanup. The WSU research team also secured a research grant in 1994–95 from Xavier University (New Orleans) to prepare a "Hanford Site Report" as part of a larger study of six nuclear weapons sites. This study examined official, public, tribal, and activist assessments of the risks to the environment and/or public health at each site.[6] A wide range of Department of Energy and contractor officials were interviewed in the course of this work.

Fifth, we collected a considerable storehouse of documents on the Hanford and Chelyabinsk facilities. A vast collection of government documents and reports is now available on Hanford, one that has continued to grow through the course of our project. Hanford's activities have generated several book-length accounts (Sanger 1989; Gerber 1992; D'Antonio 1993). Official documentation of Mayak is more difficult to acquire (among the best sources are the reports of the Natural Resources Defense Council (Cochran and Norris 1993). In addition, we assembled a database of newspaper articles on the two facilities published in the American press, and compilations of materials from the Russian press and environmental groups in Russia.

When combined, these materials enable us, in a rich and detailed way, to study the people who are directly concerned with environmental issues and attempt to use the political system to address their concerns. Moreover, the multidisciplinary nature of our project enables us to examine questions from several different research perspectives and with the benefit of evidence ranging from the mass opinion survey to the in-depth personal interview. This multidisciplinary approach provides a certain richness and robustness to our findings. The results of our collective efforts provide a unique vantage point on the nature of citizen responses to the environmental consequences of nuclear weapons production at both facilities.

The Framework of Our Analyses

At the core of our study is an interest in the process by which the public's environmental concerns become mobilized into political action, and how the political system responds to these concerns. In addressing this topic, our research touches on a range of fundamental theoretical questions about the nature of social movements, the political values and behavior of the citizenry, and the processes of governmental response. This is first a study of Hanford and Mayak, but we also see these two experiences as providing a unique setting within which to examine questions of broad relevance to social science. Therefore, we want to outline briefly the theoretical concerns that underlie our case studies, and briefly introduce the analyses that will follow in subsequent chapters.

Assessing the Damage

We begin with two chapters by John Whiteley (chapters 2 and 3) that recount the development of Hanford and Mayak, and the environmental consequences of plutonium production at the two facilities. The reader will quickly realize the exceptionality of both regions. Hanford represents the greatest concentration of nuclear and other toxic wastes in the United States. Mayak honestly deserves the title of the most polluted place on earth.

In addition to describing the historical record of Hanford and Mayak, it is equally important to compare how Americans and Russians perceive their environmental conditions, and how they think about environmental matters. Regardless of the objective environment, *perceptions* of this environment are the reality that shapes individual attitudes toward the facilities and conditions the public's willingness to engage in environmental action—and we are studying two regions with distinctive environmental conditions. The chapter by John Pierce, Russell Dalton, and Andrei Zaitsev (chapter 4) uses data from our population surveys to determine how residents of these regions perceived their own local environments in broad terms, such as the quality of the air, water, and other features of the environment. In addition, the chapter describes how residents in these two regions judged the extent of environmental damage and risk linked to the respective facilities. This chapter provides stark evidence of the environmental legacy of Hanford and Mayak as perceived by the neighboring residents.

Social Movements and the Organization of Protest

In both the Hanford and Chelyabinsk regions, significant citizen actions and organizations emerged to protest the activities of the nuclear facilities. Environmental action developed in response to widespread revelations about the nuclear sites, concerns about experimentation on the effects of radiation, and a persistent lack of candor and accuracy in the governments' accounts of those activities. In both locations, citizen mobilization involved both formal organizational development and broad-based shifts in the attitudes of the respective publics. The result has been a highly politicized environment in which to make policy about the future of these two sites.

Perhaps the most interesting and important comparison in our study involves the process by which environmental groups formed to protest the activities at Hanford and Mayak. These groups faced a daunting task: to challenge the nuclear defense complex over one of its most important and most secret facilities. Describing the evolution of the environmental movement in both regions thus provides unique insights into the political nature of both societies, and the relationship between the citizens and their governments concerning this most delicate issue.

The history of environmental action in both regions also offers the potential to examine general theories of political mobilization and social movements in these unique contexts. For example, a large portion of the current literature on social movements stresses how the characteristics of political structures and institutions affect the behavior of citizen groups (McAdam 1994; Kitschelt 1986; Jenkins and Klandermans 1995). In the context of the comparative analyses presented here, there are striking contrasts between the political structures facing environmental groups protesting Hanford and those protesting Mayak. If such factors are important, their impact should be obvious in the behavior of the respective protest groups.

Another central component of recent social movement research focuses on the capacity of the movement to mobilize effectively the appropriate resources (Zald 1992). At the level of the organization, research focuses on the effectiveness and "rationality" of the organization in mobilizing and using its resources to achieve its goals. Are movement organizations mobilizing their human and political resources in the most effective way? To what extent are groups able to engage in a calculus that heightens the probability of achieving their goals, and to what degree do their activities reflect their strengths and weaknesses?

The answers to these questions obviously are context driven. That is, they depend on what resources are available to the movement, its organizations, and leaders. Indeed, some significant theories of interest-group behavior focus on the role of the entrepreneurial leader who, much like the developer of a new business, must invest resources, effectively evaluate costs and benefits, and compete in a market both for additional resources and for policy success (Salisbury 1969). The contrasts between Hanford and Mayak in these resource terms are obviously quite extreme. Environmental groups protesting against conditions at Hanford presumably have a larger resource base to draw upon, and have available to them a more extensive network of potential allies and supporters. Moreover, there are prominent models to guide new entrepreneurs toward success within the American policy process. It is clear that groups protesting Mayak have not enjoyed a plentiful resource base for their activities, nor has the field of potential allies been broad. An environmental group protesting Hanford can turn to large and well-funded national environmental groups for advice and political entrée, they can find supportive journalists to highlight their cause, and they can work within a system that generally facilitates citizen input. Where do groups protesting Mayak find the funds for their campaigns? Where do they find allies in a Soviet or post-Soviet system? How can effective protest be channeled? The comparison of the Hanford and Mayak cases should illuminate the effects of resource availability and institutional settings in structuring the course of development and the pattern of behavior of social movements.

William Schreckhise (chapter 5) and Paula Garb and Galina Komarova (chapter 6) describe the development and evolution of environmental groups protesting the activities of the facilities at Hanford and Mayak, respectively. Both accounts also provide insights into the larger questions of social movement behavior we have outlined above.

The Public and the Environment

Our project also describes the beliefs and attitudes held by citizens residing in the area of the two nuclear weapons plants, in order to understand the basis of their environmental concerns and the thinking underlying their political action. Even more than a national study, the analysis of these two regions, where specific environmental issues have been widely discussed and debated, provides a basis for systematic comparison. This research addresses questions related specifically to the facilities, but also inquires into broader questions about the nature of environmentalism in both nations.

We want to move beyond assessments of specific environmental conditions to examine how Americans and Russians think about the environment. Past research has described Western environmental thinking in terms of a "New Environmental Paradigm" that involves a coherent set of beliefs, such as distrust of technology, acceptance of natural resource limits, and an orientation to collective values over individualism (Dunlap and Van Liere 1978; Milbrath 1984). In previous research we have argued that such environmental orientations help structure public attitudes and behavior on a broad range of environmental issues (Pierce et al. 1986; Dalton 1994). Adherents to the NEP tend to view environmental problems not just in their local context but as part of a larger conceptual framework prescribing how society should function and which values should be emphasized. More broadly, NEP values provide a model of action for many sectors of the American environmental movement and give direction to what it means to be "green" among American environmentalists (Dunlap and Mertig 1992).

It is unclear whether the same environmental paradigm exists in the formerly communist nations of Eastern Europe and the Soviet Union (Jancar-Webster 1993; DeBardeleben and Hannigan 1995). The role of private enterprise and orientations toward collective values carry a different connotation among Western environmentalists than in communist societies where values were shaped by Marxist-Leninist ideology. Thus what it means to be "green" might be different for Russians. In addition, the active conceptualization of environmental issues is a relatively new element of Russian politics. Much as might have been the case for Americans before the first Earth Day in 1970, Russians might not have thought systematically about the principles embodied in the NEP and their implications for issues such as Mayak.

We examine this question of environmental thinking at two levels. The project included opinion surveys in both regions that assess the environmental values of both publics. The chapter by Russell Dalton, Yevgeny Gontmacher, Nicholas Lovrich, and John Pierce (chapter 7) compares attitudes on the elements of the New Environmental Paradigm across nations. In addition, Paula Garb conducted extensive interviews with environmental activists in the Chelyabinsk region and Moscow. Garb (chapter 8) uses these interviews to probe more deeply into the environmental thinking of Russian activists.

Mobilization of Citizen Support

How can one explain the mobilization of citizen opinion and action in response to the widespread revelations about the Mayak and Hanford facilities? The application of theories of mobilization are especially interesting in these two instances. The stark differences in important aspects of the two sociopolitical contexts provide a unique set of circumstances for testing abstract theories of why people join forces to achieve a public good. The contrast of Russia and the United States as loci for political mobilization in response to apparently similar environmental wrongs sets up powerful comparisons of countries that differ in the historical openness of the political process, the level of information available to the citizenry, the response of the government to citizen protest, the power of the local environmental movements, the magnitude of the wrongs, and the opportunity for political organizations to act on behalf of aggrieved citizens.

We examine three major theories to explain the willingness of citizens to engage in collective action. First, social-psychological explanations of political mobilization focus on the attitudes of individuals involved in the movement (Mueller 1992; Rohrschneider 1990). Second, specific grievances can provide the link between the individual and identification with a larger social movement (Laraña, Johnston, and Gusfield 1994). Obviously, in both locations, there are significant reasons for the expression of individual and collective grievances about the environmental consequences of the facilities. Third, another body of literature argues that the individual skills and resources of citizens determines their propensity for political action, especially in noninstitutional forms of involvement, such as participation in an environmental group or environmental protests (Verba, Nie, and Kim 1978). This explanation would predict systematic differences in who might become active within each nation, as well as account for aggregate differences in political involvement across nations.

Our project provides an important context in which to examine these contending theories because we are studying areas with clear and major environmental problems where these issues are highly salient and longstanding. The chapter by John Pierce, Russell Dalton, and Ira Gluck (chapter 9) looks at public support for and involvement in environmental groups protesting both facilities. The chapter examines the correlates of environmental support in a level of detail that would not be possible in a national survey of the American and Russian publics. The results can provide valuable evidence on the sources of popular opposition to both facilities, as well as address larger questions of how political action can become mobilized.

Governmental Response

In many respects the American and Soviet governmental response to the challenge of nuclear weapons production and to its long-term consequences were remarkably similar. In terms of organizing for the task and carrying out the large-scale production of nuclear warhead materials, the United States and USSR chose comparable routes of action. In terms of the ultimate outcomes of the Cold War and the end of the nuclear arms race, however, the courses of official action taken in the two countries have been very different. Important lessons are present for scholars and decision makers in both countries. These lessons include commonalities observed in the book's two chapters on the governmental response to the environmental protests at Hanford and Mayak, and differences between these experiences.

In relating the story of the governmental response in the American and Russian cases, several questions are central to the comparison. First, who were the key institutional actors entrusted with the authority to use the secrets of the atom for military purposes? Nuclear research in the Cold War age involved a fundamental tension between scientific norms and military necessity (Groves 1962; Lens 1982; Rhodes 1986; Wheeler 1989). How did governmental institutions in both nations reconcile these tensions? What range of discretion were these agencies and their managers given to accomplish their "top secret, top priority" goals and objectives? What time frame were these agencies working under as their technical knowledge grew? Many defenders of both facilities claim that the simple pressure to produce large amounts of plutonium very quickly led to occasional resorts to safety shortcuts and the making of nonscientific administrative judgments that ultimately proved unwise. Some claim that the rapid pace of scientific breakthroughs often surpassed the existing knowledge about the environmental or health consequences of what was being undertaken. Alternatively, limited scientific knowledge led to some poor judgments; and as scientific knowledge of the atom and nuclear physics improved, so did the environmental record of the nuclear weapons facilities. It is clear that Hanford and Mayak constitute cases wherein important questions relating to both the value and limits of unquestioned faith in science and technology can be examined (Merton 1973; Wynne 1982).

Second, when things went wrong—serious off-site accidents, unintentional releases off-site, nuclear waste storage problems, or damage to local residents or the regional environment—what was the nature of the government's response? Did government officials act quickly (and openly) to address problems, or was denial and refusal of government responsibility the official response? Both facilities should be judged relative to each other but also relative to the existing political processes operant in each nation. We are interested in whether the special environmental and political problems created by these facilities resulted in deviations from the normal governmental process.

Third, what was the nature of the "organizational" cultures that arose in the agencies entrusted with the national nuclear deterrent? American democratic norms imply that there should be systems of protection for "whistleblowers" and dissident scientists, to encourage diversity of thinking and constructive criticism (Forester 1989). Were these norms followed, or were strong cultural systems developed that emphasized uniformity of thought and action? We expected less openness and diversity under the Soviet system than in the American, but we encountered more commonality than differences in this area. Thus one might ask how an uncharacteristically closed organizational system affected the operation of the Hanford and Mayak facilities, and how it influenced the environmental record of the two facilities during and after the Cold War.

Fourth, public concern over the government's ability to manage nuclear technology and protect the public's health was called into question after Three Mile Island in the United States and Chernobyl in the Soviet Union. What was the nature of the government's response to the need to reestablish public trust in its capacity to preserve and protect the public welfare? To what degree have these agencies succeeded in modifying their secret nuclear weapons production mission into a public, inclusive process of environmental cleanup? To what extent have the governmental agencies originally entrusted with production been forced to open their files and work with a broader range of governmental agencies dedicated to environmental restoration? To what extent have these efforts been met with adequate funds from central government sources?

These questions are addressed in the chapters on the governmental response to the facilities. John Whiteley (chapters 2 and 3) provides the context for our research; he describes the early history of the facilities and their environmental releases. For both the United States and Russia, the transition from the communist era of the Cold War nuclear arms race to a new phase of democratic politics has been a great challenge. The governmental response to the arms race saw similar patterns of action in the two countries, and the post–Cold War period has brought into relief the contrasts between the governmental response of a more mature democracy and that of a struggling, newly democratic land. Our study concludes with chapters focusing on the government's recent actions regarding Hanford and Mayak. The chapter by Nicholas Lovrich and John Whiteley (chapter 10) relates the American governmental response to the challenge of Hanford's cleanup. John Whiteley (chapter 11) describes the response of Russian agencies to Mayak.

In conclusion, the seemingly obvious explanation of citizen protest against the actions of officials responsible for the Hanford and Mayak nuclear weapons complexes—that people were angry about being placed at risk and having been systematically misled by their governments—conceals many other important lessons for both Americans and Russians. As the above review suggests, the actual dynamics of citizen action (and policy influence) concerning these facilities involve a variety of important policy lessons and considerations for social science theory (Rose 1993). The imposition of the Russian and American contexts on these two case studies helps us account for the variations in the character of citizens' reactions and evaluations of the facilities, both between the two countries and across locations within each country. These frameworks will also help us understand the consequences of actions by the citizenry and the government for the furtherance of democratic governance in both the United States and Russia. The chapters in this book do not directly confront all of the suggested explanations and the testable hypotheses that derive from the relevant social science theories. Yet much of what we present has obvious implications for the central tenets of how citizens can influence their own government and how democratic governments need to accommodate the transition from Cold War expediency to a more open and citizen-inclusive form of political life.

Notes

1. The production facilities were first identified as Chelyabinsk-40, the number designating the post office box for Mayak. This later changed to the Chelyabinsk-65 designation, and the secret city associated with Mayak is currently known as Ozersk. The separate secret city of Chelyabinsk-70 housed nuclear research institutes and other related enterprises; it is now known as Snezhinsk. We will

use the hyphenated post office box designations throughout this book, since these were the standard reference at the time of our data collection.

2. In recognition of the unprecedented engineering feat of rapid construction and immediate successful operation, B Reactor is listed on the National Register of Historic Places.

3. Another parallel between the two facilities is their link to domestic nuclear power. The Hanford facility was the site of the ill-fated Washington Public Power Supply System (WPPSS) that defaulted on its bonds in 1983. The largest nonmilitary atomic project in the world, WPPSS planned on the construction of five nuclear power plants, including three at Hanford. Greatly overestimated power demands led to the financial debacle that occurred in the WPPSS case (Impact Assessment 1987).

4. See, for example, Edward Walsh's study of citizen protest against nuclear power in the wake of the Three Mile Island accident (Walsh 1988).

5. One of the grand jurors empaneled to hear charges about environmental cover-up at the Rocky Flats facility stated: "The big story is that they all knowingly violated environmental laws. Congress passes laws it doesn't intend to see enforced. The government has a lack of will to enforce the law. The government is lying to us. It's a breach of public trust. That is the biggest crime of all" (*Los Angeles Times Magazine,* August 15, 1993, p. 44).

6. The Consortium for Environmental Risk Evaluation (CERE) project was mandated by Public Law 103-126 to discern the risks and concerns associated with the cleanup of six sites formerly associated with nuclear munitions production. The U.S. Department of Energy contracted with Tulane University to conduct the CERE study, and Tulane contracted with Xavier University to conduct the "Public and Tribal Concerns" portion of the risk assessment work. Xavier University contracted with Washington State University to conduct research on and prepare a site report for the Hanford facility (O'Connor et al. 1995).

II

The Historical Legacy

2

The Hanford Nuclear Reservation: The Old Realities and the New

John M. Whiteley

Just 50 years ago several Indian tribes wandered and foraged this land, and 6,000 farmers from the towns of Hanford, Richland, and White Bluffs grew fruit in orchards irrigated from the Columbia. But after the Manhattan Project expropriated 570 square miles of land in 1943, plutonium and its lasting legacy, nuclear waste, became Hanford's crop, forever altering the land.
—Karen Dorn Steele

In 1905, Judge Cornelius Hanford of Seattle gave his name to a town site six miles down river from White Bluffs, Washington, on land he and some fellow investors purchased through the Priest Rapids Irrigation and Power Company. In the years immediately following the passage of the 1902 Newlands Reclamation Act, irrigation projects proliferated, land prices appreciated, and by 1909 three railways served the region. Occasional floods, omnipresent dust storms, and the porous soil, however, combined to make some portions of the Columbia River basin less of a bonanza than Judge Hanford and his fellow investors had hoped. Years after the time of Judge Hanford, his name is best known throughout the world as a leading site of plutonium production to fuel the nuclear weapons program of the United States.

This chapter traces the significant developments at the Hanford site through its history as a nuclear weapons facility.[1] This examination of the old reality of Hanford covers the beginning and the development of the plutonium production mission, and the contamination of the environment that resulted from this mission. The discussion of the new realities at Hanford describes the ending of production, and the beginning of public discourse about the facility and its legacy. This includes the release by the Department of Energy in 1986 of previously secret documents, which was followed by a series of formal agreements on the Hanford cleanup and the mobilization of the public concerned with the health and environmental impact of the facilities.

The Beginning of the Old Reality at Hanford

In December of 1942 and January of 1943, a small group of men, acting under the cover of wartime secrecy, selected Hanford as the site for the Manhattan Engineer District. One of those men, Lieutenant General Leslie R. Groves, reportedly observed that the farms did not appear of much value. The survey team from the DuPont Corporation recommended the Hanford site as "being far more favorable in virtually all respects than any other" (Matthias quoted in Gerber 1992a, 12).

Hanford's remoteness benefited the secrecy of the Manhattan Project. It was a very large site (570 square miles) located along the Columbia River, a considerable distance from the state's major metropolitan areas of Seattle and Spokane. Because it was located in a remote region of the country, few Americans were aware of Hanford, and even the local residents were unsure about the activities occurring at the site. Hanford's location along the Columbia River also provided an abundant source of water and electric power for use in the nuclear reactor program.

Until the formation of the Atomic Energy Commission (AEC) after World War II, General Groves personally made many of the critical decisions about installations and practices. The DuPont Corporation was to serve as the primary industrial contractor, and one of the company's conditions, which Groves accepted, was to agree that DuPont would be indemnified by the U.S. government from "any damages . . . incurred in the course of the work" (Groves 1962, 51–57). One concern was that, because of the need for wartime security, employees could be exposed to radiation risks about which they had not been fully informed. The extension of the principle of corporate contractor indemnification has been one of the recurring obstacles to accomplishing sound environmental practices in the nuclear weapons complex in the United States, and to the associated issue of protecting the health of the surrounding population. Groves acknowledged that significant risks to health and environment were possible: "We knew, too, that in the separation of plutonium we might release into the atmosphere other highly toxic fumes which would constitute a distinct hazard . . . I was more than a little uneasy myself about the possible dangers to the surrounding population" (Groves 1962, 69).

The Army Corps of Engineers began construction of the plutonium production facilities in March of 1943. At the peak, 50,000 workers labored on the construction; most of them lived in temporary barracks and were relocated after the work was completed. The B reactor, which became the world's first functioning large scale reactor, began operation in September 1944.[2] Between 1943 and 1945, Hanford constructed two additional reactors (D and F) to produce plutonium, and additional facilities for chemically extracting plutonium from irradiated fuel (the T and P plants). The reactors were graphite-moderated and single-pass, water-cooled reactors that drew cooling water from the Columbia River, then returned it. Hanford-produced plutonium went into the first nuclear weapon exploded at Alamogordo, New Mexico, in July of 1945, and fueled the bomb dropped on Nagasaki in August.

After World War II ended, there was a temporary lull in activities; in 1946, the B reactor was placed on standby status. But the heating up of the Cold War led to a dramatic expansion of Hanford, particularly during the years 1947–1955. The H reactor went into production in 1949, followed by the DR reactor in 1950; both used variants of the graphite-moderated, water-cooled design of the first reactors. After the Soviet Union's first atomic bomb test in September 1949, the activities at Hanford took on the character of a race to ensure America's defense. From 1950 to 1953, there was a doubling of the plutonium production facilities, and new reprocessing plants were constructed. The REDOX (reduction oxidation) facility, which began operation in 1952, extracted uranium and plutonium from the irradiated reactor fuel. Additional reactors were constructed in the early 1950s in reaction to national security concerns spawned by the Korean War (C reactor in 1952, KW and KE reactors in 1955). Soon after, Hanford built the PUREX (plutonium uranium extraction) plant, which used a new solvent-based process to extract plutonium and other nuclear elements. The final (N) reactor—a thermal reactor for producing plutonium for nuclear weapons and breeder reactors—was completed in 1964.

By the 1960s, Hanford had largely assumed its full structure (figure 2.1). The reactors were located in 100 Area—strung along the Columbia River. Indeed, adjacent land a bit further downriver, but still on the Hanford Reservation, was used in the mid-1960s for the ill-fated construction of a set of commercial nuclear power reactors.[3] 200 Area (West and East) contains the chemical processing and waste storage operations; this is where the PUREX facility is located. In addition, this area holds 149 single-shell tanks constructed between 1944 and 1964 for the storage of high-level waste. The tanks range up to 1 million gallons in capacity, and in total, hold roughly 40 million gallons of liquid wastes produced in plutonium extraction processes. Between 1968 and 1986, an additional twenty-eight double-shelled storage tanks were built. Each of the tanks has a capacity of between 1 and 1.4 million gallons, and their double-shelled construction is designed to minimize the possibility of leakage. Currently they contain over 24 million gallons of mostly liquids, with

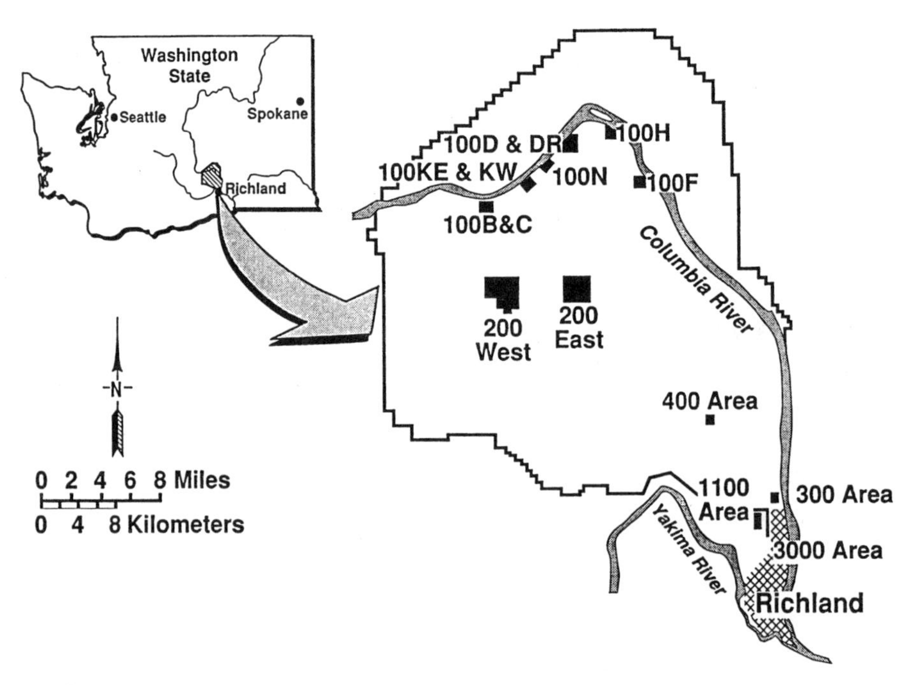

Figure 2.1
U.S. DOE Hanford site

some sludges and salts. 300 Area contains laboratories, fuel fabrication, and reactor design facilities. 400 Area houses the Fast Flux Test Facility (FFTF), an experimental, fast-neutron reactor used for irradiating breeder reactor fuels. Administrative and support services for Hanford are located in 1100 Area and 3000 Area.[4]

The American Nuclear Weapons Complex

Throughout its nuclear history, Hanford has been first and foremost an integral contributor to the nuclear weapons complex of the United States.[5] It is part of an industrial empire that is a "collection of enormous factories devoted to metal fabrication, chemical separation processes, and electronic assembly. Like most industrial operations, these factories have generated waste, much of it toxic" (Office of Technology Assessment 1991, 3–4). These facilities are scattered throughout the country, from the West Coast to the East Coast of the United States, and have included nuclear testing sites ranging from the Aleutian Islands to Bikini Atoll in the South Pacific (figure 2.2). By most accounts, the creation of nuclear weapons and the development of the American nuclear arsenal was one of the most expensive projects ever undertaken by mankind.

The Manhattan Project unleashed the nuclear genie and created the foundations for the American nuclear weapons complex. Hanford was to be the main supply source for plutonium and enriched uranium. The Oak Ridge Reservation in Tennessee also produced plutonium and enriched uranium for nuclear warheads; the bomb dropped on Hiroshima in August 1945 was constructed from materials produced at Oak Ridge. The research and development work, and the actual construction of the earliest nuclear weapons, was done at Los Alamos National Laboratory in New Mexico. It was at Los Alamos that Robert Oppenheimer, Edward Teller, Hans Bethe, and other scientists solved the theoretical and practical problems of building nuclear weapons (Rhodes 1986; Nichols 1987). All three of these facilities were key components of the Manhattan Project, and as the U.S. nuclear weapons program grew over the years, so did these facilities (table 2.1).

The primary research and development site for the nuclear weapons complex was the Los Alamos National Laboratory. As the nuclear weapons program expanded, a second design center was established in

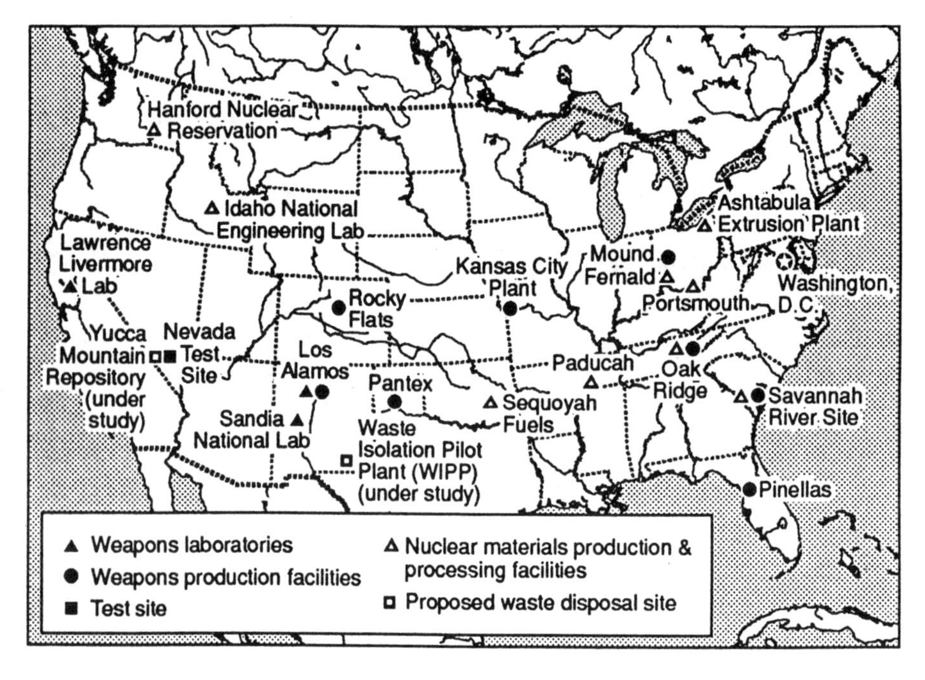

Figure 2.2
Nuclear weapons testing and production sites in the United States
Source: Makhijani, Hu, and Yih 1995; reprinted with permission.

1952 at the Lawrence Livermore National Laboratory in northern California. Lawrence Livermore's role focused on warhead design and development. Both Los Alamos and Lawrence Livermore are administered by the University of California.[6] In 1945, the Sandia National Laboratory was established in Albuquerque, New Mexico; its primary mission was to design non-nuclear components of nuclear weapons, such as electrical systems and firing systems. All three laboratories continue to function, though they have reoriented a portion of their research to non-weapons applications.

In addition to Hanford and Oak Ridge, a third facility was established in the 1950s at Savannah River, South Carolina, to produce plutonium and tritium, and to provide additional research and development activities. Yet another source of plutonium and other materials was the Idaho National Engineering Laboratory, where the PUREX process was used to recover plutonium from spent reactor fuel. The initial source of enriched

Table 2.1
Major American nuclear weapons facilities

Facility	Activity	Location
Ashtabula Extrusion Plant	Uranium Processing	Ashtabula, OH
Buffalo Works	Weapons Production	Buffalo, NY
Feed Materials Procession Plant	Uranium Production	Fernald, OH
Hanford Reservation	Plutonium Production	Hanford, WA
Idaho National Engineering Laboratory	Reprocessing Facility	Idaho Falls, ID
Lawrence Livermore National Laboratory	Design Laboratory	Livermore, CA
Los Alamos National Laboratory	Design Laboratory	Los Alamos, NM
Nevada Test Site	Test Site	Nevada
Oak Ridge Reservation	Uranium Enrichment	Oak Ridge, TN
Paducah Gaseous Diffusion Plant	Uranium Enrichment	Paducah, KY
Pantex Plant	Weapons Production	Amarillo, TX
Portsmouth Gaseous Diffusion Plant	Uranium Enrichment	Piketon, OH
Rocky Flats Plant	Fabrication of Plutonium Components	Golden, Co
Sandia National Laboratory	Design Laboratory	Albuquerque, NM
Savannah River	Plutonium Production	Savannah River, SC

uranium for the construction of nuclear weapons was the Oak Ridge facility. In short order, additional production facilities were developed at Paducah, Kentucky, and at the Portsmouth Uranium Enrichment Complex in Ohio.

The actual production of nuclear materials into weapons involved several different facilities with redundant specialities. The Rocky Flats plant in Colorado fabricated plutonium; the Feed Material Production Center at Fernald, Ohio (near Cincinnati), converted uranium materials into bomb products. The various weapons parts were shipped to the Pantex plant in Texas for assembly; the disassembly of retired weapons also occurred at Pantex.

As the American nuclear weapons program scaled back in the 1980s and 1990s, many of these facilities have been significantly downsized, closed or been converted to other applications. As noted elsewhere in this chapter, Hanford's plutonium production reactors were closed down by 1988, and plutonium production activities have now ceased completely. By the 1980s, the United States had amassed a large surplus of enriched uranium. Consequently, the Oak Ridge enrichment facilities closed down production in 1987, and production ceased at Portsmouth in 1992. The remaining enrichment capabilities at Paducah are now used for civilian purposes. The Fernald plant was closed in 1989.

Even though the production of weapons-related material has been significantly scaled back, and the number of nuclear weapons in the American arsenal is being reduced, the environmental legacy of the nuclear weapons program is staggering (Office of Technology Assessment 1991; National Research Council 1995). Virtually all of the weapons production facilities have accumulated a legacy of accidental releases that have created severe environmental consequences. The combined cost estimates of environmental cleanup are staggering. A recent Department of Energy budget allocates $5.7 billion for environmental remediation and waste control at the various weapons facilities (Department of Energy 1998a), and the total expense of environmental cleanup may exceed the cost of producing these weapons in the first place.

Each of the different sites of the United States nuclear weapons complex had its distinct function, but they are all linked by common factors. The nature of the manufacturing processes were inherently waste-producing. The facilities also share a long history of emphasizing the urgency of weapons production in the interest of national security, but too often to the neglect of environmental considerations. There has also been a lack of knowledge about, or proper attention to, the consequences of environmental contamination. Finally, the nuclear weapons complex has operated in official secrecy for decades, with little independent oversight or meaningful public scrutiny. Thus the history we recount for Hanford echos an experience shared by the other facilities and by their neighbors (Makhijani, Hu, and Yih 1995).

Immediate Consequences of the Old Reality at Hanford

The expansion at Hanford coincided with the development of the Cold War and the successes of the Soviet bomb programs. The pressure to produce plutonium (and enriched uranium) was the driving force behind Hanford's expansion. Reports from the period and firsthand accounts are replete with examples of how feelings of urgency guided decision making (Gerber 1992a, ch. 2; Groves 1962). Moreover, it is clear that perceived production needs often led to decisions that minimized safety and/or environmental concerns. For example, plutonium production could not be restricted to days when weather was favorable for dissipating pollutants, and production was not always halted when concerns arose about the safety of plant workers or the environment. The centrality of Hanford to the production of plutonium during the Cold War, along with the secrecy and lack of knowledge about the consequences of nuclear production for human health and the environment, provided the root causes of the enormity of the pollution problems that were created.

The overall secrecy surrounding Hanford extended to worker health and safety. In order to keep the mission secret, the Manhattan Engineer District retained its own health files during World War II, and General Groves was the official responsible for "project health matters" (Jones 1985). During World War II, strict security regulations prevented disclosure of the existence of radioactivity to most employees (Gerber 1992a, 49).

Following World War II, various officials spoke out in behalf of the safety and health record of the Hanford facility. The Atomic Energy Act of 1946 obligated the Atomic Energy Commission to take the "necessary steps to protect life and property from hazards arising out of its work" (quoted in Gerber 1992a, 67). However, even though researchers extensively studied radiation-related issues, early public reports tended to downplay the health and environmental impact of the facility. In 1949, for example, the AEC issued a public report that assured readers that Hanford's discharge standards were such that ". . . no damage to plants, animals or humans has resulted" (Atomic Energy Commission 1949, 12–15). In 1957, the AEC praised Hanford's environmental monitoring programs by stating: "Minute quantities of radioactive contamination in air, vegetation, soil, surface water, and groundwater are detected by radiochemical means . . . All radioactive materials routinely detected beyond the plant perimeter are at or below one-tenth of the maximum permissible limits" (Atomic Energy Commission 1957, 197, 206–207).

The assertions by the Manhattan Engineer District and the AEC during and after World War II were false on both the protection of the environment and the protection of human health (Office of Technology Assessment 1991; Heeb 1994; Heeb and Bates 1994). Ironically, the stated government policy was one of open disclosure. The AEC called for "the dissemination of scientific and technical information relating to atomic energy . . . and [to] provide that free interchange of ideas and criticisms which is essential to scientific progress" (Atomic Energy Commission 1947, 15). In practice, however, relevant documents were routinely classified as "top secret" and subject to national security protection and legal status. In 1948, for example, only 2 of 158 Hanford reports were considered "publishable" (Gerber 1992a, 69). Moreover, the secret documents told a different story than the public record, especially as research evidence grew to discount early forecasts of the environmental impact of nuclear weapons production.[7] As early as 1944, Hanford scientists knew that the Columbia River and its drainage basin were being contaminated in ways that would enter the food chain from both air and water. There was a comprehensive program of monitoring milk, water, air, plants, wild and domestic animals, fish, and insects. But in the absence of public access to primary environmental records, and without open citizen discussion of issues of scientific uncertainty and environmental risk, the environmental impact of Hanford did not reflect in either its image in the press or in official, public testimony before the U.S. Congress.

With the release by the Department of Energy of previously secret documents (which it began in 1986), and from other recent research, it is possible to reconstruct at least some of the production practices, as well as the deliberate "coverup" at Hanford, which resulted in monumental harm to the environment. It is useful in understanding the scope of contamination to identify categories of deleterious activity: airborne contaminants, river contaminants, and groundwater contaminants.

Airborne Contaminants

One source of radiation release was the processing of nuclear fuels. When the irradiated fuel elements from the nuclear reactors were placed in acid solutions as part of either the REDOX or PUREX process, airborne releases, such as iodine, strontium, cerium, ruthenium, and other elements resulted. During the early production years, measurable amounts of uranium and plutonium also escaped from Hanford. Some of the "hot particles" were highly radioactive and long-lived: ruthenium-106, strontium-90, plutonium-239, and cerium-144. Current estimates gauge that more than several thousand curies of these particles were released into the air during Hanford's production period, and some traveled beyond the Hanford Reservation boundaries.[8]

The greatest volume of radiation, and thus the greatest source of concern, involved iodine-131. The Hanford Environmental Dose Reconstruction Project (HEDR) now estimates that over the length of Hanford's plutonium production, between 630,000 and 980,000 curies of iodine (I-131) were released into the air between 1944 and 1972 (Hanford Health Information Network 1997).[9] To understand these statistics in context, the reactor accident at Three Mile Island was considered a major accident in the U.S. nuclear power industry—but it released only 14 curies of radiation.

The Hanford Environmental Dose Reconstruction Project used a detailed account of Hanford's production records to estimate the yearly iodine releases from 1944 until 1972. Releases were quite high in the earlier years, but as new technologies were developed for the reprocessing facilities, and the REDOX (1952) and PUREX (1956) facilities came online, these new plutonium processing procedures minimized airborne releases (figure 2.3). The airborne releases are considered a major potential health threat to the downwind populations inhabiting the area during these years because I-131 can concentrate in human thyroid glands, giving rise to cancer, hypothyroidism, thyroid nodules, and other health risks. Hanford researchers repeatedly found much higher than allowed levels of I-131 in the flora and fauna surrounding the facility (Gerber 1992a, 86–87, 89–90).[10]

In 1999, the Hanford Thyroid Disease Study reported the results from their nine-year study conducted by the Fred Hutchinson Cancer Research Center. The study evaluated whether thyroid disease was increased in a sample of 3,441 people who had been exposed as children in the 1940s and 1950s to radioactive iodine from the Hanford Nuclear Site. Dr. Scott Davis, the director of the study, reported that "We found no evidence

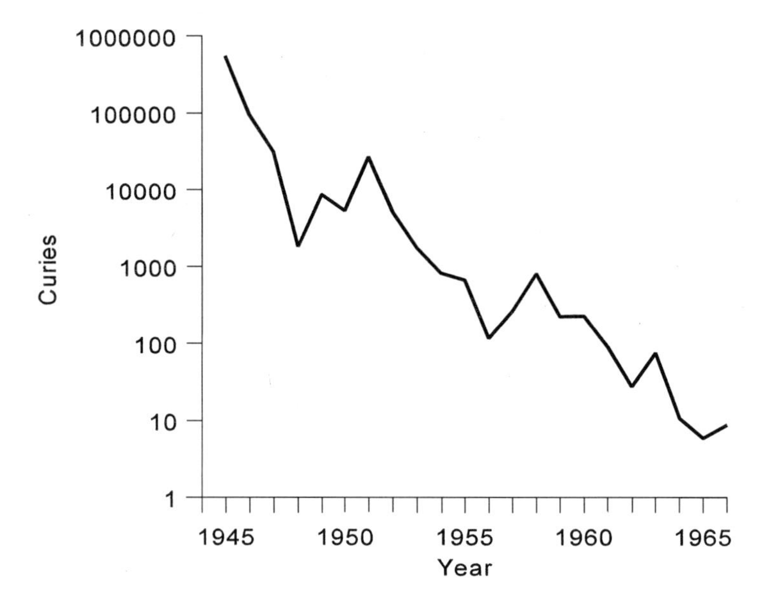

Figure 2.3
The amount of I-131 in airborne releases (logarithmic scale), 1944–1966
Source: Heeb (1994, table S1).

that the number of people with disease increased significantly in relation to radiation dose" (Hanford Thyroid Disease Study 1999, 1).

More ominous was the infamous Green Run of 1949. This event was a planned release into the atmosphere of 7,800 curies of I-131 along with 4,750 curies of Xe-133 on December 2–3, 1949 (Gerber 1992a, 91). The U.S. government now acknowledges that one of the purposes of the Green Run was to simulate the production process at Mayak, so the U.S. intelligence services could better use aerial surveillance to estimate Soviet plutonium production.[11] Instead of cooling fuel for an extended period before processing, as was the American practice, the plutonium was processed "green" after only a nineteen- to twenty-day cooling period. This meant a larger amount of I-131 and Xe-133 was released into the atmosphere. In addition, pollution control devices (scrubbers) on the processing plants were not operated during the Green Run.[12]

The release did not go as planned (Harlan, Jenne, and Healy 1950; Faden 1995, ch. 11). Rain, snow, and wind conditions concentrated the fallout on surrounding communities and farms. Instead of dissipating across a wide area of the Pacific Northwest, most of the radiation settled in areas immediately downwind of the facility. The radioactive iodine released by the Green Run covered extensive portions of eastern Washington and central and eastern Oregon.[13] Even worse than the failed experiment was Hanford's response. Despite the known risk that this accident posed, and high radiation readings obtained from the test (Faden 1995, ch. 11), Hanford did not notify local residents or local government agencies of the planned release or the adverse effects that might result from the adverse weather conditions.

River Contaminants

The eight original nuclear reactors built along the Columbia River featured "single-pass" cooling systems; water from the Columbia ran straight through the reactor cores as the principal means of temperature control. Once "passed through" a reactor, the river water settled in huge concrete holding basins so that radioisotopes of short half-lives would decay. Despite this precaution, significant quantities of radionuclides and nuclear debris reached the river. For instance, when a fuel rod leaked, radioactive material would escape into the river via the cooling process; Heeb and Bates (1994) report nearly 2,000 fuel element failures at Hanford. In addition, heavy elements contained in the river water, such as phosphorus, were irradiated as the water passed through the reactor, then were discharged into the Columbia. As the pressure for plutonium production increased, the holding time for cooling water was shortened, and thus the amount of radioactive releases increased. Finally, the reactors' cooling systems were periodically purged, which discharged accumulated radioactive wastes from the systems.[14]

These activities resulted in significant releases of radioactivity into the river. By the mid-1950s, the "single-pass" reactor cooling system was releasing 7,000 curies a day into the Columbia River even after impounding the water (Gerber 1992a, 123). By 1963, Gerber reports (1992a, 125), the average daily release of radionuclides exceeded 14,500 curies per day. An estimated "all-time high" for release of radioactivity was reached in April 1959 at 20,300 curies, though the daily release normally ran around 18,500 curies for the period between August 1957 and November 1959 (Gerber 1992a, 125). (This particular form of radioactive pollution of the Columbia River ceased by 1971 with the closure of the last of the eight "single-pass" reactors.)

In 1997, there was an updated report by the Hanford Environmental Dose Reconstruction Project on estimates of releases into the Columbia River from the Hanford Reservation during the period 1944–71 (Hanford Health Information Network 1997). Table 2.2 reports the amount of curies released and the half-life of each radionuclide. The HEDR Project indicates that eating fish and shellfish was the main pathway of human radiation exposure from Hanford's discharges into the Columbia.

Much of this radioactive material was composed of elements with short half-lives or minimal environmental consequences. However, the accumulated releases of five key radionuclides were staggering, totaling more than 20 million curies (Heeb and Bates 1994: Table S1). In a claim that was made in 1945 and found subsequently to be incorrect, Herbert Parker, head of the Health Instruments Section of the Manhattan Engineer District, declared that radioactivity in the Columbia River was limited "to the local area near the plant" (Parker 1945). A substantial portion of the releases involved elements that could accumulate in the river or the plants and animals of the river. Fish studies conducted in 1946 found that radioactivity became concentrated six to thirty times higher in fish

Table 2.2
HEDR estimates of radiation releases into the Columbia River, 1944–1971

Radionuclide	Amount released (Curies)	Half-life
Sodium-24	13,000,000	15 hours
Phosphorus-32	230,000	14 days
Scandium-46	120,000	84 days
Chromium-51	7,200,000	28 days
Manganese-56	80,000,000	—
Zinc-65	490,000	245 days
Gallium-72	3,700,000	14 hours
Arsenic-76	2,500,000	26 hours
Yttrium-90	450,000	64 hours
Iodine-131	48,000	8 days
Neptunium-239	6,300,000	2.4 days

Source: Hanford Health Information Network (1997).

bodies than in the water (Gerber 1992a, 117). Hanford officials periodically considered bans on fishing, hunting along the river, and the use of untreated river water because of health concerns—but no action was ever taken.

From the perspective of hindsight, the introduction to this previously secret memorandum is instructive. It indicates how the Health Instruments Section leadership understood the process of releases of radioactivity into the river. Parker's report read in part:

> The process waste waters become temporarily radioactive, and are held up in the Retention Basins until the activity is mainly exhausted. The activity of the water is measured at three points in each basin—the inlet end, an intermediate point, and the exit end. The reading at the exit end defines the safety of the water, and the inlet and intermediate readings are used only to detect a possible flow of excessively active water before it reaches the exit, so that the process could be stopped and the water held up longer. This has never been necessary in practice. The water released from the basins has always been considered safe for humans or fish to swim in. It is further diluted before return to the Columbia River by mixing it with other wastewater free from activity. The water returns to the river by underwater pipes to be released in mid-river and have the maximum chance to mix with the river water to further reduce the already safe concentration of radioactivity (Parker 1945, 1).

In this same secret memorandum, Parker observed: "It is understood that published evidence indicates that fish are less sensitive to radiation [than people]" (Parker 1945, 1). A series of other fish, algae, and animal studies confirmed that living organisms in the food chain concentrate radioactivity to a great extent, some 400 to 3,700 times greater than that of the river water.

Reports such as these were kept secret in the interest of national security, and because the researchers did not believe that the radiation levels in these animals posed an immediate threat to human health. In addition, environmental impact assessments were not always reconsidered when Hanford's waste production increased as a consequence of the nuclear arms race of the 1950s and 1960s.

The environmental impact assessments at Hanford also did not take into account the special situation of the Native American populations in the region. Three tribal governments, and their populations, are located in the region and directly involved in Hanford issues: the Confederated Tribes and Bands of the Yakama Indian Nation, the Confederated Tribes of the Umatilla Indian Reservation (CTUIR), and the Nez Percé Tribe. Until the mid-1800s, the aboriginal territory of these tribes included land in eastern Washington State including Hanford. In 1855, the U.S. government entered into a treaty that established reservations for each tribe. The Yakama Reservation is on 1.4 million acres west of Hanford along the Yakima River; the Umatilla Reservation is 172,000 acres in southeastern Washington and northeastern Oregon; the Nez Percé Reservation of 138,000 acres is near the Idaho-Washington border. These tribes still retain fishing, hunting, gathering, grazing, and other rights in the Hanford region under the 1855 Treaty.

The resources of the Columbia River were a central feature of self-definition for these Native American tribes. The Hanford facility lies within their traditional homelands, and these lands hold special cultural and religious significance to them.[15] Furthermore, many Native Americans were dependent on salmon and other fish from the Columbia River as a basic food source, so increased concentrations of contaminants in the river and its fish would pose greater risks to them. All three Native American nations in the region have been highly critical of the Columbia River Impact Evaluation Plan that assessed the environmental risks to the river and proposed actions to address these problems (O'Connor et al. 1995).

Groundwater Contaminants

The production of plutonium involves the creation of large quantities of low-level, liquid radioactive wastes as by-products of the plutonium extraction process. Hanford chose to discharge these liquid wastes directly into the permeable soil (high-level wastes, discussed later, went into underground storage tanks). In 1948, Herbert Parker characterized the decision to dispose of low-level liquid wastes in the ground in Hanford's 200 Area (location of the chemical separation complexes) to be a "temporary but necessary expedient . . . pending the development of intrinsically safer and more effective means of ultimate disposal" (Parker 1948, 9). Parker indicated that "provisional studies during the war indicated that it was feasible to discharge less active wastes to ground to avoid absurd costs on tank storage, evaporation equipment, or equivalent" (Parker 1948, 3). The report concluded with the observation: "Present disposal procedures may be continued, in light of present knowledge, with the assurance of safety for a period of perhaps fifty years" (Parker 1948, 9).

Parker's study reported some data that might have led to calls for further research (and, perhaps, to alternative conclusions) if they had been publicly released at the time. For example, Parker reported that plutonium contamination "has nowhere penetrated to a depth of more than 25 feet from the point of entry, although it has traveled laterally to about 200 feet" (Parker 1948, 3). Fission product contamination (unspecified) was found to have penetrated downward as much as 100 feet in two years in one case; lateral penetration up to nearly 300 feet had occurred. Perhaps the most disturbing finding was the following: "In one special case of uranium disposal, rapid percolation to groundwater was noted" (Parker 1948, 3). This finding was not further explained as to circumstances or implications.

The planners of the Manhattan Engineer District apparently were not able to factor into their calculations the extreme variability in the water table lying under the Hanford Reservation in their strategy for disposal of low-level liquid wastes. While the average level is 250 feet below the surface, the pattern fluctuates from a few feet to over 300 feet (Gerber 1992a, 146), with the aquifers flowing generally toward either the Columbia or the Yakima River. Therefore, projecting the statistics from the 1948 Parker report to the long-term practice of dumping nuclear waste into the soil, one might have forecast that the migration of the waste plume toward the adjacent rivers would become a real problem.

There were other potential approaches to disposing of low-level, liquid radioactive waste: dumping it into trenches, or into dry shafts in the ground (the so-called "reverse wells"). Extensive tests were done on soil absorption and the movement of liquids underground, particularly in the period between 1947 and 1950. But the studies were kept secret like those of air and river contamination. Moreover, the magnitude of waste discharge was expanding rapidly. By March of 1951, 5.5 billion gallons had been discharged into the ground. Between 1951 and 1955, an additional 20 billion gallons were added. Between 1956 and 1960, the volume dumped in Hanford soils was 32 billion gallons. Hanford was floating on a sea of liquid wastewater.

The peak of release of low-level liquid wastes into the Hanford soil occurred between 1959 and 1965, with the total exceeding 32 billion gallons. From 1966 through 1970, an additional 27 billion gallons was discharged into the soil (Gerber 1992a, 159, 162). Despite the high volume of discharge, a series of AEC reports in the 1960s and 1970s generally endorsed this method of disposal.

In one early report, Hanford scientists indicated that the time of movement of groundwater from the 200 Area (chemical reprocessing facilities) to the Columbia River was "estimated as varying from 50 years to 1,500 years, depending upon the initial location of the water" (Healy 1953, 3). This study by J. W. Healy from the General Electric Company addressed the important long-term question of the impact of the release of radioactive wastes directly into the soil. Healy was either tentative in his conclusions or withheld judgment on important matters, such as identifying key areas where data was lacking, acknowledging that information about the content of past wastes was uncertain, and calling for additional research. Nevertheless, he concluded that "some increase in the levels cribbed may be allowed" (Healy 1953, 2).

The Healy study was completed in 1953. The disposal of low-level waste into the soil significantly increased in the years immediately following even though Healy had clearly indicated that such action would be premature. It is likely that without the government's policy of total secrecy, independent scientists would have recommended further areas for data collection and analysis, and possibly alternatives to disposal into the soil.

Another declassified report indicates that William Bierschenk also conducted a study of waste disposal and groundwater in 1958. He estimated that since 1944, 3.8 billion gallons of intermediate-level wastes had been discharged into the ground containing a total of 2.5 million curies of beta emitters. His estimate of travel time to the Columbia River was 175–180 years (Bierschenk 1959).

As the period of soil storage of low-level waste came to an end, the overall cumulative estimates were staggering: over 120 billion gallons of liquid, over 3.2 million curies of beta emitters, over 280,000 grams of plutonium, and over 117,000 kilograms of uranium (Essig 1971). The report by T. H. Essig of the Battelle Pacific Northwest Laboratories was compiled from secret or private (de facto secret) reports by four Hanford contractors: Atlantic Richfield Hanford Company, Battelle-Northwest, Douglas-United Nuclear, and the Westinghouse Advanced Development Company (WADCO). By way of summary, Essig reported that by 1970, "Releases of radionuclides to air and water from 100 and 200 Area facilities showed a continuing downward trend, reflecting the effects of production facility retirement and the results of programs aimed at reducing releases" (Essig 1971, 2). Even in 1970, after a series of procedures had been introduced to limit water pollution, the total introduction of radioactive wastes into the Columbia was considerable. Had the Essig report and its sources not been secret at the time, requests would have been forthcoming for a fuller explanation of the meaning of these releases for human health and the environment.

For the first time, however, other agencies of the U.S. government began to question the soundness of the strategy of ground disposal (Office of Technology Assessment 1991; National Research Council 1996). More recent research has documented the current extent of the problem. For example, according to the Office of Technology Assessment of the U.S. Congress, "Tritium and nitrate contamination has been found in plumes covering 122 square miles. Other pollutants have been detected in more localized groundwater areas at levels that exceed drinking water standards" (Office of Technology Assessment 1991, 150).

In 1998 the General Accounting Office released a report (GAO 1998) that indicated that the Department of Energy's understanding of waste migration was inadequate for making key policy decisions. At issue was the migration of highly radioactive waste to the groundwater from leaking underground storage tanks. It had been previously believed that the area above the water table—called the vadose zone—was an effective barrier between tank waste (and waste dumped into the soil) and the groundwater. In a letter of transmittal (Jones 1998), the GAO concluded that "The Department's understanding of how wastes move through the vadose zone to the groundwater is inadequate to make key technical decisions on how to clean up the wastes at the Hanford Site in an environmentally sound and cost-effective manner. For many years, DOE assumed that wastes would move slowly, if at all, through the vadose zone." The report questioned this assumption.

Storage Tanks for High-Level Waste

One hundred and forty-nine single-shell waste storage tanks were constructed at Hanford between 1944 and 1964. These tanks were designed to hold high-level radioactive wastes that were too dangerous for ground disposal. Confirmed tank leaks began in 1956. The high heat and caustic nature of the compounds proved too corrosive for the materials used in the original tanks, primarily carbon steel shells surrounded by outer, concrete envelopes. Through 1998, at least 67 of the 149 single-shelled tanks were documented as leakers (IPPNW 1992, 63). The mixture leaking into the soil from these tanks consists of radioactive constituents (e.g., cesium-137 and strontium-90) and chemical solutions including various nitrates, nitrites, and sodium compounds. Ferrocyanide was added to about twenty tanks in the 1950s to make cesium-137 settle to the bottom.

The largest reported leaks by volume in the 1950s consisted of 55,000 gallons from Tank 104-U in 1956. The largest leak by amount of radioactivity contained 20,000 curies from Tank 101-U in 1959. In the ten years beginning in 1962, eight significant tank leaks occurred (Gerber 1992a, 166). In 1971, 70,000 gallons containing 50,000 curies escaped into the environment from Tank 102-BX. In 1973, 115,000 gallons containing 40,000 curies of cesium-137, 14,000 curies of strontium-90, 4 curies of plutonium, and various fission products escaped from Tank 106-T (Fleming 1973, 1).

Tank 106-T is representative of the single-shell tanks in terms of the potential danger to the environment, and the problems of monitoring leakage from them.[16] A report on the leak of Tank 106-T illustrates the type of problem that affects the entire program:

> The exact cause of the leak is still unknown, and may never be determinable but it most probably resulted from corrosion of the aging (29-30 years old) carbon steel tank by the caustic waste solutions to which it had been exposed during its lifetime. . . . The immediate cause of the failure to recognize the leak earlier and minimize the quantities of fluid lost was the failure of ARHCO [Atlantic Richfield Hanford Company] employees to promptly review liquid level and radiation level data. (Fleming 1973, 4–5)

The 149 single-shelled tanks range up to 1 million gallons in capacity. As noted before, at least 67 of them are assumed to have leaked. If the single-shelled tanks to which ferrocyanide was added between 1954 and 1959 were to become extremely hot, an explosion or runaway chemical reaction could occur. As of November of 1997, about 5.7 million gallons of pumpable liquid remains in 30 tanks.

The introduction of 28 double-shelled tanks between 1968 and 1986 provided an additional storage capacity for high-level wastes; so far, these tanks have not produced any known leakage. Currently they contain 24 million gallons of mostly liquids with some sludges and salts. The waste in these tanks is estimated to contain 110 million curies of radiation.

Several thousand Hanford employees work on the problems associated with operating the tanks safely and preparing for the permanent disposal of the wastes. The Tank Waste Remediation System will continue to monitor and ensure the stability of the waste tanks. Liquids are gradually being removed from the single-shelled tanks; the long-range plans call for the solidification of the wastes in all these tanks and their disposal to another storage site. The storage of these high-level wastes will be a long-term concern at the Hanford site.

Another waste storage problem concerns the fuel from the N-reactor. Two water-filled basins were initially designed for interim storage of fuel from the K-reactor; these basins now hold nearly 2,100 metric tons of irradiated reactor fuel. During the 1970s, the basins leaked about 15 million gallons, containing an estimated 2,500 curies (Makhijani, Hu, and Yih 1995, 223). Similar problems exist with the fuel from the FFTF reactor. Hanford has constructed a new canister storage building to provide an alternative storage facility for this fuel. The fuel will be removed from the storage basins and placed in canisters, which will be moved to the new facility for dry storage. The severity of the environmental problems in the storage basins has delayed implementation of this plan until at least November of 2000 (Niles 1998). A permanent solution to the storage of these fuel elements has not been found, and continued use of temporary storage arrangements invites further accidents.

End of Production

As the American nuclear weapons complex expanded and modernized, and the American arsenal reached its upper plateau, the pressure to produce plutonium decreased. This was reflected as early as 1964 in President Lyndon Johnson's executive order authorizing the gradual shutdown of

Hanford's production reactors. In April of 1971, the last of Hanford's eight graphite reactors was closed. The PUREX chemical separation plant went on a standby mode in June of 1972 because of a lack of irradiated fuel to reprocess.

This slowdown process was reversed in the Fall of 1983 when the nuclear weapons build-up of the Reagan administration increased plutonium production. The N-reactor switched to plutonium production, and the PUREX plant resumed operation. This proved to be a temporary development, however. Plutonium production again came to a halt in 1988, and the PUREX facility was placed on standby status.[17] The end of the Cold War subsequently formalized this change; it marked the end of Hanford's era of production. The N-reactor and the PUREX plant have been closed, and the military plutonium production mission has ended.

The official mission of Hanford has changed from producing nuclear materials for the American weapons program to addressing the consequences of the accumulated environmental damage done to the area. This mission represents both a monumental redirection of efforts and a monumental challenge. Hanford contains more than 1 million cubic yards of waste from plutonium production. This is about two-thirds, by volume, of all the high-level nuclear wastes in the United States, and about half of the nation's total transuranic waste (Office of Technology Assessment 1991). Environmental remediation activities at Hanford will last well into the next century. Whether the technology and the funding will be there to resolve these threats to the environment and public health are major uncertainties of Hanford's new reality.

The New Realities at Hanford

The process of change from the old to the new realities at Hanford was not abrupt—either in time or in the perceptions of nuclear experts, the environmental community, or the general public. There are also multiple definitions of the new realities. The following discussion will identify the new realities and the multiple perceptions of them.

The decade between 1983 and 1993 was extraordinarily important for Hanford due to specific events, disclosures, and developments. Local environmental activists petitioned under the Freedom of Information Act for secret documents concerning Hanford's release of radioactive materials and for research on the environmental impact of Hanford's activities. In February of 1986, the Department of Energy released an initial 19,000 pages of previously secret documents. The disclosures had a dramatic impact as people learned that the chemical and radioactive wastes that had been released into the Columbia River ecosystem far exceeded previous disclosures.

The news galvanized the environmental community, the general public, and local, state, and national governmental officials (see chapter 5). The State of Washington formed a Hanford Historical Documents Review Committee, which, among other tasks, prepared abstracts of the released documents. The Hanford Health Effects Review Panel (formed by the Indian Health Service, the Yakama, Umatilla, and Nez Perce Nations, and the states of Oregon and Washington) mobilized resources from the U.S. Centers for Disease Control and Prevention and the Washington State Department of Social and Health Services to calculate the doses of radiation potentially received by the affected populations. These efforts led to the previously discussed Hanford Thyroid Disease Study and the Hanford Environmental Dose Reconstruction Project. Another new reality is the Hanford Federal Facility Agreement and Consent Order, known as the Tri-Party Agreement (TPA) (Department of Ecology 1989), which was concluded in May 1989 among the U.S. Department of Energy, the State of Washington's Department of Ecology, and the U.S. Environmental Protection Agency. It committed the signatories to a scheduled cleanup of the Hanford site, assigned a major oversight role and joint decision making to the Department of Ecology, and required continued involvement by the general public. In May 1991, the parties agreed on modifications, but the TPA's structure still provides direction for the cleanup work at Hanford.

The Tri-Party Agreement represented a dramatic departure from how Hanford had been governed previously. The generation, treatment, storage, and disposal of hazardous waste would no longer be secret and would be administered publicly by the government. Hanford's environmental cleanup would be regulated by the Department of Ecology of the State of Washington, and the DOE was mandated to comply with state and federal law. Before beginning any cleanup work, the DOE must notify both the EPA and the Department of Ecology of the scope of work to be conducted. The Hanford site was also subject to the Resource Conservation and Recovery Act (RCRA) of 1976 and the Comprehensive Environmental Response, Compensation, and Liability Act of 1980 (also known as Superfund). Together these laws focus on reducing the generation of hazardous waste and conserving energy and natural resources (RCRA), and on assessing contaminant releases (both radioactive and hazardous) from abandoned waste sites (Department of Ecology 1989, 1–7). The TPA includes a provision that if the DOE believes it cannot comply with some provision of the TPA based on some alleged inconsistency with the Atomic Energy Act of 1954, the Department of Ecology can seek judicial review in the U.S. District Court in Spokane.

The general purpose of the Tri-Party Agreement was to "ensure that the environmental impacts associated with past and present activities at the Hanford site are thoroughly investigated and appropriate response action taken as necessary to protect the public health, welfare and the environment" (Department of Ecology 1989, 7). This is a fundamental departure from the rationale employed for past activities. It removes the secrecy provisions that previously shielded Hanford's actions and makes the facilities subject to state oversight and applicable state and federal environmental regulations. The TPA also provides for multiple channels of public involvement in agency decisions regarding Hanford's cleanup.

Another new reality at Hanford is that in 1989, the then-Secretary of Energy, James D. Watkins, released a five-year plan for environmental restoration and waste management at DOE's nuclear-related waste sites, as well as a priority list for steps to be taken for achieving compliance with applicable environmental regulations (Department of Energy 1989). Several points made by Secretary Watkins in his introduction are important to understanding more recent activities at Hanford. There was an expressed willingness to develop a formal agreed-upon methodology for placing into priority its remedial activities (Department of Energy 1989). At the same time, the statement cautioned that the technology available for environmental restoration is not sufficiently mature or cost-effective to assure meeting the Department's goals or to efficiently use public resources.

The five-year plan itself is 441 pages long. Two elements deserve comment. First is the commitment to change the DOE's organizational culture, and the specific characterization of environmental contamination at Hanford. The report described the DOE culture as having grown out of a "strong AEC culture devoted in large part to the national defense mission to produce nuclear materials for nuclear weapons" (Department of Energy 1989, 6). The characterization of the culture is noteworthy, and the commitment to its modifications unprecedented. Second is the commitment to fundamentally address environmental restoration and waste management issues in order to minimize waste generation and to conduct efficient disposal in a spirit of environmental stewardship. One of the factors determining Hanford's future is whether this "commitment to changing DOE's culture" can be achieved.[18] We will return to this theme later (see chapter 10).

Another key factor for Hanford's future was the plan for environmental cleanup. Each of the sites at Hanford were evaluated in terms of eight factors: extent and type of contamination, regulatory drivers (such as the Tri-Party Agreement or the Resource Conservation and Recovery Act), the regulatory authorities, health risks, milestones, funding by program, and funding by priority level. The regular updates of this document are important components of the new reality at Hanford. Particularly relevant are the characterization of health risks and the assignment of priorities to known problems, and the assessment of progress.

Finally, the new reality of Hanford also includes new relations with the Native American tribes in the region. The 1855 Treaty provided the Yakama, Umatilla, and Nez Perce with continuing rights of usage of the Hanford area for fishing, hunting, gathering, and other traditional activities. The 1982 Nuclear Waste Policy Act established the DOE precedent of formal consultation with these tribes over the possible use of Hanford as a nuclear waste repository. Subsequent legislation on Hanford's waste management and environmental restoration, and DOE's American Indian policy, have institutionalized the tribes' consultation role.[19] Representatives of each tribe participate in various Hanford advisory groups, such as the Hanford Advisory Board, the Hanford Natural Resources Trustee Council, and the Hanford Cultural and Historic Resources Program. Furthermore, as federally recognized tribes, each can claim government-to-government rights to consultation and consent on the activities that occur at Hanford, including the processes of environmental cleanup (Liebow, Younger, and Broyles 1987). Thus the Yakama, Umatilla, and Nez Perce have become significant actors in discussions and decisions about Hanford cleanup activities.

Toward the Future

This change in its official mission is far from the end of Hanford's history. Indeed, we have described in this chapter just the beginning of Hanford's new reality—and the challenges it faces in achieving this reality. Subsequent chapters will continue this story, discussing how citizens responded to the new information on Hanford's environmental impact and how the government responded to the challenges of Hanford's new reality.

If Judge Hanford were to return to the land he and his fellow investors in the Priest Rapids Irrigation and Power Company had purchased in the early years of the twentieth century, he would find that land prices have appreciated and major railways still serve the region. A considerable portion of the $1 trillion dollars invested by the federal government in the nuclear weapons complex at Hanford has produced a product that he would not have recognized. At the dawn of the twenty-first century, he would find a debate not about what the next crop should be, but how to protect the health, safety, and environment for thousands of years into the future from what was the last crop on Hanford land: plutonium and its lethal by-product, nuclear waste.

If General Leslie Groves were to return to Hanford, he would find that what he worked so hard to start, the production of plutonium, had ceased forever. In addition, the Cold War is also over forever. It would no doubt surprise him that environmental matters deferred fifty years ago probably would cost more to contain and stabilize than the original production cost of plutonium.

The descendants of the jack rabbits, burrowing owls, and rattlesnakes that are radioactive today will still be radioactive in thousands of years. There is ample evidence that the 100 and 200 areas at Hanford, for example, can never be made to return to the condition that existed before World War II. And it is highly likely that neither can between 150 and

200 square miles of contaminated groundwater. Evidence available to nuclear experts has coalesced around three long-term threats to safety, health, and environment: contaminated groundwater; volatile, high-level waste in leaking storage tanks; and corroding, spent nuclear fuel in the K-Basins along the Columbia River.

It is clear that the challenges for the coming decades are primarily social and political in nature—not just technical. The problem now is to devise new frameworks for building cooperation that will surmount decades of secrecy and of mistrust, all in the face of declining resources and daunting technical obstacles.

Nuclear waste has forever changed Hanford and its surrounding environment. What has not changed since 1942 is the vulnerability of the surrounding population to a contaminated nuclear environment. With a more complete understanding of the risks now, the next great challenge is summoning up the social and political will to make concrete progress on all the areas of environmental management that require attention at Hanford.

Notes

1. The conflicting claims by various information sources is part of Hanford's legacy. To the extent possible, we have attempted to rely on current information sources and generally accepted authoritative sources for information on the facility. One of the standard works is Gerber (1992a), who was the staff historian for the Westinghouse Hanford Company. Cochran et al. (1987) provides a detailed history about Hanford and the other U.S. weapons sites. More popular, and critical, accounts have been written by D'Antonio (1993), Sanger (1989), and Leslie (1993). Finally, in recent years, a number of new U.S. government studies that are cited below and in chapter 10 have attempted to provide accurate, nonclassified information about the facility and its past practices.

2. The A-reactor was never built because its design was not approved during the design phase.

3. The Washington Public Power Supply System (WPPSS) was to be a commercial nuclear power facility located on the Hanford Reservation. One reactor is now in operation. In 1983, WPPSS defaulted on its bonds, and two more reactors on the site remain unfinished. The WPPSS default is the largest American municipal bond failure in history, entailing $6.7 billion in principal and $23.8 billion in interest.

4. As a symbol of defense conversion, in 1997, the 70-acre Hanford 3000 area and buildings were transferred to the Port of Benton for economic development.

5. A substantial part of this material was drawn from the Office of Technology Assessment (1991); Makhijani, Hu, and Yih (1995); and Cochran et al. (1987).

6. Part of the University of California's contractual arrangement for managing the sites calls for a portion of the overhead to be used for research on peace and conflict research, administered through the Institute on Global Cooperation and Conflict (IGCC). The IGCC funded a portion of this research.

7. As we discuss below, early reports frequently made optimistic assumptions about the production parameters of the facility, or about the environmental impact of production processes. Subsequent research often yielded new knowledge that corrected these inaccuracies. Nuclear health and environmental research was an emerging field, so incorrect assumptions were to be expected. However, the implications of the new findings were often overlooked. For instance, decisions about disposal of liquid wastes were based on assumptions that later proved highly inaccurate, but the same disposal processes continued, despite new information that raised doubts about them.

8. There is a "range of uncertainty" in the estimates of air releases from Hanford according to the Hanford Environmental Dose Reconstruction Project which is now under the Centers for Disease Control and Prevention (Hanford Health Information Network 1997). HEDR estimates a range in radiation (in curies) released into the air by Hanford from 1944–1972 as follows:

Radionuclide	Low	Average	High
Iodine-131	630,000	740,000	980,000
Ruthenium-103	330	1,200	4,100
Ruthenium-106	110	390	1,400
Strontium-90	23	64	180
Cerium-144	1,400	3,800	11,000
Plutonium-239	0.1	1.8	31.0

9. These statistics are substantially greater than Hanford implied before the 1986 release of documents, and have been gradually adjusted upward. For example, Gerber (1992a, 78) first estimated that 420,000 curies of I-131 were released by 1955; the HEDR estimates in 1991 were very similar; revised HEDR estimates in 1994 raised the total to 685,000 curies (Heeb 1994). In 1997, a further HEDR estimate produced a "range of uncertainty" (Hanford Health Information Network 1997).

10. The farming and Native American populations near Hanford were especially vulnerable because their diets included produce, milk, and livestock products that may have concentrated their exposure to radioactive particulates (Gerber 1992a, ch. 4).

11. An April 4, 1986, letter from Michael Lawrence, then manager of the Richland Office of the Department of Energy, to Representative John Dingell, chairman of the U.S. House of Representatives Committee on Energy and Commerce, stated that the Green Run was "related to development of a monitoring methodology for intelligence efforts regarding the emerging Soviet nuclear program"

(quoted in Thomas 1990, 6). This was also the finding of the Advisory Committee on Human Radiation Experiments (Faden 1995).

12. The Green Run was one of Hanford's worst radiation releases. It was an attempt to simulate the "normal" operating procedures of Mayak. This observation has ominous implications for the health effects of the radiation releases that were regularly occurring at Mayak.

13. Hanford's environmental reports often did not include the Native American populations. Even the first phase of the HEDR project did not include indigenous populations in its study. Moreover, because of their reliance on subsistence farming and cattle grazing, many Native Americans may have had higher exposure to radiation contaminants.

14. In addition, occasional accidents released other pollutants into the Columbia River. For instance, the Hanford Health Information Network reports that in 1948 a breach in a waste pond released 28 pounds of uranium into the Columbia. Gerber (1992a, 133–138) discusses other episodic releases into the Columbia.

15. In 1996 a 9,200-year-old skull was uncovered in Kennewick. Since then it has generated conflict over disposition of the remains, with the Umatilla tribe pressing for its reburial. It also underscored the Native Americans' early presence in the region.

16. The surface level of the waste in tank 106-T was 183.7 inches from the bottom of the tank on April 25, 1973. Despite the fact that regular monitoring of the surface level was done between April 25 and June 7, 1973, and the reports of that monitoring indicated that the level of the surface had dropped below 149.2 inches, there was no recognition in that critical time frame that the tank had leaked over one-fifth of its capacity. Prompt identification of the leak could have minimized the environmental damage by pumping the contents into other tanks.

17. There was also a period in the late 1980s when Hanford was evaluated for consideration as a potential national repository for commercial, high-level nuclear waste (Dunlap, Kraft, and Rosa 1993). Hanford was removed from consideration in 1987.

18. Hugh Gusterson (1996) provides an insightful anthropological study of this corporate culture based on his research at the Lawrence Livermore National Laboratory.

19. DOE's American Indian Policy mandates a special government-to-government relationship with the tribes. Specific legislation, such as the Native American Graves Protection and Repatriation Act or the Archeological Resources Protection Act, require consultation with the tribes regarding the many tribal sites within the Hanford facility before DOE can take action.

3

The Compelling Realities of Mayak

John M. Whiteley

In the center of a rural town called Kyshtym, a digital sign flashes the time, the temperature and one more piece of information: the radiation level. Two years ago, residents saw a rose-colored cloud float through their village, which is about 10 miles from a nuclear-weapons plant [Mayak]. The digital sign flashed numbers above 200 microroentgens per hour—10 times the norm—then "broke."

The official response: Nothing untoward occurred. The dosimeter was fixed and now shows normal background radiation. And to this day, no one knows what really happened. This is Chelyabinsk, a region that for over 30 years has suffered through secret nuclear dumping and two deadly accidents . . . and now claims the unwilling distinction of being one of the most polluted spots on Earth.

Alina Tugend
Orange County Register
June 6, 1993

This chapter describes a region of Russia at the edge of Siberia east of the Ural Mountains, which has long been shrouded in the secrecy imposed by the Cold War nuclear arms race. It reports the history and developments spanning over half a century of the Soviet Union's first sourcing of plutonium for thermonuclear weapons. The incredible beauty of birch forests, mountains, rolling farmland, and lakes has been as hidden from outside view as the nuclear pollution from undisclosed nuclear accidents, explosions, and direct discharge of nuclear waste into the environment.

With the end of the Cold War, the region has been in a period of major economic depression. After describing the region and recounting the development of the enterprise Mayak—variously known in its history as Chelyabinsk-40 and Chelyabinsk-65 (after its post office box addresses)—this chapter will introduce the principal legacies of the environmental consequences of nuclear development.

Chapter 3 is organized into seven sections. Section one provides an overview of nuclear weapons production facilities in the USSR, and Mayak's place in that system. Section two characterizes the Chelyabinsk region of Russia, home to the Mayak facility. Section three describes the nuclear facilities of the Mayak Chemical Combine. Although the plutonium production reactors are closed, Mayak is still very much in the nuclear business, and has a business plan with many "operational units." Section four details the series of nuclear waste tragedies and accidents at Mayak that have made it so infamous with environmentalists and antinuclear activists. Section five presents what is publicly known about the plan that was formulated in 1993 (and earlier) for Mayak to continue as an active nuclear enterprise. Information in section five is unevenly known within Russia and highly controversial within the environmental community. Section six discusses the central elements of Mayak's problematic future, and section seven details the changes that the principal founder of Mayak, Igor Kurchatov, would confront if he returned today.

The Soviet Nuclear Weapons Complex

Since the waning days of World War II, the USSR had made the development of nuclear weapons a national priority.[1] Nuclear weapons were one of the bases for the Soviet Union's superpower status, and the guarantors of the USSR's military security.

The initial development work was conducted at the All-Union Scientific Research Institute of Experimental Physics (commonly known as Arzamas-16) under the direction of Igor V. Kurchatov. Arzamas-16 was the functional equivalent of the Los Alamos National Laboratory in the United States. The first Soviet atomic bomb, which exploded on August 29, 1949, was designed and constructed at Arzamas-16, as was the first Soviet hydrogen bomb, which was detonated on November 22, 1955.

To produce plutonium and other nuclear elements for the weapons program, the USSR created the Mayak Chemical Combine (Chelyabinsk-65) in the Southern Urals, about 60 miles from the industrial city of Chelyabinsk. Mayak was the initial and primary producer of nuclear material for the Soviet weapons program—the Soviets' equivalent of Hanford. Construction started in 1945, and Mayak's first plutonium production reactor became operational in 1948. Close to the Mayak facility, the USSR established the All-Union Scientific Research Institute of Technical Physics (called Chelyabinsk-70). Chelyabinsk-70 was a weapons research and design institute; it also houses facilities for high-energy testing.

The facilities at Chelyabinsk-65 and Chelyabinsk-70 were just two elements of a large complex devoted to nuclear weapons production in the former Soviet Union. As the USSR's nuclear weapons program grew, other facilities were established (figure 3.1). In all, at least a dozen major facilities were devoted to different aspects of the Soviet nuclear weapons program. The convention was to name a facility according to the nearest city, using a post office box number or some other numeral to signify it. Most of these facilities were associated with a "closed city" that housed employees and their families. The residents of the closed cities had amenities that were significantly better than those enjoyed by the Soviet population at large.

Each facility had a specialized function in the weapons production process (table 3.1). Nuclear weapons design was done at Arzamas-16 and

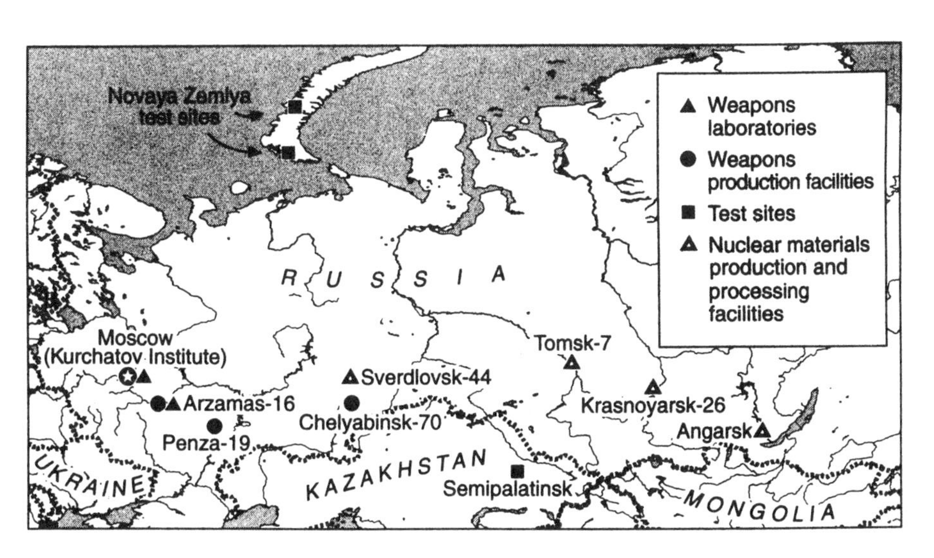

Figure 3.1
Major Soviet nuclear weapons production and testing sites
Source: Makhijani, Hu, and Yih 1995; reprinted with permission.

Table 3.1
Major Soviet nuclear weapons facilities

Facility	Activity	Closed city (1990 pop.)	Region
Angarsk	Uranium Enrichment	—	Siberia
Arzamas-16	Design Laboratory Weapons Production	Kremler (80,300)	Mordovia
Chelyabinsk-65	Plutonium Production Fabrication of Plutonium Components	Ozersk (83,500)	Urals
Chelyabinsk-70	Design Laboratory	Snezhinsk (46,300)	Urals
Krasnoyarsk-26	Plutonium Production	Zhelenogorsk (90,300)	Siberia
Krasnoyarsk-45	Uranium Enrichment	Zelnogorsk (63,300)	Siberia
Novaya Zemlya	Test Site	—	Arctic
Penza-19	Weapons Production	Zarechnoye (61,400)	
Semipalatinsk-21	Test Site	—	Kazakhstan
Sverdlovsk-44	Uranium Enrichment	Novouralsk (88,500)	Siberia
Sverdlovsk-45	Weapons Production	Rusnoy (54,700)	Siberia
Tomsk-7	Uranium Enrichment Plutonium Production Fabrication of Plutonium Components	Seversk (107,700)	Siberia
Zlatoust-36	Weapons Production	Torifuginuy (29,800)	Urals

Source: Cochran and Norris (1993b); Donnay et al. (1995).

Chelyabinsk-70. An initial step in the weapons production process is the enrichment of uranium. In the United States, the primary facility is at Oak Ridge and the Paducah, Kentucky plant. The major Soviet centers for the production of enriched uranium were Krasnoyarsk-45, Sverdlovsk-44, and Tomsk-7. As production needs increased, Krasnoyarsk-26 and Tomsk-7 were developed in addition to Chelyabinsk-65 as centers producing plutonium and other weapons-related materials. The major testing site for nuclear weapons was in Semipalatinsk-21; this is equivalent to the Nevada test site in the United States. Semipalatinsk-21 was the site of the first Soviet nuclear explosion, and much of the later testing activity took place at this facility (it was closed in 1991, when Kazakhstan declared its independence). Novaya Zemlya in the Arctic Sea was the other major nuclear weapons testing site.

As figure 3.1 shows, these facilities are spread across the Soviet Union. (Within only a decade, most of these facilities and their closed cities were missing from official maps of the Soviet Union.) Several facilities are located in the Southern Urals, in part because of Mayak's initial placement in this region and in part because of the concentration of heavy industry there. The map also shows how weapons plants are often clustered together because of their interrelated activities and their dependence on a common scientific and technical expertise. Two facilities are located near Chelyabinsk, two near Krasnoyarsk, and two near Sverdlovsk, while complexes at Arzamas-16 and Tomsk-7 have multiple functions.

Precise statistics on the total extent of the nuclear weapons program for the Soviet era are understandably difficult to obtain. By one account, approximately 1 million people were directly employed in nuclear weapons production at the end of the 1980s. A total of population estimates for ten secret cities in table 3.1 alone accounts for almost 800,000 people. Add to that individuals employed in uranium mining and processing, waste storage facilities, government ministries, and the military—a large number needed to maintain the USSR's big nuclear weapons program. It is estimated that the Soviet Union had more than 45,000 nuclear weapons at the peak of its inventory, a significantly larger stockpile than the United States both in terms of numbers and total nuclear material.

The Chelyabinsk Region of Russia

Chelyabinsk Oblast is a region the size of the State of Indiana. Located a thousand miles from Moscow east toward Siberia, it was closed to foreigners until early 1992. It has a population of over three million people with a major metropolitan center, the city of Chelyabinsk, that has over 1.2 million inhabitants located on the Trans-Siberian Railway.

Looking at a map of the Chelyabinsk region, the striking physical feature is the number of lakes, some 3,000, dotted throughout the landscape. That beauty was well captured in an announcement of an international children's ecology camp:

> The Urals mountain chain stretches along the border between Europe and Asia in the very heart of Russia. The Urals region is not only a geographical reference point, it is a special region of our country with a rich history and unique culture

and traditions. . . . The Urals region is beautiful in any season: in winter when the ground is covered with a white veil of snow, in summer when the depth of the azure sky competes with the unruffled surface of more than 3000 lakes, in spring when the first greenery appears and forest meadows are strewn with flowers, and in autumn when the frost flames with brilliant foliage in crimson and gold.

The invitation went on to describe the broader Southern Urals region as an important scientific, industrial, and cultural center for Russia. But it was beauty to which the text of the invitation returned:

. . . a picturesque part of the Southern Urals surrounded by ancient mountains covered with beautiful primordial forests. Pine forests and birch groves create not only a sense of inimitable beauty, but also a unique climate. Our (the camp's) guests will enjoy clean, fresh air, gather herbs, mushrooms, and berries, and swim in clear waters of Sungul Lake.

By way of sharp contrast, the industrial character of Chelyabinsk City has a very different environmental quality. Daniel Sneider of the *Christian Science Monitor* described the urban environment as follows: "Beyond the worn slopes of the Southern Ural mountains, the horizon of empty snow-covered Siberian lands is broken only by forests of smokestacks, belching grey smoke into the perpetual twilight. Here, close to the deposits of coal and iron ore and beyond the reach of invaders, Joseph Stalin built the sprawling steel mills and arms plants that were the bedrock of Soviet industrialization. . . . The pre-revolutional settlement of 75,000 hardy souls in Chelyabinsk has grown into a city of 1.2 million, its broader avenues and stolid structures indistinguishable from the other products of assembly-line Soviet urbanism" (Sneider 1992).

The Yekaterinburg (formerly Sverdlovsk) region of the Southern Urals to the immediate north of Chelyabinsk was Boris Yeltsin's original base of support in the Soviet era. Three of four voters in the Chelyabinsk region supported him in the 1990 election. With the decline in production in the military-industrial sector, however, and the breakup of the supply agreements associated with the disappearance of the former Soviet Union's command economy, many factories are idle either for lack of defense orders or because of the absence of spare parts and raw materials. During World War II, Chelyabinsk City had the nickname "Tankograd" since one of every two Soviet tanks came from the area's assembly lines.

There were no tanks produced in 1991 at the giant Chelyabinsk Tractor Works. Budget subsidies from Moscow have declined precipitously since the dissolution of the Soviet Union.

The general economy of the region benefited disproportionately from the buildup of the immense military strength of the Soviet Union. In a society where half of all manufacturing jobs were for the military, Chelyabinsk City was one of the leading centers of productivity. Now the giant enterprises that turned out ordinary tanks and munitions must face a difficult choice: either shut down or find new sources of financing for civilian production. The worsening economy will be a long-term source of distress for the people of the Chelyabinsk region, and it has had a disquieting ripple effect. It impedes efforts to stem environmental pollution. There is no money available to close down environmentally damaging industries. Moreover, people of the region lack the financial resources to move away from the chemical and bacterial pollution associated with life in Chelyabinsk City and the surrounding area.

A symbolic contrast to the industrial pollution and to the economy in crisis is a visible manifestation of the Soviet Union's most enduring legacy in the Chelyabinsk region: a dramatic statue in the center of the city commemorating nuclear development. Erected toward the end of the Soviet era, the statue is an imposing representation of the figure of nuclear physicist Igor Kurchatov with a metallic model of the atom above his head. At night, the movement of light animates the model of the atom. Kurchatov had come to the Chelyabinsk region at the end of World War II at the direction of Joseph Stalin and under the supervision of Lavrenti Beria, head of the secret police. Kurchatov's secret mission was to found and run a production plant for plutonium.

The presence of the dramatic sculpture is highly ironic: for most of its existence it was the only public acknowledgment of one of the Soviet Union's most tightly kept secrets. The secret cities and production facilities that were created under Kurchatov's direction did not appear on the region's maps until nearly half a century later. Equally invisible on a map was what those cities and facilities produced: "nearly every known radioactive and hazardous waste and type of contamination known to man" (Laverov 1993, 1).

Mayak Chemical Combine and Its Nuclear Facilities

The centerpiece of Kurchatov and his associates' work is the Mayak Chemical Combine, 60 miles northwest of Chelyabinsk City and 9 miles east of the village of Kyshtym. Mayak means "lighthouse" or "beacon." The closed city where Mayak workers live is now called Ozersk. Construction began on the northeast shore of Lake Kyzyltash in November of 1945 using forced labor by prisoners. The first of five graphite-moderated reactors (now closed) and two water-moderated military production reactors (still operating) started working in June of 1948, with the last of the graphite-moderated reactors ceasing operation in November of 1990 (Cochran and Norris 1992). Up until 1978, plutonium was separated from spent fuel from the military production reactors. From about 1978, the irradiated fuel elements from the military production reactors at Mayak were sent to Tomsk-7 for reprocessing, until about 1991 (presumably three to six months after the November 1990 shutdown of the last graphite reactor).[2] The chemical separation facility at Mayak has been converted to the reprocessing of spent fuel from naval reactors and civilian power plants. One purpose of the recovered plutonium is to serve as the fuel source for the civilian fast-breeder reactor program.[3]

The plutonium output from Mayak remains a national security secret, as does the essential descriptive information on the amount of radioactive wastes that have resulted from plutonium production. Nonetheless, a considerable body of declassified information has emerged since 1989 about plutonium production and nuclear waste at Mayak.

Some of the information released about plutonium production and nuclear waste was a product of the post-Chernobyl reaction against nuclear power within the Soviet Union. Some was released to Russian environmental activists and the media as part of the policies of glasnost and perestroika of the Gorbachev government (Commission for Investigation of the Ecological Situation in the Chelyabinsk Region, 1991). Some was released by Russian doctors after a decision by the then Soviet Ministry of Atomic Power and Industry to declassify some medical records.[4] Some is the product of systematic independent scholarship (Medvedev 1976, 1977). Since 1994, some is the product of cooperative research by the

United States Department of Energy and the National Laboratories of the United States in association with the Russian Ministry of Atomic Energy and the Russian Academy of Sciences (Bradley, Frank, and Mikerin 1996).

The scope of plutonium production and nuclear waste generation at Mayak is still being understood; details remain largely classified. The broad outlines appear to be agreed upon; under contention, however, is what can be done to stabilize (and restore where possible) the environment surrounding Mayak. That has been terribly contaminated by four decades of plutonium production, uncontrolled and controlled radioactive waste disposal directly into the environment, a major nuclear waste explosion, and an unseasonable drought followed by strong winds that dispersed radionuclides over a wide area.

The history of Mayak as a plutonium production site dates from November of 1945. Seventy thousand prisoners from twelve labor camps were used in the construction of the city (Soran and Stillman 1982). In contrast to the U.S. plutonium production site at Hanford, where workers live in cities outside the facility gates, at Mayak, roughly 100,000 inhabitants live within the reservation.

Cochran, Norris, and Bukharin (1995, 117) called the environs of Mayak "arguably the most polluted spot on the planet—certainly in terms of radioactivity." Among the circumstances they offered in support of this characterization were: 1) The discharge of 123 million curies of strontium-90 and cesium-137 into the environment (an estimated 15 percent of what had been produced at Mayak); 2) the exposure to elevated levels of radioactivity of more than 437,000 people; 3) the presence of 340 million cubic meters of radioactive water in on-site open reservoirs; and 4) over 200 solid waste burial sites at Mayak containing over half a million tons of solid radioactive waste. These waste disposal sites were located to minimize their distance from the production centers (Cochran, Norris, and Bukharin 1995).

Mayak is a large complex, occupying more than 200 square kilometers. Within its fences are a number of different types of facilities:

Production Reactors: There are two general types of nuclear reactors: five (now closed) graphite-moderated reactors for plutonium production, and two operational light-water-moderated reactors for the production of special isotopes and tritium.

Chemical Separation Plants: The plants for separating plutonium from nuclear fuel removed from the production reactors are closed. The original chemical separation facilities took irradiated fuel elements from the plutonium production reactor cores and separated, by chemical means, the plutonium and uranium from the other highly radioactive fission materials.

This chemical separation process produces a vast amount of toxic waste, and remains a major source of Mayak's contamination of the surrounding environment. Based on data from the Worldwatch Institute, the manufacture of 1 pound of weapons-grade plutonium produces: 1) 300 liters of highly radioactive wastes; 2) 60,000 liters of low- and medium-level wastes; and 3) 2.5 million liters of contaminated cooling water. The radioisotope plant remains operational, as does the plant for reprocessing naval and spent fuel from nuclear power plants (although it is shut down periodically for repairs).

Fuel Fabrication Plants: Some of the plants in this category are closed, and others of recent origin (1980s) remain open to produce mixed-oxide fuel. A plant identified as "Complex 300" is only partially completed. Its purpose would be to manufacture fuel for the as yet unbuilt BN 800 fast-breeder reactors that would be part of the South Urals Nuclear Power Station (discussed later in this chapter).

Storage Facilities for Nuclear Materials: There are a variety of storage facilities for nuclear materials: a partially completed facility for interim storage of spent fuel from nuclear power reactors, a completed facility for storage of plutonium recovered from naval and nuclear power reactors, and a facility under construction for the storage of weapons materials from dismantled nuclear warheads.

Nuclear Industry Facilities: These facilities, which support the operation of Mayak as an industrial enterprise, are concerned with the design and manufacture of instrumentation, and research on scientific, technical, and engineering problems.

Nuclear Power Generation: Known as the South Urals Nuclear Power (or Atomic Energy) Station, this facility was planned to include three liquid-metal, fast-breeder reactors to generate electricity. Construction began in 1984. This is a strongly contested initiative, the construction of which was halted by opposition in the late 1980s from both the environmental and the antinuclear communities, and more recently because of lack of funds. Various arguments have been advanced by the Ministry of Atomic Energy and by government officials of the Russian Federation for why fast-breeder reactors should be built, but they have not persuaded either most members of the Western scientific community, or the Russian environmental community.

Nuclear Waste Facilities: Perhaps the most infamous undertaking at Mayak, this activity consists of waste storage tanks for intermediate- and high-level waste (one of which exploded in 1957), a vitrification plant, and a series of trenches, reservoirs, marshes, and lakes without direct outlets. The Techa River, part of a river system that drains into the Arctic Ocean, was the choice for direct disposal of nuclear waste in the late 1940s and early 1950s.

The vitrification plant incorporates radioactive waste within phosphate glass. The glass blocks that result are then placed in interim storage. Viktor Fetisov, current director of Mayak, indicated on January 5, 1993, that, for the first time in its history, Mayak had begun to reprocess more nuclear waste products than it had produced (Kachenko 1993). Subsequently he argued that if Mayak treated and buried nuclear waste imported from German nuclear power plants, it could earn "millions of German marks" (Agence France Press, March 1993).

Waste storage practices at Mayak have changed over the years, but remain highly controversial. Also, they are the source of deep antagonism in relationships with the broad environmental movement. The Russian Ministry of Atomic Energy is continuing the tradition begun by the Soviet Ministry of Atomic Power and Industry of keeping secret basic data on the disposition of radioactive waste from plutonium production and on other waste-producing aspects of the Mayak nuclear enterprise (Broad 1994).

Nuclear Waste Tragedies at Mayak

With the advent of glasnost, researchers have begun to piece together an understanding of waste storage practices. The available record has been augmented by three environmental and health tragedies that resulted from waste disposal practices gone awry at Mayak: direct dumping into the Techa River of nuclear waste starting in 1949, a nuclear waste storage tank explosion in 1957, and a radioactive dust storm in 1967.

Direct Dumping of Nuclear Waste into the Techa River

According to Nikolai Yegorev, Deputy Minister of Atomic Energy, the first waste disposal method for high-level radioactive waste at Mayak was treating it like ordinary waste. That determination led to a decision in late 1948 to dilute the high-level waste from the chemical separation process essential to producing plutonium and dump it into the Techa River. A corollary of that decision—in keeping with the strict secrecy surrounding national security—was not to tell the residents downstream.

Information on the Techa River releases is now available from several sources. The most extensive public report of the dumping and its consequences was commissioned by Soviet President Mikhail Gorbachev with presidential decree #RP-1283 of January 3, 1991. The public report, known in English as the "Proceedings of the Commission on Studying the Ecological Situation in Chelyabinsk Oblast," appeared in April of 1991 in two volumes. It is a vitally important glasnost-era document for understanding how nuclear development affected the environment and human health. Another was a paper released at the 1991 UC Irvine Conference on the Environmental Consequences of Nuclear Weapons Development. Nuclear experts primarily from the Soviet Academy of Sciences and the Supreme Soviet (parliament) collaborated in the preparation of the presentation (Chukanov et al. 1991). Some Soviet nuclear environmentalists criticized the report as only a partial portrayal of the technical history of Mayak. But it did serve as one of a beginning series of reports from Russian scientists, physicians, and government representatives. Another source of important information about Mayak's environmental history in the glasnost era was the International Radiological Conference held in Chelyabinsk, Russia in May 1992 (see chapter 6). One of the most informative conference papers was by G.I. Romanov (1992). This paper provided new information on the relationship of distance downstream from Mayak of different villages and the effective dose equivalent for inhabitants of those villages. Another report relevant to the health consequences of the disposal of nuclear waste by direct discharge into the Techa River was by Kossenko, Degteva, and Petrushova (1992).[5] Since the collapse of the Soviet Union, a number of other studies have updated and refined these early reports.[6]

According to Kossenko, Degteva, and Petrushova (1992), the discharge of high-level waste into the Techa River is estimated at 2.75 to 3.0 million curies—which is now the officially accepted estimate. This is an amount of radiation roughly comparable to the amount released by the Hiroshima bomb. Seventy-six million cubic meters of water contaminated by nuclear waste were dumped into the river, with daily releases of curies ranging from 4,300 to over 100,000 on some days.[7] The heavy dumping into the Techa River continued until after September of 1951 when radioisotopes were found in the Arctic Ocean over 1,000 kilometers away. Figure 3.2 from Kossenko and her collaborators reports the isotopic composition and the average daily curies released from 1949 until 1956.

As a consequence of living along the banks of the Techa River, 124,000 people were exposed to radiation (Akleyev and Lyubchansky 1994). Cochran, Norris, and Bukharin (1995, 102–103) documented the population centers along the Techa that were evacuated and those that were not (figure 3.3). This record has been augmented by Kossenko et al. (1997), who added village population and ethnicity data (table 3.2). The impact on the villages that were immediately downstream from the Mayak facility was devastating. For instance, the 1,242 inhabitants of the village of Metlino did not begin evacuation until 1953, and the evacuation was not completed until 1956. All of the thirteen villages in the first 75 kilometers downstream were evacuated; the village of Muslyumovo at 78 kilometers was inexplicably not evacuated.[8] In fact, four villages further downstream from Muslyumovo were evacuated. The flood plain of the Techa River in the center of Muslyumovo was fenced off in 1956 (Kossenko, Degteva, and Petrushova 1992). At that time, the water supply for the village was shifted to underground sources.

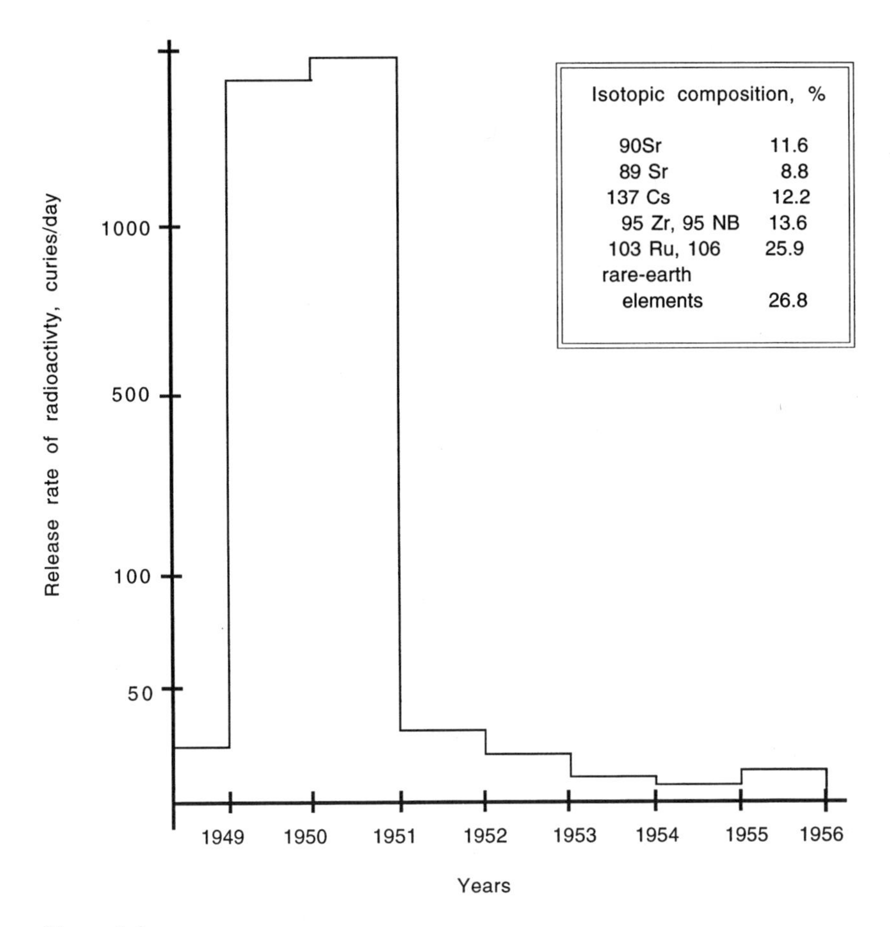

Figure 3.2
Average amount of radioactivity released per day into the Techa River, 1949–1956
Source: Adapted from Kossenko et al. (1992).

In the early 1950s, the Techa River was diverted into artificial channels within the grounds of Mayak. The area of the former riverbed was converted into a series of artificial reservoirs with earthen dams. Construction began in 1951 for four reservoirs to isolate contaminated water that had already been released into the environment. The first dam was completed in 1951, with subsequent ones completed in 1956, 1963, and 1964.

In addition to the Techa releases, low-level radioactive waste was released into Lake Staroe Boloto (3 million curies), and medium-level radio-

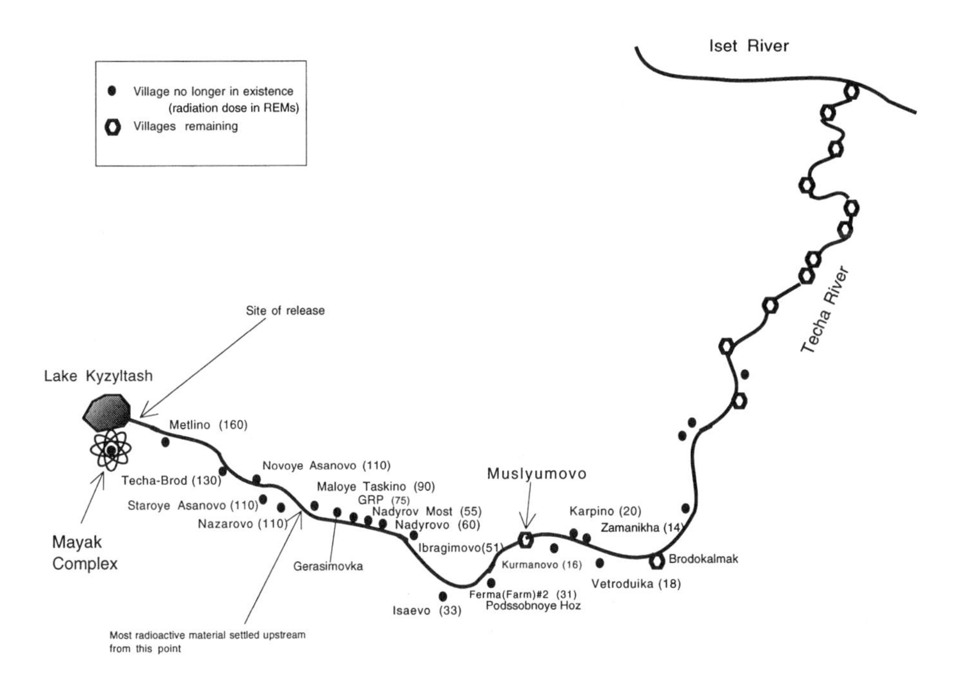

Note: The thirteen villages upstream from Muslyumovo and the first four villages downstream were evacuated between 1952 and 1956.

Figure 3.3
Schematic map of Mayak complex and Techa River villages (approximate scale)

active waste was diverted into Lake Karachay (120 million curies). Neither Lake Staroe Boloto nor Lake Karachay have direct outlets into the stream or river systems. Figure 3.4 is a schematic representation of the lake, river, and reservoir system within Mayak.

The use of Lakes Boloto and Karachay as an interim solution to the storage of radioactive waste in the early 1950s carried long-term problems of its own. Radioisotopes have migrated in the groundwater over 3 kilometers from Lake Karachay into the flood plain of the Mishelyak River (Cochran and Norris 1993b, 71). There is no proven technology to stabilize or reverse this groundwater migration. Officials at Mayak have undertaken an approach to stabilization that consists of filling Lake Karachay with concrete, rock, and soil. This is to prevent the effect that windy days have of lifting mist from the surface of the water along with radionuclides and sending it into the countryside.

Table 3.2
Techa River villages downstream from Mayak

Village name	Population in 1950	Distance from Mayak in kilometers	Radiation dose in REMS	Ethnicity
Metlino	1,242	7	160	Russian
Techa-Brod	75	18	130	Russian
Novoye Asanovo	157	27	110	Tatar/Bashkir
Staroye Asanovo	637		110	Tatar/Bashkir
Nazarovo	98		110	Tatar/Bashkir
Maloye Taskino	147		90	Tatar/Bashkir
Gerasimovka	357		NR*	Russian
GRP	260		75	Russian
Nadirov Most	240		55	Russian
Nadyrovo	184		60	Tatar/Bashkir
Ibragimovo	184	48	51	Tatar/Bashkir
Isaevo	434		33	Tatar/Bashkir
Ferma(Farm)#2	487		31	Tatar/Bashkir
Muslyumovo	3,230	78	IR**	Tatar/Bashkir
Kurmanovo	1,046		16	Tatar/Bashkir
Karpino	195		20	Russian
Zamanikha	338		14	Russian
Vetroduika	163	109	18	Russian
Brodokalmak	4,102	115	NR*	Russian

* NR indicates no record available to researchers.
** IR indicates inconsistent record available to researchers for this village, which was not evacuated: Muslyumovo at 78 km. Brodokalmak at 115 km (est.) was also not evacuated.
Note: Historical records available to researchers differ slightly. Kossenko et al. (1997, 55) present population and ethnicity data for villages. Cochran, Norris, and Bukharin (1995, 102) report radiation dose in REMS. The village of Gerasimovka appears in Kossenko et al. (1997, 55). Ferma (Farm) #2 appears on a map at the May 1992 conference in Chelyabinsk and in Cochran, Norris, and Bukharin (1995, 102). It is identified in Kossenko et al. (1997) as Podssobnoye Hoz. Some villages like Muslyumovo were on both sides of the river.

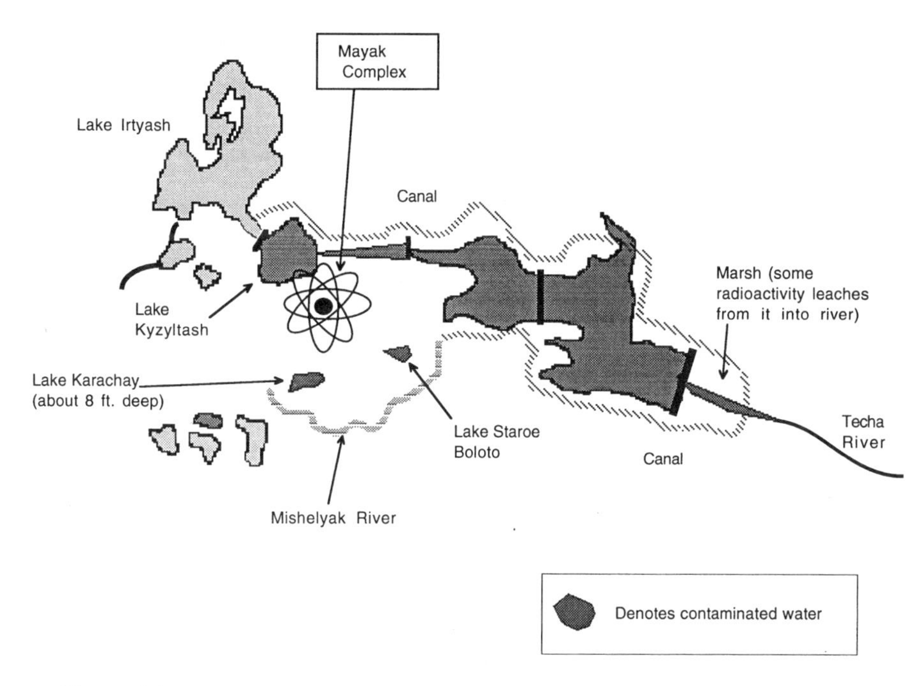

Figure 3.4
Schematic representation of the lakes, rivers, and reservoir system within Mayak

The high levels of radiation along the shore make this a most difficult task in and of itself. Soviet officials reported that the radiation exposure rate near the discharge line into the lake is 600 roentgens per hour (Nikipelov et al. 1990), an amount Cochran and Norris estimate is sufficient to provide an individual with a lethal dose within an hour (Cochran and Norris 1991, 20).

Another problem with this 1950s interim solution to nuclear waste storage is that the earthen dam reservoirs constructed in the riverbed of the diverted Techa River are close to overflowing. If they were either to overflow or to burst, there would be a further radioactive contamination downstream. Mironenko et al. (1995) estimated that if Dam 11 failed in a sudden rupture, the 230 million cubic meters of contaminated water that currently covers 44 square kilometers would flow downstream in a flood. The flood would reach Muslyumovo fifteen hours later at a height of 12.3 meters, with a velocity of 10.9 kilometers an hour. Sergerstahl,

Akleyev, and Novikov (1997) estimated that the Techa Valley would be uninhabitable for three years.

The difficulty of understanding the migration of radionuclides into the groundwater below Lake Karachay is found in the work of Petrov and his colleagues (1994). They focused on developing a numerical model of groundwater flow and contaminant transport from the surface radioactive waste reservoirs of Mayak. There were three prediction tasks undertaken by the investigators: 1) to predict contaminant plume behavior (shape, concentrations in time); then 2) to predict the discharge of the contaminated groundwater below Lake Karachay into the Mishelyak River; and 3) to predict the efficacy of safety engineering measures (Petrov et al. 1994, 4).

These authors indicate that since 1957 there has been significant data collected on both the hydrogeological situation (groundwater flow) and the containment transport process. At the same time, although there is a large volume of input data, "their quantity and quality may be insufficient to realize the complex modeling that is necessary to reflect the phenomena of interest. In this case, the modeling process must provide the basis to design subsequent tests and additional observations" (Petrov et al. 1994, 10).

The researchers encountered a number of practical difficulties. For example, the natural groundwater flow structure has been changed by such Mayak initiatives as raising the water level of Lake Karachay by artificial filling, or by taking a former marsh in 1954 and creating in its place a reservoir (Reservoir 17), with a water level six meters higher. The main features they encountered were the nonuniformity of the aquifer in the fractured rock, and the high density of the contaminated water solutions. In a fiscal climate of severe restraint, the problem of groundwater migration from Lake Karachay seems beyond practical remediation for lack of finances as well as of known technology.

Nuclear Waste Explosion in 1957—"The Kyshtym Disaster"

The second nuclear waste tragedy associated with Mayak occurred in the tank storage system for nuclear waste. Some of the high-level wastes produced at Mayak have been stored in stainless steel, single-walled tanks. The number of such tanks are variously reported to be from 60 (Chukanov et al. 1991) to 99 (Cochran and Norris 1993b, 81). The tank storage system used at Mayak in the 1950s consisted of rectangular concrete structures housing stainless steel tanks that were cooled by water flowing between the outer concrete structure and the inner tanks.

On the afternoon of September 29, 1957, an explosion occurred in one of those waste storage tanks. The force of the explosion was sufficiently powerful (estimates range from 70 to 100 tons equivalent of TNT) that it blew 70 to 80 tons of high-level radioactive waste containing 20 million curies into the surrounding countryside.[9] A total of 18 million curies landed within Mayak, and the remaining 2 million formed a radioactive cloud that covered 15,000 to 23,000 square kilometers with at least .1 curies of strontium-90 per square kilometer (Cochran and Norris 1993b, 72–80).

Estimates vary on the impact of the 1957 explosion on the inhabitants of the region. The range of evacuees listed by various sources is from 10,100 to 11,000. Figure 3.5 identifies the trace from the 1957 explosion and the area evacuated in relationship to the Mayak complex.

In retrospect, there are several historical anomalies that merit comment. The Soviet authorities elected to wait thirty-two years to inform the world about the accident—even their own people, who were affected by it. The United States knew about radioactivity being present in the Techa River as early as 1954, and at least by 1959 knew that some radiation accident had occurred at the Mayak complex. Despite the anti-Soviet stance of the United States government at the time, what was known about the accident was not disclosed until a nonprofit public interest group, Critical Mass, filed a request under the Freedom of Information Act in 1977 (IPPNW/IEER 1992). The first public disclosure of this waste disposal explosion, popularly known as the Kyshtym 1957 disaster, was by a dissident Soviet biologist Zhores Medvedev (Medvedev 1976). At the time, his disclosures were met with some degree of skepticism by elements of the Western nuclear establishment (IPPNW/IEER 1992, 81).

Soviet waste disposal policy changed after the 1957 disaster to greater use of deep-well injection. Laverov, Omelianeako, and Velichkin (1994, 2) presented the three official rationales for Soviet-era decisions on nuclear waste disposal after the Kyshtym explosion. The first rationale was that nuclear weapons production had to be continued, and a slowdown or cessation of production was impossible on national security grounds. The second rationale was that any solution to nuclear waste disposal had

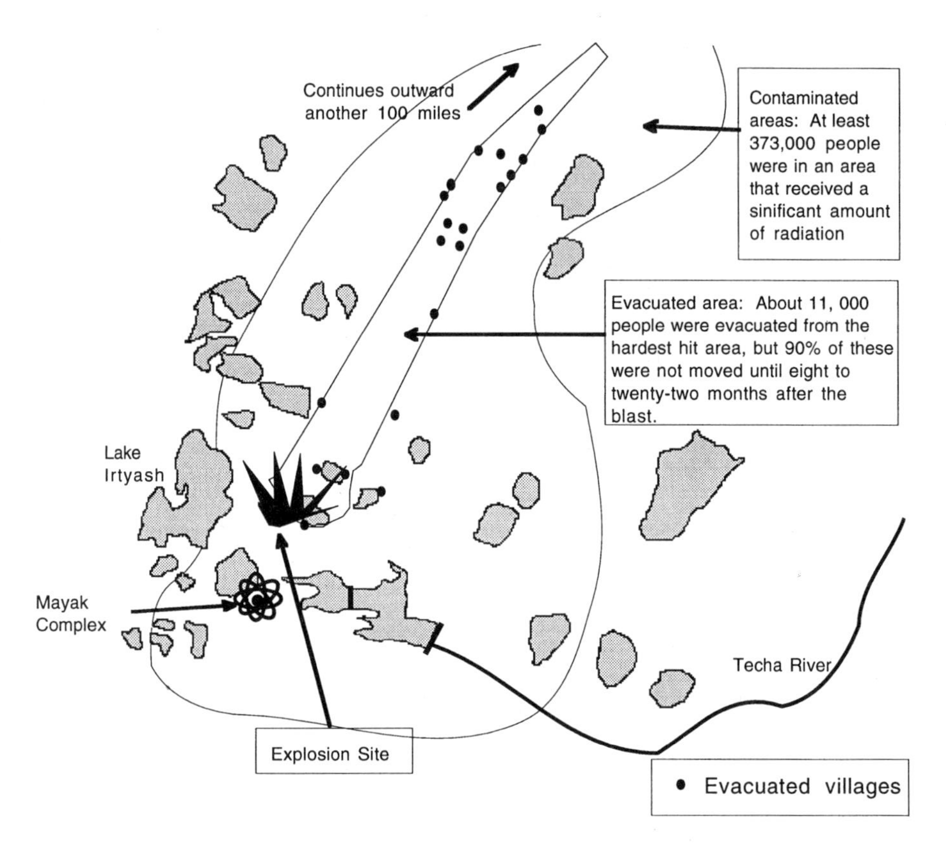

Figure 3.5
1957 Waste dump explosion at Kyshtym

to be inexpensive (which excluded expensive foreign technologies). The third rationale was that the hazards associated with deep-well injection were less than hazards associated with disposal on the surface.

Specifically applied to Mayak, however, the decision to dispose of high-level waste in water-bearing underground sites did not work: "Unfortunately, the technology of disposing of liquid NW [nuclear waste] in water-bearing underground locations was unfit for the "Mayak" region. In that region, waste was discharged into open reservoirs (Laverov, Omelianeako, and Velichkin 1994, 2). They report that the Soviet-era nuclear waste disposal decision at Mayak led to severe pollution, and violated International Atomic Energy Agency standards:

> Like the disposal of NW in underground water-bearing sites, storage of NW in open reservoirs is a serious violation of IAEA standards. Scattering of aerosols during strong winds, the almost-certain underground water contamination, risk of waterspouts, and other factors contribute to the high probability of radio-ecological catastrophes. Before much passage of time the theoretical hazards developed into actual events. After saturation by radionuclides, the bottom muds and underlying foams of the "Karachai" lakes actually prevented the penetration of radionuclides into underground waters. But particularly severe contamination was caused by direct penetration of radioactive water from "Karachai" to underground water through fractures of crystalline rocks in 1961, when considerable expansion of the reservoir occurred during the rainy summer (Laverov, Omelianeako, and Velichkin 1994, 3).

As of 1994, the authors reported that 5 million cubic meters of polluted water had migrated from Lake Karachay, and was spreading to the south and north at 80 meters a year, "threatening to enter water intakes and rivers" (Laverov, Omelianeako, and Velichkin 1994, 3). In a classic understatement, the authors observed that "theoretical hazards developed into actual events" (Laverov, Omelianeako, and Velichkin 1994, 3).

In November of 1994, officials of the Russian Ministry of Atomic Energy disclosed that Soviet officials, after the Kyshtym disaster of 1957, began a process that has resulted in the injection of 3 billion curies of high-level nuclear waste into deep wells at three other sites around the country. They are near major rivers in what were thought to be stable and impervious geological formations (under layers of shale and clay). All three injection sites have experienced unexpected migration of radionuclides into the surrounding groundwater (Broad 1994).

Lake Karachay's 1967 Dust Storm

The third nuclear waste tragedy associated with Mayak was a consequence of the decision to store nuclear waste in Lake Karachay. A drought in the area in 1967 caused extensive evaporation of the water in the lake, a condition that exposed expanses of dried sediment. While the details of the cause are unclear (windstorm; tornado; strong, hot, dry winds), some wind condition blew 600 curies of radioactive isotopes of cesium-137 and strontium-90 over 1,800 square kilometers affecting an estimated 41,000 people. Figure 3.6 identifies the footprint of the 1967 dust storm or storms.

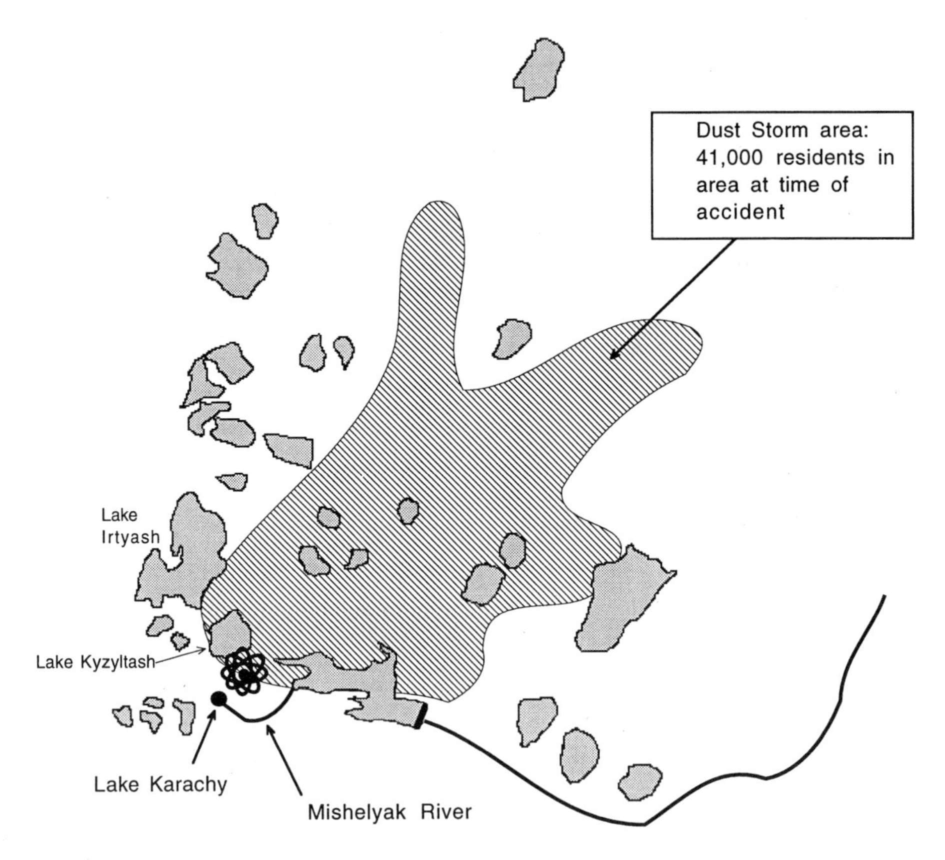

Figure 3.6
1967 radioactive dust storm

There was overlap between the footprint of the dispersal of radioactivity from the Kyshtym 1957 explosion and from the 1967 Karachay dust storms. There are several glasnost-era reports of the 1967 disaster at Lake Karachay, including *Proceedings of the Commission on Studying the Ecological Situation in Chelyabinsk Oblast* (Commission for Investigation of the Ecological Situation in the Chelyabinsk Region 1991). One problem was that the lake was only 8 feet deep at its deepest point, so the evaporation from the unseasonable heat exposed an expanse of contaminated dried mud around the rim of the lake. The external radiation dose to 4,800 of the residents closest to Mayak was about 1.3 rem. For the residents more removed from Mayak, but within the trace, the external radiation dose was 0.7 rem.

Additional Nuclear Problems at Mayak

The "normal" operation of the facilities at Mayak also produced inevitable radiation releases. For instance, the chemical processing of nuclear fuel to extract plutonium releases iodine-131 and other airborne radionuclides. Chapter 2 reported that over the length of Hanford's plutonium production, large amounts of iodine-131 were released into the air; and the worst release came when Hanford was simulating Mayak's production processes (Heeb 1994). Official Russian statistics now estimate that 560,000 curies were released by Mayak between 1949 and the halt of plutonium production (Degteva et al. 1996a, 1996b). This statistic seems to be a significant underestimate of releases, based on the production methods used at Mayak, the estimates of total plutonium production, and the Soviets' limited pollution technology.[10] But even the low current estimate of I-131 releases represents a significant health risk for the residents of the closed city of Chelyabinsk-65 and for residents in nearby rural areas. Early research suggests that there have been elevated levels of thyroid cancer and related diseases in the Mayak area that can be attributed to past radiation releases (Kossenko 1996; Kossenko et al. 1997).

Similarly, the plutonium production process inevitably generated large amounts of radioactive and toxic substances. Safety provisions for workers at the facilities also appear to have been less stringent than American standards. Thus the early studies, based on preliminary and severely limited data, find evidence of increased cancer deaths among Mayak workers who were exposed to high radiation levels (Koshurnikova et al. 1996; Hohryakov, Syslova, and Skryabin 1994). Indeed, in addition to the major nuclear accidents and nuclear waste tragedies at Mayak identified above, there have also been smaller accidents. The newspaper *Komsomolskaya Pravda* (November 11–14, 1994) and Cochran, Norris, and Bukharin (1995, 98–99) document more than a dozen such problems, ranging from several critical events to a series of small explosions that involved radioactivity releases.

Finally, and perhaps most ominously, the lasting legacy of plutonium production is the accumulated wastes that still reside at Mayak (Cochran, Norris, and Bukharin 1995). From accounts in the mid-1990s, the Techa River reservoirs and the deposits in lakes within the Mayak facility held between 350 and 400 million curies. Liquid storage tanks at Mayak hold 250 million curies (Bradley, Frank, and Mikerin 1996, 43). Managing these nuclear wastes, and their eventual disposal, presents an even greater challenge than that at the Hanford facility.

Mayak's Business Plan Circa 1993 (and Earlier)

As with Hanford, the end of the Cold War has generated extended discussions about the fate of Russia's nuclear weapons complex, and specifically, the environmental legacy of Mayak. Gradually, an increasing amount of information about Mayak's past activities has become public. This both invigorated the environmental movement (see chapter 6), and provided a basis for the beginnings of a reasoned scientific discussion of Mayak's legacy. Since the mid-1990s, American and Russian scientists have collaborated to begin studies similar to the dose reconstruction project under way at Hanford.

At the same time, within the confines of the Russian government, the Russian Ministry of Atomic Energy released a plan in January of 1993 that was supposed to chart Mayak's future (Bukharin, 1997). This classified plan, "Program of Structural Reconstruction of the Russian Nuclear Industry," has not been made available for public scrutiny. Nine elements of that program of structural reconstruction for Mayak are apparent, however, from either publicly reported actions, or from publicly announced plans that have been disclosed to the media. The elements of this plan, which are discussed below, provide a framework for judging how in 1993 officials at Mayak and the Ministry of Atomic Energy viewed Mayak's future (see the discussion of Mayak's 1999 plan in chapter 11). The goals were maintaining Mayak as a functioning institution, supporting the scientists employed at the facility, ensuring the security of the nuclear arsenal and technology, and developing a revised role for the facility in a post–Cold War world. This reflects official thinking at the beginning of the processes discussed in later chapters.

One continuing mission was that Mayak would still produce and process tritium for nuclear weapons. Tritium is produced in the Ruslan and Ludmilla reactors at Mayak by neutron irradiation of lithium-6 targets. Since tritium decays at a rate of 5.5 percent a year, a dependable long-term supply is needed in order to maintain the Russian nuclear arsenal. Tritium enhances the explosive power of the fission trigger of thermonuclear weapons. As a business initiative, tritium production is a national security project of the Russian Federation. Given the requirements for secrecy, and the limited, if any, commercial application, it is highly unlikely that the sources of payment to Mayak for this mission will extend beyond the financial capabilities of the Russian government.

Another mission for Mayak was to become a generator of commercial nuclear power. This activity would require new nuclear reactors. Since the 1970s, the Ministry of Atomic Energy and its predecessor organizations have advocated expansion of reactor capability at Mayak, such as the South Urals Nuclear Power Station, to generate nuclear power for commercial purposes. This proposed project has been the source of unbridgeable conflict between the Ministry of Atomic Energy and the Chelyabinsk environmental community; it was halted in the late 1980s. At that time, the ministry had completed work on foundations for some basic structures and for utility connections for the administration buildings. The first support girders had been raised at two of the reactor sites.

In June of 1989, Boris N. Nikipelov, first deputy minister of Medium Machine Building (later the Ministry of Atomic Power and Industry at the end of the Soviet era) held a news conference at which he stated that the choice of the reactor site had been made specifically to enhance the environment. The ministry argued that the reactors could be used to evaporate water from the wastewater holding ponds to prevent overflowing the dams (which would send water-borne radionuclides downstream), or to prevent an even worse situation of an earthen dam bursting and sending a torrent of contaminated water down the Techa River. Cochran and Norris (1991, 17) noted that the Ministry of Atomic Power and Industry also had argued for continued construction of the power plant on the grounds that "the facility is needed to provide employment for the skilled workers who have lost or will lose their jobs as a result of the shutdown of the (plutonium) production reactors. . . ."

However, environmentalists and nuclear experts have questioned the practicality of the breeder reactor plan. For example, Cochran and Norris (1991) cited several, general nuclear safety considerations about Russian breeder reactors since that is what is proposed for the South Urals Nuclear Power Station: They pointed out that the breeder program is plagued by safety concerns—leaks in the sodium-water heat exchangers and the possibility of a runaway chain reaction during an overheating accident—and by problems encountered in the development of "mixed-oxide" (MOX) plutonium fuel. The BN-600 breeder at Beloyarskiy continues to operate at half-power, and until recently operated with highly enriched uranium rather than plutonium. The Soviet breeder reactor program is increasingly vulnerable to charges that it is uneconomical.[11] In 1994, a fire at another Russian breeder reactor at Yekaterinburg renewed safety concerns (*Los Angeles Times,* May 7, 1994).

However, it is clear that irrespective of documented problems and cost factors, the Ministry of Atomic Energy considers breeder reactors to be safe, and is unswerving in its support for the South Urals Nuclear Power Station. Since the Soviet era, there has been consistent support for the South Urals Nuclear Power Station. For example, included on the "list of the most important construction projects for 1992, the financing of which will be carried out with funds from Russia's federal budget, was continuation of construction work on the South Urals nuclear plant" (Belyanivov 1992). Authority for the resumption of construction was in a decree (EeG-P11-11639) signed by Yegor Gaidar, first deputy prime minister of the Russian Federation on March 26, 1992. The ministry has also obtained the support of local scientists and government officials in the Chelyabinsk region. In November of 1994, the Chelyabinsk regional parliament voted to support resumption of construction.

Furthermore, confirmation of the commitment of the Russian government to construct the South Urals Nuclear Power Station was provided by Oleg Soskovets, Russian first deputy prime minister on May 15, 1996, during a visit to Ozersk (*Nuclear News* 1996). He reported that the government intended to allocate 15 billion rubles (about $3 million) in 1996. Soskovets also said that the Cabinet of Ministers had instructed the ministries for economics and finance to find the additional 50 billion rubles needed for construction. Soskovets claimed that the South Urals Nuclear Power Station would make two contributions to improving the ecological situation in the region. First, it would use plutonium already in storage at Mayak for fuel. Second, it would reduce the water level in already overfull reservoirs by using the irradiated water in them as cooling fluid (Tanner 1997).

Another perspective on the future of the South Urals Nuclear Power Station was provided by Evgeny Mikerin, head of fuel cycle operations for the Ministry of Atomic Energy. According to Hibbs (1996), Mikerin indicated that 28 metric tons of plutonium from reprocessing, spent fuel from VVER reactors by RT-1 is stored at Mayak in anticipation of the operation of the BN-800 fast reactors. Mikerin indicated, however, that not enough funding had been allocated for BN-800 construction, and the units will not be built until some time in the twenty-first century.

A third mission for Mayak involves the as yet unfinished Complex 300. This facility would fabricate plutonium into MOX (mixed-oxide) fuel for use in the one to three BN-800 fast-neutron reactors of the South Urals Nuclear Power Station, which, in turn, would generate electricity and more plutonium. As already noted, at present Mayak has neither the fast-neutron reactors nor the funds to build them.

A fourth, continuing mission for Mayak would be to encase excess plutonium in molten glass, a process known as vitrification. Bradley, Frank, and Mikerin (1996) report that Mayak's work on vitrification began in 1967. Between 1969 and 1971, several small-scale facilities were developed for "direct electric heating" melters. Between 1973 and 1977, simulated wasters and radionuclide tracers were tested in pilot projects. The first ceramic melter was put into operation in 1986–87. This had to be shut down because of electrode problems. A second ceramic melter began operation in June 1991.

In what Bradley, Frank, and Mikerin (1996) call one of Mayak's major accomplishments, an amount of nuclear material equivalent to the total high-level-waste-tank inventory currently at Hanford has been vitrified. This amount is 25 percent of Mayak's total tank inventory of 250 million curies (Bradley, Frank, and Mikerin 1996, 43). The problem for the future is that the vitrification unit has been operating for twice its expected period of use, and there are no funds to replace it.

Fifth, in contrast to the environmental stabilization and restoration problem that exists at Hanford, Mayak has an even more extensive environmental cleanup problem. Bradley, Frank, and Mikerin (1996) estimate that plutonium production operations and nuclear accidents released more than 1.7 billion curies of radiation into the environment surrounding Soviet-era plutonium production locations, including Mayak. A mission statement intended to guide decontamination, decommissioning, and cleanup, is known as the "Passport of the Federal Program 'Management of Radioactive Waste and Spent Nuclear Materials, Their Utilization and Disposal for 1996–2005' " (Chernomyrdin 1995). Viktor Danilov-Danilyan, head of the state committee for the environment, estimated that the problem of radioactive waste would cost 700 billion rubles at a minimum. With respect to the problem of nuclear waste around Mayak, in Danilov-Danilyan's assessment, "as far as potential threats to the economy and people's health the situation with nuclear waste is very threatening as it is not kept in proper conditions" (Tanner 1997).

It was not clear from the interview what Danilov-Danilyan thought could be accomplished for 700 billion rubles. Elsewhere in the interview he said that the surface water supply, which for 60 percent of Russians was unsatisfactory, would cost 200 billion rubles to clean up. The basic problem, according to Danilov-Danilyan, is that "the amount of pollution dumped during communism was so great that to free ourselves from it, we need a much longer amount of time and more measures than have been adopted" (Tanner 1997).

In 1997, Interfax News Agency quoted the then head of the Ministry of Atomic Energy, Viktor Mikhailov, as saying that the budget for the Russian Federation provided only $3 million in funding for nuclear waste disposal at Mayak, or only 15 percent of the need, which he estimated at $50 million annually (Interfax News Agency 1997). (The annual expenditures at Hanford are in excess of $1 billion). The $3 million in 1997 from the Russian Federation was supplemented by $12 million from the Ministry of Atomic Energy's export earnings and $5 million from the Mayak Production Association according to Minister Mikhailov.

Mikhailov linked the problems with nuclear waste clean-up at Mayak and elsewhere to a shortage of funds resulting from debts of energy customers to the Ministry of Atomic Energy. In addition, another obstacle to waste clean-up was the failure of the government of the Russian Federation to remit to the Ministry of Atomic Energy another 1,600 billion rubles which had been budgeted for it (Interfax News Agency 1997).

Sixth, another mission for Mayak is commercial reprocessing of spent nuclear fuel. The RT-1 facility at Mayak reprocesses highly enriched uranium fuel from plutonium and tritium production reactors, as well as spent fuel from the nuclear fleet of the Russian Navy and several power reactors (VVER-440 and BN-600). Mayak has in the past reprocessed spent fuel from Hungary and Finland.

The future of reprocessing at the RT-1 chemical separation plant is far from assured. Mark Hibbs (1996) focused on the problems of identifying future customers and the escalation of the costs associated with actual reprocessing. Spent nuclear fuel from the Loviisa nuclear power station in Finland had been reprocessed at Mayak under the terms of a nuclear fuel supply contract for Soviet-designed VVER-440 reactors—a contract that Mayak officials had honored after the dissolution of the Soviet Union. In 1994, however, the Finnish Nuclear Energy Act was amended to halt shipments by the end of 1996 (*Nuclear News* 1997).

Evgeny Dzekun, chief engineer at RT-1, blamed interference by the United States government for hurting demand from foreign countries for reprocessing services at Mayak. The nature of the interference, according to Dzekun, was persuading operators of VVER reactors outside of Russia to adopt another method of disposal for their spent fuel, rather than choosing reprocessing at Mayak. According to Dzekun, this pressure from Washington, D.C., was directed at officials in Finland, Hungary, Slovakia, and the Czech Republic.

Another factor that is influencing foreign nuclear plant operators not to send their spent nuclear fuel to Mayak is a provision of Russian nuclear legislation passed in 1995 that requires foreign clients to take back in vitrified form high-level nuclear waste from reprocessing. Dzekun stated that foreign clients, specifically from Finland, were "intimidated" by this new requirement of Russian law (Hibbs 1996).

Russian inflation and market price reforms were another factor identified by Dzekun as hurting the reprocessing business at Mayak. After noting that prices now approach Western levels for the cost of such inputs for RT-1 reprocessing operations as transportation, equipment, chemicals, and electricity, Dzekun said, "We will have to raise the price of reprocessing for Russian reactors to continue justifying operation of the plant" (Hibbs 1996, 10). There are two constraints on raising the prices for reprocessing. The first constraint is that there are preexisting contracts tied to premarket conditions (noninflationary conditions) that preclude passing on to customers the increased costs. The second is that nuclear utilities are being paid only unevenly by their own customers for the electricity they provide; therefore they may not be in a position to pay for reprocessing services under the existing contracts, let alone at a new, inflated price.

An official of the Ministry of Atomic Energy, identified as an aide to then head, Viktor Mikhailov, speculated that a strategy for RT-1 management would be to work on the inventory of spent fuel from the Russian Navy from both the submarine and icebreaker fleets. Hibbs (1996) quoted him as saying, however, that neither the Ministry of Defense nor the Ministry of Atomic Energy could fund "major management" of the reactor fuel inventory.

An added source of difficulty was noted by Evgeny Mikerin. Hibbs quoted Mikerin as saying that the large inventory of icebreaker fuel in temporary storage in Murmansk would be "difficult to transport" (Hibbs 1996, 10). Further, according to Mikerin, uranium-zirconium fuels used in some submarines would be difficult to reprocess at Mayak. Yet another long-term problem for the reprocessing business for RT-1 at Mayak is that the Ministry of Atomic Energy is not including VVER-440 reactors in its future development plans. According to someone Hibbs identified as a senior Russian fuel-cycle official, in the future "the real issue for RT-1 is who its clients will be" (1996, p. 10).

Seventh, a possible post–Cold War mission is commercial application of Mayak's nuclear technology for export. Included in this category are uranium enrichment and radioisotope production. Mayak currently produces radioisotopes as a by-product of the operation of the RT-1 reprocessing facility and the tritium production reactors. The Department of Energy, for example, contracted with the Ministry of Atomic Energy to produce at Mayak plutonium-238 as a power source for civilian space applications (Department of Energy 1994). But even this development would require additional investment in the transportation infrastructure around Mayak.[12]

Eighth, another potential mission for Mayak is to apply its technology sophistication to commercial products of both low and high technology for the domestic and international markets. There are numerous obstacles to be overcome in attempting a transition from a core nuclear weapons production mission to a series of commercial initiatives.

Ninth, beyond the role of tritium production mentioned earlier, a continuing mission for Mayak and the other plutonium production cities is contributing to nuclear stewardship for the Russian nuclear weapons stockpile. With the end of the Cold War nuclear arms race, and as long as nuclear weapons have a role in the national defense for Russia, it is still necessary to maintain the nuclear arsenal, albeit at lower levels. Two arms control agreements that have the most relevance to the reduction in warheads are the INF and START treaties. Under these agreements, and based upon some unilateral Russian decisions as well, the number of warheads that must be maintained in ready condition number about 10,000 (Bukharin 1997). Maintenance of this number of warheads is well below the combined capacities of Mayak and Tomsk-7. Therefore, some restructuring and downsizing must occur, which will negatively impact one or both of the facilities.

In summary, in the circa 1993 business plan, defense conversion, and structural reconstruction at Mayak have both nuclear and nonnuclear elements. As plans for these proposed activities have evolved, many are either highly controversial or lack reliable funding. Important new activities are also supposed to be undertaken in response to initiatives that evolved during the 1990s (see chapter 11).

Future Realities of the Chelyabinsk Region

Several realities of the Chelyabinsk region are past. One is the use of terror by Joseph Stalin and the head of the secret police, Lavrenti Beria. They used terror to exert pressure for the production of plutonium as part of a strategy for success against America and the West in the Cold War nuclear arms race. Perhaps the story is apocryphal, but it has been reported that Beria had a list drawn up of who would be shot if the explosion of the first Soviet bomb was not successful (Barwich 1977, 116; Holloway 1994, 215).

A second reality, which was not fully understood at the start of the development of the Soviet nuclear weapons program, is the long-term negative impact of nuclear waste on the environment and, through the environment, on human health. The problems of health and environment for the Chelyabinsk region, and the interrelationship with economics extends beyond the nuclear arena at Mayak. For example, the closure of the copper smelter in the copper-industry town of Karabash has left a legacy of pollution and underemployment. The smelter had operated for seventy years. A report on Russian television indicated that it had discharged 12 million tons of harmful substances into the atmosphere, and the maximum permissible levels of arsenic, lead, and mercury in the soil have been exceeded hundreds of times over.[13] Cadmium and lead are both present in children's hair. For the smelter's former workforce of several thousand, only seven positions were advertised through the Karabash employment service in the first ten months of 1996. Unemployment benefits have run out.[14]

Closer to Mayak than Karabash, Muslyumovo, the main village remaining along the upper reaches of the Techa River, faces an uncertain future. Paula Garb (1997) reports that the average life span of the women of Muslyumovo is forty-seven (versus seventy-two for Russia as a whole) and the average life span of the men is forty-five (versus fifty-nine). Sterility among males who went through puberty in the late 1940s and early 1950s is also reported to be commonplace. Numerous other studies document the impact that Mayak's activities have had on the health of downriver populations (e.g., Kossenko 1996; Kossenko et al. 1997). Addressing these problems seems increasingly difficult as Muslyumovo seems stuck in the downward spiral of decreasing support from Russian government resources (see chapter 11). Andrei Ivanov (1997) reported that the Chelyabinsk regional administration signed a decree in August of 1997 authorizing the resettlement of Muslyumovo and adjacent areas along the Techa. The resettlement is to take place when funds are available.

Despite the uncertainty of the time frame (if ever) for actual resettlement, and the uncertain prospects of funding for it, Ivanov quoted a spokesperson for a local ecological protest group, Techa, on the symbolic significance of the resettlement authorization by the Chelyabinsk regional government: "Nevertheless, we still see it as a significant victory. . . . The decree will be of huge symbolic value for the environmental groups of

Chelyabinsk. After more than seven years of fighting, we have gained a solid lever to demand that the county administration should now fulfill its obligations" (Ivanov 1997).

A third reality not fully understood at the start of systematic nuclear development in both the Soviet Union and the West is that if catastrophic nuclear accidents were to happen, trust in the judgment and capability of the nuclear expert community would be extremely difficult, if not impossible, to rekindle. In the context of the legacy of obsessive secrecy on the part of nuclear experts during the Soviet era, little has developed since to build trust among the general public.

A fourth reality is the difficult future of the closed city of Ozersk. Several years ago, inhabitants indicated in an opinion poll that they preferred to remain a closed city out of concern for the level of crime that has become such a problem in other regions of Russia. The actual living conditions within Ozersk are not publicly reported. The 1996 suicide of Director Vladimir Nechai of Chelyabinsk-70 lifted the secrecy briefly on the region's closed cities. There are undoubtedly parallels to Mayak. One insight into the circumstances behind the suicide was provided by Cragg (1996), who noted that there was ample reason for Nechai's depression: his 16,000 technical and support workers had not been paid for four months, his organization was $12 million in arrears on wages, and Chelyabinsk-70's bank accounts had been frozen because it had debts of $32 million.

Another problem facing Nechai at the time of his suicide—directly applicable to Mayak—is the relatively unchanged Soviet-era structure of life in a closed city. Life in isolated military-industrial cities of the Soviet era was: "the deeply developed Russian perception of communality. The idea that 'enterprises' should provide shops, hospitals, schools, roads and towns is not purely communist inspired, but goes back to the pre-Revolutionary era" (Cragg 1996, 1).

The conflict for places like Mayak is rooted in the transition from a communist state to a free market. One path is to confine the boundary of Mayak as a business to the (more) narrow task of becoming a profitable entity. This narrow boundary would be an advantage for competing in the international marketplace. Another path is to define the boundary broadly to include responsibility for the social welfare safety net of the closed city. Cragg placed these alternate paths in a context such that private enterprises cannot have it both ways: "a license to take the road to profit by abandoning the notion of the enterprises' obligation to the work force and low levels of corporate taxation. Yet this is precisely what a great many Russian entrepreneurs are having and increasingly expect" (Cragg 1996, 1). For closed cities these changes intensify the downward economic spiral.

As the Chelyabinsk region confronts the implications of the end of the Cold War, a series of daunting problems face everyone involved. With an economy closely tied to the military-industrial complex and to the nuclear arms race, the region will have a particularly difficult adjustment. With the worst nuclear contamination in the world, there are pressing issues of environmental stabilization. At Hanford in the State of Washington, there are more employees now working on complex issues of cleanup than were employed at the height of plutonium production. It is unlikely that the Russian government will provide the massive financial support necessary for a comparable undertaking in the Chelyabinsk region.

It is premature to forecast how the Chelyabinsk region will mobilize itself to confront the nuclear problems of the past and the economic challenges of the immediate future. Quite separate realities divide the three communities—the nuclear expert community of the Ministry of Atomic Energy, the depleted environmental activist community, and the general public.

Epilogue

If Igor Kurchatov returned to the Chelyabinsk region today, to the scene of so much of his life's significant work on nuclear weapons development, he would be immediately astounded. He and his associates had labored in strict secrecy; even the towns they had built did not appear on maps. First, he would see the larger-than-life statue of himself located in the middle of Chelyabinsk City.

He would also be surprised to find that he did not return to either the Soviet Union or the Cold War rivalry with the West. Rather, he would find in the place of his nuclear empire a crushing lack of financial resources and, following the disaster at Chernobyl, a global scrutiny beyond his wildest fears. Even the three previously secret nuclear disasters associated with the crown jewel of his creation, Mayak, would be known worldwide.

Moreover, instead of spies or traitors having leaked secrets about his beloved enterprise, Kurchatov would learn it was a Ministry of Atomic Energy senior official who, for example, told the world of the 1 billion curies of radionuclides that had been deposited irretrievably in a deep injection well in Dimitrovgrad. And this startling disclosure did not occur in Moscow, but in one of the national laboratories of the United States Department of Energy, the heartland of America's nuclear brain trust.

Instead of living well in the Soviet Union as he and his contemporaries had done, he would find in his successors in Russia a dispirited workforce, many of whom receive very low wages that frequently are not paid on time. Instead of a steady infusion of money for new projects, he would find that plans to double electric power generated from nuclear sources have been approved by the government, but construction has not proceeded in any substantial manner. Kurchatov would also be astounded by the role of the national press in Russia. Nothing is sacred anymore, or exempt from scrutiny, including those who labor on behalf of the country's national security, or the projects they undertake, such as the construction of the South Urals Nuclear Power Station.

But he would also find some things in present-day Chelyabinsk and Russia much the same. The foothills of the Ural Mountains remain a national treasure with their gorgeous foliage and white birch trees and many unspoiled lakes. He would discover that glasnost and perestroika and democratization have not penetrated into how ministries conduct their business, especially concerning the environment. While a minister no longer needs the sanction of the Politburo or the staff of the Central Committee of the Communist Party (neither of which exists anymore) the sanction of the elected Duma is also not necessary. Rather, most real business is conducted in secret between representatives of the Ministry of Atomic Energy and representatives of the appointed prime minister and his staff. One exception is that the most important business is discussed in secret with the staff of the elected president of Russia, then issued in the form of a presidential decree without public commentary or debate in the open press or in the Duma.

He would find compelling evidence that his most noteworthy legacy is not celebrated by the statue in the center of Chelyabinsk City, but in the enduring radioactive pollution in the Techa River Valley and surrounding environs, and in the bones and soft tissue of the individuals who lived downstream and downwind from Mayak. Those people were never told of the nuclear disasters that negatively impacted their health and that of their descendants for generations to come.

Igor Kurchatov would soon learn that America is not going to solve its own environmental pollution problems, which were created in the nuclear weapons complex of the United States during the Cold War. If he were to examine the comparable experience in America with the environmental consequences of nuclear weapons development, he would find that $350 billion will not be enough for the environmental management effort in the United States, that technology does not exist to solve key problems such as contaminated groundwater, and that whatever environmental stabilization and restoration is undertaken in the United States will extend well into the twenty-second century. The challenges in America are not only technical in nature, but also social and political. The difficulties include surmounting decades of secrecy and generations of mistrust in order to devise new frameworks for building cooperation among the various stakeholders—at a time when technical obstacles are not solved and resources are declining.

Kurchatov would observe at Mayak some close parallels with the American experience, and he would be confronted with apparent differences. His successors have nowhere near the necessary funds for environmental cleanup, and if they did, investing in environmental restoration is still not an accepted priority, either for solving problems from the past or for preventing them in the future. Tilting against the environment remains business as usual, an enduring legacy from the former Soviet Union.

If Igor Kurchatov were to return to Mayak today, he would find that he had to pass through a new electronic security system financed by the United States government. As he left, he and his vehicle would be assessed by nuclear materials detectors. Among the residents of the still closed city of Ozersk he would find representatives of the Departments of Defense and Energy of the United States government, as well as American building contractors and construction employees.

He would undoubtedly be skeptical of the claim of the Kabirov family of Muslyumovo that one of their sons had died prematurely as a result of unknowing exposure to radioactive waste in the Techa River. The family's history is compelling. The father was one of the first guards hired to keep people away from the river's banks. He died in his early forties. The mother was responsible for taking three water samples from the river on a daily basis, and keeping them in her home until authorities from Mayak could collect them. The children helped her by wading into the river. The boy, who in the family's opinion died prematurely, was not yet a teenager when exposure to radioactive waste occurred unknowingly. The suspicions about Mayak on the part of its surviving neighbors cannot be erased by the magnitude of its industrial accomplishments or by appeals for compensated sacrifice on behalf of national security.

Surely Kurchatov would find it ironic that Mayak's greatest accomplishment, its contributions to a cornerstone of Russia's national security, the 1,350 metric tons of fissile materials for nuclear weapons, has become the source of its greatest notoriety as the most polluted place on the planet.

Notes

1. A substantial part of this material was drawn from Cochran and Norris (1993b), Donnay et al. (1995), and Holloway (1994).

2. Cochran, Thomas B., Personal Communication, August 13, 1993. Both water-moderated reactors are still operating. They were probably not used for plutonium production, but for the production of tritium and other special isotopes. The spent fuel and target elements of the water-moderated reactors are processed at a separate chemical separation area at Mayak.

3. Cochran, T.B., Personal Communication, August 13, 1993. One of the chemical separation facilities, RT-1, is for spent fuel from naval reactors and from VVER-440 civilian power reactors. The other facility is for "special isotope production."

4. The background of this report is that on April 26, 1990, Boris V. Nikipelov, deputy minister of atomic power and industry, signed a document officially declassifying the medical records of those exposed to radiation from Mayak.

5. Dr. Kossenko was a longtime physician and researcher for the Institute of Biophysics, Branch 4, of the Soviet Ministry of Health. Prior to the action of Boris Nikipelov in declassifying medical records, she had been bound by an oath of silence that she had signed as a condition of employment with the Institute of Biophysics.

6. See, for example, the issue of *Health Physics* (July 1996) devoted to the environmental impact of Mayak (e.g., Kossenko 1996; Hohryakov, Syslova, and Skryabin, 1996). In addition, a joint Russian-American research project has begun to release its findings (Degteva et al. 1996) and a Russian-Norwegian study has also begun to publish its findings (Christensen et al. 1997).

7. Russian military plant's waste irradiated half-a-million. Kyodo News Service, June 7, 1991.

8. In related research, Paula Garb (1997) discusses the special situation of Muslyumovo and its minority populations. She states that the local Bashkirs and Tatars did not regard their situation as the result of policies aimed particularly at their ethnic groups. Most activists have stressed that Soviet nuclear policies were anti-human, and did not target non-Russians.

9. There are now a number of very interesting accounts of what happened that afternoon in September of 1957. The reader is referred to: Medvedev (1979); Cochran and Norris (1991); Nikipelov et al. (1990); IPPNW (1992); and "An explosion in Southern Urals," *Priroda,* no. 5 (897). English translation for Lawrence Livermore National Laboratory, Livermore, California, pp. 4–11.

10. For example, Cochran and Norris (1993b, 150–151) estimate that Mayak produced at least as much plutonium as the Hanford complex; and Mayak generally used procedures that increased environmental releases.

11. Kyshtym and Soviet nuclear materials production, *Science and Global Security,* vol. 1, nos. 1–2 (1989), p. 174 (a fact sheet containing technical information collected during a visit to Chelyabinsk-40 by an NRDC/Soviet Academy of Sciences delegation July 7–8, 1989).

12. For example, the ITAR-TASS News Agency reported that a drunken tractor driver collided his vehicle with a truck transporting radioisotopes from Mayak. The shipment of twenty-two containers fell to the ground as a result of the collision. The outer casings of containers of cobalt-60 and iridium-192 became detached, and background readings of radiation were twenty-five times higher than normal at the point of impact—but normal 10 to 15 meters away (ITAR-TASS News Agency, 1997). No radiation leak after Urals road accident, September 17, 1997. British Broadcasting Corporation Monitoring International Reports, record number: 00805*19970917*00358. The major access routes to Mayak are by rail, and an improvement of highway access would require major new construction.

13. British Broadcasting Corporation (1996). Urals copper town suffers pollution and unemployment. British Broadcasting Corporation Summary of World Broadcasts, December 7, 1996. Russia TV Channel, Moscow record number: 00805*19961207*0025.

14. See note 11.

4

Public Perceptions of Environmental Conditions

John C. Pierce, Russell J. Dalton, and Andrei Zaitsev

They have been concealing the truth about radiation pollution from the people for thirty years. There has been silence. How many people have suffered? I hope we know better now.
—Boris Yeltsin

There are two realities to the environmental conditions found at Hanford and Mayak. The first is the history of radiological releases and environmental damage as described in the prior two chapters. The other involves public perceptions of these conditions. In both settings a broad array of government officials and citizen advocate groups have presented frequently contradictory evidence on the seriousness of the environmental conditions and the risk to public health at Hanford and Mayak. Public assessments of local environmental conditions, and of the dangers associated with the facilities themselves, reveal how the citizens of these two areas judge these competing claims. Their perceptions and environmental concerns also provide a measuring stick for evaluating claims that environmental conditions have changed for the better over the past two decades.

In addition, public concerns about environmental conditions can motivate environmental action (McCormick 1993). Public perceptions of the environmental problems associated with the plutonium production facilities can structure how individuals define the context for potential political action (Pearce 1991). The extent of perceived environmental problems associated with Hanford and Mayak may also generate a feeling of victimization or grievance that may be important in mobilizing support for environmental groups more generally (see chapter 10).

This chapter compares citizen perceptions of the environment in the Hanford and Chelyabinsk regions based on representative public opinion surveys (for more information on the surveys see the appendix). This comparison addresses the following three questions. First, do the respondents in our American and Russian surveys have similar perceptions of the general and specific environmental conditions in which they live, or are those perceptions substantially different? Second, how are Russian and American perceptions of their respective environmental conditions related to residence—namely, does geographic proximity to the plutonium production sites affect public perceptions and attitudes in the same way at Hanford and Mayak? Third, are the correlates of environmental perceptions and attitudes similar or different among American and Russian citizens in the Hanford and Mayak areas?

Our analytic strategy presumes that even while the two publics may show similarities in their overall environmental perceptions, these views still may differ across political or socioeconomic groups. If there are different social and/or political foundations for environmental perceptions, that would have obvious implications for this comparative analysis, both in how the facilities are perceived and in the potential political and social bases for environmental action in the two areas.

Perceptions of Environmental Conditions

We wanted to examine citizens' views about Hanford and Mayak in the context of the broader environmental concerns of these two publics. This approach would allow us to isolate environmental perceptions of the two facilities from more general feelings about the quality of the local environment. Similarly, we wanted to determine whether the correlates of site-specific perceptions are unique to those views, or whether they are similar for perceptions of other environmental conditions. Consequently, this chapter incorporates a rather broad range of perceptions of environmental conditions. The first items measure general evaluations, such as the quality of the environment at the local, national, and global levels. The second set of evaluations involves potential sources of local environmental concern. The third focuses on specific environmental issues pertaining to the production of nuclear weapons and the storage of nuclear wastes at the Hanford and Mayak complexes.

Views of the Environment

We expected that environmental perceptions of Hanford and Mayak would differ substantially by an individual's relationship to the facilities. Therefore, we used a stratified sampling design to collect data from four distinct areas in both regions (see methodological appendix). Our goal in this sampling design is not to make claims about public perceptions that are representative of the region as a whole or of the national publics in both nations. Rather, we designed a contextual analysis that focuses on the distinctive relationships of individuals to the facilities in specific communities. We wanted to isolate the effects of context in relation to the two nuclear facilities, then determine whether the larger political environments surrounding the specific communities mitigated contextual patterns. Thus our sampling strategy maximizes the contextual differences within each country in order to strengthen the analyses of contextual patterns. A larger, more geographically diverse regional sample would gain representativeness to a larger population, but at the loss of the contextual specificity that underlies our research.

First, we surveyed the "host city" that is economically and professionally connected to the nuclear weapons sites. In the United States we surveyed residents of Richland, the town immediately adjacent to the facilities, where a large proportion of Hanford employees live.[1] In Russia we surveyed residents of Chelyabinsk-70, a closed and formerly secret city that houses research and technical facilities related to Mayak.[2] Both host areas' survey sites are quite different from the surrounding areas as well as from the other areas in their region. In the two sampled communities the residents are highly dependent on the facilities for employment and their economic well-being. They generally are highly educated and well paid. The host-city residents are likely to be more knowledgeable about the facilities' activities, and more tolerant of the risks and environmental impacts associated with them. For example, Dunlap and his colleagues (1993) found that compared to the state overall, Tri-Cities residents were more positive about locating a nuclear waste repository at Hanford and more likely to discount pollution risks associated with such a facility. In a real sense, the nuclear complexes are the rationale for the presence of these two host communities, unlike any other location in the region. Thus the host-city population provides a unique perspective from which to view perceptions of the facilities.

Second, we surveyed the populations of the major urban population centers most affected as downwind from the sites—in the United States, Spokane, in Russia, Chelyabinsk. "Downwind" refers to people who unknowingly were exposed to radioactive contaminants from Hanford and Mayak. Spokane is directly in the path of the major weather patterns passing over the Hanford Reservation. It is also the largest city and the largest media market in the region. The major television and newspaper outlets in Spokane were the primary vehicles through which much of the critical information about Hanford, its history, and its secrecy was first disseminated publicly. Chelyabinsk is a major metropolitan area and media center located approximately 40 miles from Mayak. The city of Chelyabinsk was indirectly affected by the Mayak releases, since most radiation distribution patterns were away from the city; more immediate were the pollution problems resulting from the chemical and metallurgy industries in the city. In any case, Chelyabinsk served as the major urban rallying point for environmentalists and was the setting where public criticisms and discussion of Mayak and its environmental history first occurred. Because of their relationship to the facilities, we expect the residents of the downwind cities will be more critical of the nuclear facilities than will host-city residents.

Third, we studied opinions in the rural communities in the vicinity of the sites that had a large proportion of indigenous and minority residents. In the United States we surveyed residents of three small communities on the Yakama Indian Reservation (White Swan, Toppenish, and Wapato) that include a large share of Native Americans or Hispanics. In Russia we surveyed residents of Kyshtym and Muslyumovo, communities with significant populations of Tatars and Bashkirs (Garb 1997). While the indigenous populations are not the same, and do not share the same histories and cultures, there are important similarities among these groups in relation to the facilities. In both countries these populations historically have been relatively powerless in comparison with host-city and downwind-city residents. Moreover, in both cases these indigenous residents may have been disproportionately exposed to radiation releases. In the United States, for example, many Native Americans were dependent on fishing from the Columbia River and subsistence farming as food sources.[3] This potentially increased their exposure to Hanford contaminants released into the air and water. In Russia the ethnic minority populations in Kyshtym, Muslyumovo, and other rural areas surrounding Mayak were seriously affected by the facilities. Kyshtym is closely adjacent to the facilities; Muslyumovo is located roughly 60 miles downstream on the Techa River. The river remains officially off-limits because of high residual radiation levels in the river bottom, and Muslyumovo was one of the few riverfront villages that was not relocated after the early detection of the river's excessive radiation. Its population is predominately Tatar and Bashkir, and these groups also are politically disadvantaged relative to the dominant Russian population (Garb 1997).

Fourth, we surveyed the population in two "control" cities in the same general areas as the sites, but with no obvious economic dependence or downwind effects directly associated with the nuclear weapons plants. In the United States we interviewed residents of the city of Wenatchee. In Russia we surveyed residents of Chebarkul. We refer the reader to chapter one or the appendix for methodological details on the collection of the survey data.

To tap public judgments about the environment in these different communities, our surveys began by asking about perceptions of environmental quality in the respondents' own community, in their nation, and in the world as a whole. Figure 4.1 displays the Americans' and Russians' environmental perceptions, pairing results in each of the four types of communities. The figure entries are the percentage giving a "very good" or "fairly good" evaluation on a five-point Likert-type scale.[4]

Figure 4.1 yields several noteworthy patterns. In an overall sense, Americans perceived their environment to be in much better condition than the Russians did theirs. Comparing the two weighted samples, for instance, more than 80 percent of Americans gave a favorable rating to their local environment, as compared with fewer than 20 percent of the Russians. This same pattern is repeated in our paired comparison of equivalent communities as well in national survey data reported in the Gallup *Health of the Planet Survey* (Dunlap, Gallup, and Gallup 1993a;

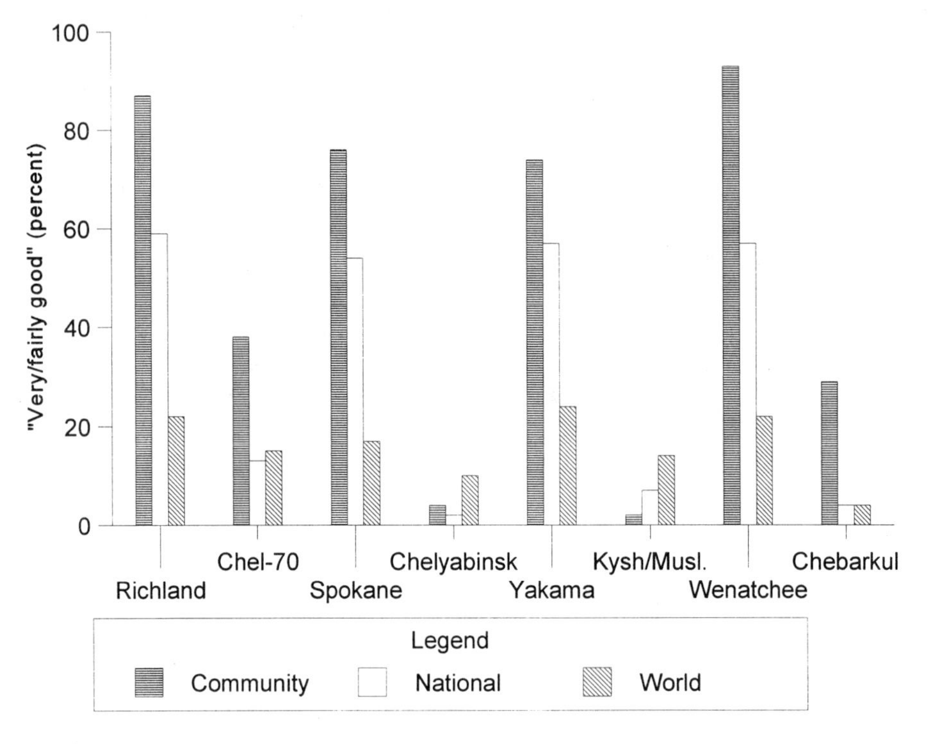

Figure 4.1
Perceptions of community, national, and world environmental conditions

Dunlap 1994).[5] In point of fact, residents in the Chelyabinsk region were more negative toward their local environment than were Russians nationally, and residents in the Hanford region were more positive about their local environment than were Americans as a whole.

These broad differences between the Chelyabinsk and Hanford surveys validate what we know about the overall objective environmental conditions in both regions. Spokane and the eastern Washington area is a region with low population density and a small industrial sector. It has a largely agricultural and natural-resource oriented (forestry and tourist industries) economy, with primarily small-town settlement patterns. The Superfund cleanup site at Hanford (prominent on the National Priority List [NPL] of the Environmental Protection Agency) stands out in glaring contrast to the otherwise relatively healthy-looking surrounding environment in eastern Washington. The city of Chelyabinsk, however, is a major industrial center for Russia, containing a high concentration of metallurgy plants and related industries. D.J. Peterson (1993, ch. 2) lists Chelyabinsk among the Russian cities with the worst air quality; in addition, water quality in the area is much lower than the Russian average (Zaitsev 1993). Added to those environmental issues are Mayak's monumental nuclear contamination problems.

Within each nation, there were similar marked differences across the four types of communities. For instance, 38 percent of Chelyabinsk-70 residents were positive about their local environment, compared with only 4 percent in the downwind city of Chelyabinsk and 2 percent in the communities of Kyshtym and Muslyumovo, which are adjacent to Mayak. Chelyabinsk-70 residents, perhaps reflecting their privileged status as a closed city, gave even more positive environmental ratings than residents of Chebarkul, our control site.[6] This same overall pattern of community differences existed for perceptions of national environmental conditions.

Among the eastern Washington residents, the cross-locale differences were proportionally smaller and were much more positive overall than among the Russian counterpart location. Residents of the host city of Richland and the control city of Wenatchee had more favorable views of the community environment (87 percent and 93 percent, respectively) than did citizens of the downwind city of Spokane (76 percent) and the rural towns on the Yakama Reservation (e.g., White Swan at 76 percent).

Perceptions of differences in national environmental conditions were sharply divided between the two publics, and this mirrors what we know from objective environmental statistics at the time (Peterson 1993; Feshbach and Friendly 1992). Only global environmental perceptions were somewhat similar across the two publics. In this case, the cross-national similarity is primarily a consequence of the American public's negative assessments dropping to near the level of those held by the Russians.

In summary, then, the Americans of eastern Washington had a much more favorable view of their living space environment than did their Russian opposites, especially with regard to their immediate community and with respect to their nation as a whole.[7]

Local Environmental Conditions

Our survey also asked for specific evaluations of potential environmental problems in both regions. We asked respondents to indicate the seriousness of ten particular environmental problems in their area (see appendix B). As a first reference point, Figure 4.2 compares the national data from the 1992 Gallup *Health of the Planet Survey* with the weighted results from our Chelyabinsk and Hanford surveys (Dunlap 1994).

At the national level, Russians were much more likely than Americans to perceive a variety of local environmental problems as being very serious. For example, 39 percent of the total Russian public saw water quality as a very serious problem compared with only 22 percent of Americans. Similarly, 40 percent of Russians saw air quality as a serious problem compared with 18 percent of Americans. As we have seen previously, the two regions surrounding the Russian and American nuclear production facilities accentuated these national contrasts. Residents of the Chelyabinsk region were more concerned about air, water, and other environmental quality problems than were most Russians. The polluting smokestacks of the manufacturing plants, the waste products of the iron and steel industries, and the pollution of the local water supplies were damaging the Chelyabinsk environment severely even before the legacy of plutonium production at Mayak. In comparison, although there were environmental quality issues involving the agricultural economy of eastern Washington, residents there were much less concerned about local environmental conditions such as water quality or the destruction of nature.

The most striking cross-national contrast involves an area that was not included in the Gallup national survey—namely radiological pollution. In the region surrounding Hanford a modest number of our survey respondents expressed very serious concern about radiation-related pollution (11 percent). This figure is worrisome because of the severe risks associated with radiation exposure, and this percentage was undoubtedly much higher than would be found if national data for the United States were available. However, worries about radiation pollution were many times more severe among residents of the Chelyabinsk region. A full two-thirds of those surveyed expressed very serious concern about problems stemming from radiation pollution. Government authorities and officials

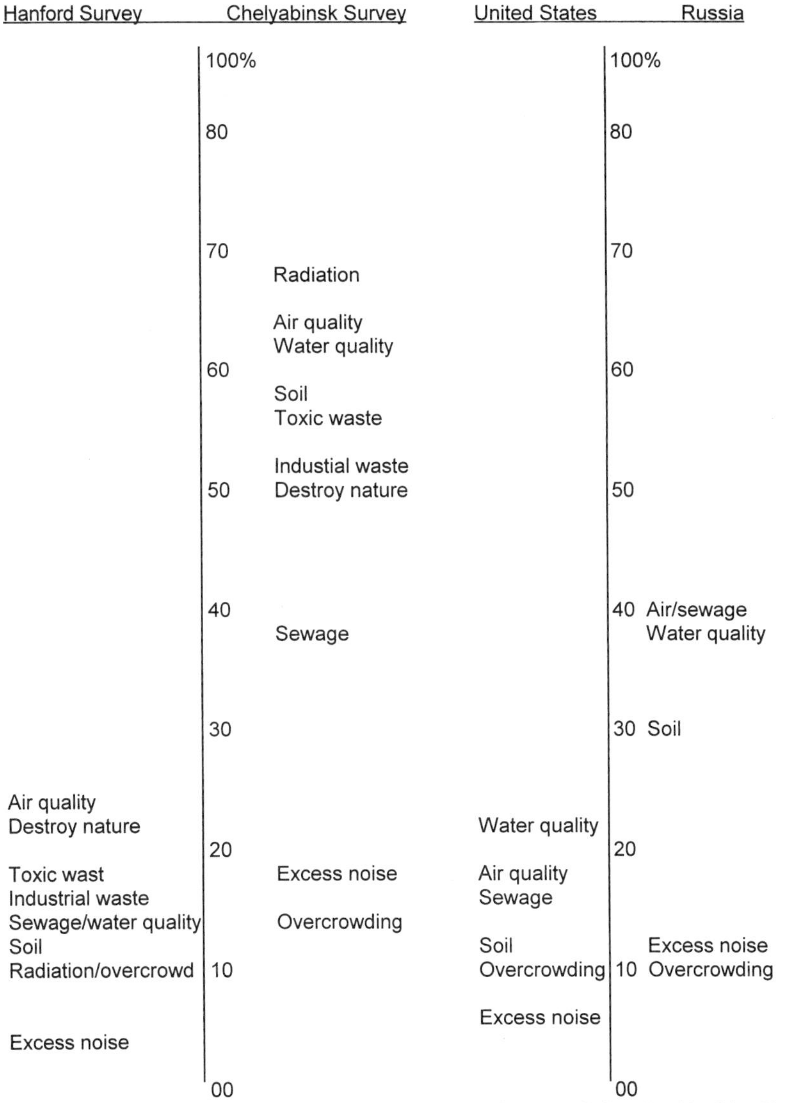

Hanford Survey		Chelyabinsk Survey	United States		Russia
	100%			100%	
	80			80	
	70			70	
		Radiation			
		Air quality			
		Water quality			
	60			60	
		Soil			
		Toxic waste			
		Industial waste			
	50	Destroy nature		50	
	40			40	Air/sewage
		Sewage			Water quality
	30			30	Soil
Air quality					
Destroy nature			Water quality		
	20			20	
Toxic wast		Excess noise	Air quality		
Industrial waste			Sewage		
Sewage/water quality		Overcrowding			
Soil			Soil		Excess noise
Radiation/overcrowd	10		Overcrowding	10	Overcrowding
			Excess noise		
Excess noise					
	00			00	

Note: National data for the United States and Russia is from the Gallup Health of the Planet Survey Dunlap (1994).

Figure 4.2
Perceptions of local environmental problems (percent "very serious")

from Chelyabinsk-65 have repeatedly attempted to reassure the population of the region that no significant radiation risks existed (see chapters 2, 11). These authorities have clearly been unsuccessful in their efforts, however. The troublesome legacy of Mayak has created serious, ongoing worries among the facility's neighbors.

There are also several significant patterns to highlight in environmental concerns across paired sets of communities (table 4.1). Starting with the close-in communities, Richland respondents saw contaminated soil, and toxic wastes, and radiation as the most serious concerns, but at a rather low level of anxiety. In sharp contrast, Chelyabinsk-70 residents rated radiation contamination, soil pollution, and toxic wastes as highly significant problems. In overall terms, Chelyabinsk-70 residents expressed nearly four times higher rates of concern than did Richland residents.[8]

Residents of the downwind city of Chelyabinsk were very concerned with the serious urban environmental problems created by the heavy industry located in the region, in addition to being concerned about the radiation problems associated with the Mayak facility. Roughly two-thirds of the region's residents rated air quality, radiation, toxic wastes, water quality, and industrial wastes as serious environmental problems. Only three conditions were seen as "very serious" by fewer than half of those surveyed—namely, sanitation (46 percent), noise (27 percent), and overcrowding (20 percent).

The residents of the downwind city of Spokane considered air quality, sanitation, water quality, and industrial waste as their most serious environmental concerns. A variety of groups in Spokane have worked to place these local environmental issues on the public agenda (League of Women Voters of Spokane, 1985). Spokane has been under EPA special regulations for nearly a decade in response to air quality problems in the downtown area. There has also been considerable controversy over the advent of major urban growth immediately to the east of the city in advance of proper sewers and storm drain infrastructure, endangering an EPA-designated sole source aquifer (the Rathdrum Prairie Aquifer), which supplies the area's water (Soden, Lovrich, and Pierce 1985). Finally, Spokane hosts the last "waste-to-energy" plant (waste incinerator) to be built in the United States; its licensing and construction caused great public controversy. The levels of environmental concern in Spokane

Table 4.1
Perceptions of local environmental problems by community

Status	Richland	Chelyabinsk-70	Spokane	Chelyabinsk	Yakama Reservation	Kyshtym/ Muslyumovo	Wenatchee	Chebarkul
Water quality	12	37	18	62	10	79	24	56
Air quality	11	35	29	75	20	76	11	36
Contaminated soil	16	43	14	58	22	78	11	41
Sewage, sanitation	6	23	18	46	20	31	9	52
Overcrowding	7	6	12	20	18	10	10	9
Noise	3	8	67	27	6	14	6	9
Radiation	13	49	11	65	19	89	8	48
Toxic wastes	16	38	16	65	23	68	11	34
Industrial wastes	12	35	17	61	23	53	11	41
Natural areas	12	48	25	52	24	50	17	49
Average number rated "very serious"	1.0	3.7	1.6	5.7	1.9	5.6	.9	4.0

Note: Table entries are the percentage rating each problem area as "very serious." Missing data were included in the calculation of percentages.

appear modest—but this may only be because we are comparing perception in Spokane with those of the horrific environmental problems in Chelyabinsk.

A similar contrasting pattern emerges for the rural communities near both facilities. Residents on the Yakama Indian Reservation displayed high levels of environmental concern relative to the other American sites. However, these environmental concerns pale in comparison to those documented among the citizens of Kyshtym and Muslyumovo. Seven of the ten local environmental conditions were rated as very serious by half or more of the residents in these two Russian communities.

For the problems that are more directly related to the nuclear weapons facilities—for example, radiation and toxic waste—host and downwind communities differed relatively little in the Hanford region. For instance, 13 percent of Richland residents rated radiation pollution as a very serious concern, compared with 11 percent of Spokane residents. Comparable cross-community differences were much sharper in Russia. Radiation concerns were much greater in the City of Chelyabinsk (65 percent) than in Chelyabinsk-70 (49 percent). Perceptions of Mayak-related problems reached their peak in the communities of Kyshtym and Muslyumovo, which lie in Mayak's shadow; there nearly nine-in-ten of the respondents perceived very serious problems of radiation pollution!

In summary, Russians expressed a much more negative view of the quality of the environment than did Americans, both in general terms and in the comparison of paired cities. Within both countries, respondents in the close-in host sites and the control sites were less worried about the environment than were the downwind cities and the closeby rural communities. By all accounts (see chapters 2 and 3), these differences reflect both the reality of the situations and the respondents' relative economic dependence on (or familiarity with) the nuclear weapons facilities (Benson and Shook 1985).

Images of the Facilities

Although there are broad differences in the general environmental conditions of eastern Washington State and the Chelyabinsk Oblast, both regions contain a complex that was primarily responsible for the large-scale

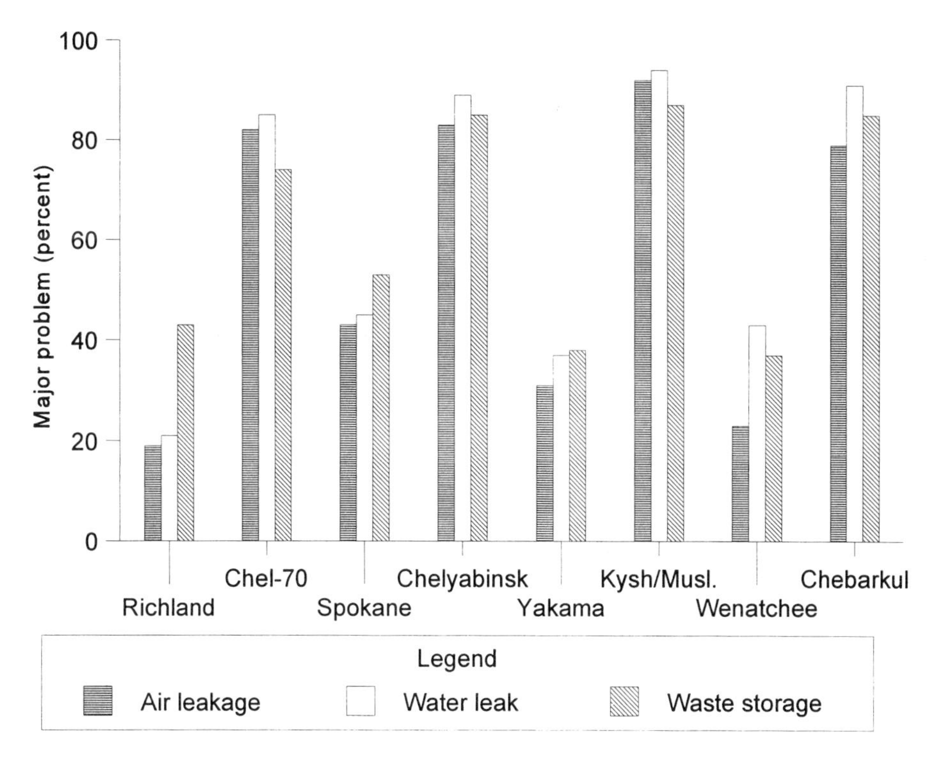

Figure 4.3
Perceptions of past environmental problems at the facilities

production of plutonium and related radionuclides for nuclear armaments—and the public image of these two facilities is a central concern of this study.

Our survey included a variety of measures intended to tap public perceptions of Hanford and Mayak, and their respective environmental legacies. One battery of items asked about the public's awareness of past environmental problems at the facilities (figure 4.3).[9] On the whole, most residents of the Chelyabinsk region were aware that the leakage of radioactive gases into the atmosphere, the migration of radioactive waste into the groundwater, and the safety of long-term nuclear waste storage have been major problems in the past. These sentiments probably reflect the information disseminated by the media, the Movement for Nuclear Safety and other environmental groups. Glasnost and perestroika created a new awareness about Mayak among residents in the region; increased access to information from abroad confirmed the environmentalists' claims that environmental conditions could damage the public health and safety.

Concerns about Hanford's past environmental problems among residents of the various neighboring communities were also quite high, again likely as a result of the activities of environmental groups, the 1989 release of Freedom of Information documents, the dissemination of information by the tribal governments, and the attention Hanford has received from critical news media in Spokane. Nevertheless, American concern over the Hanford facilities was much lower than in the Russian study. In each paired comparison of Russian and American communities, the Russian respondents were substantially more likely than the Americans to see major problems associated with nuclear contamination and the storage of nuclear waste at their respective facilities.

We also find significant differences in perceptions of the facilities' past problems across communities. For example, residents of Richland tended to discount Hanford's past problems; only one-fifth conceded to major airborne releases or leakage into the groundwater in the past, although two-fifths acknowledged major problems with the long-term storage of nuclear wastes.[10] Richland residents often have the most direct familiarity with the activities at Hanford, but Richland is largely a company town dependent on the Hanford facility, and individuals associated with the nuclear industry tend to discount environmental claims against a facility. These sentiments were expressed by one of our Richland respondents: "The storage vaults at Hanford, with very few exceptions, pose no problem to the public. Do not equate change with destruction. Just because the environment changes does not mean it's destroyed." Another Richland resident was even more explicit: "I think that Hanford, in years past, had environmental problems, but as they have been made aware of them, I think they have tried to handle them. With Hanford being mostly cleanup, I think it should be highly respected. I also think Seattle and Spokane people should stay out of our business." Indeed, many Richland respondents were openly critical about questions that implied that Hanford had damaged the environment. This mix of experience and corporate culture generally diminishes environmental concerns.

In a separate meeting with members of the Native American Tribal Nations that are located near the Hanford facility (Yakama, Umatilla, and Nez Perce), we uncovered deep levels of concern that illuminate the general pattern in figure 4. 3. The tribal representatives expressed major worries about the effects of hazardous materials from Hanford on the tribal lands adjacent and downwind (O'Connor et al. 1995). The Yakama, for example, charged that Hanford had transgressed the Yakama Nation's territory through its waste disposal practices. All of the tribes were troubled about the contamination of local rivers and water supplies and the possibility that harmful sediment was endangering fish and wildlife. Moreover, there were concerns that the DOE had not recognized the special status of the Indian tribal governments and had not sufficiently involved them in decisions on Hanford's cleanup.

Russian communities were fairly uniform in their awareness of Mayak's past problems. Even in the closed city of Chelyabinsk-70, nearly 80 percent acknowledged that Mayak had major radiation pollution problems in the past.[11] Starting from this high baseline, residents of the downwind communities scored only slightly higher. Indeed, because of the magnitude of Mayak's past difficulties, and because of the open (albeit grudging) acknowledgment of this past by the government, few could credibly discount this tragic legacy.

Concerns about past problems are worrisome enough, but it is even more troubling if the public expected these problems to continue into the future. Accordingly, we asked respondents in both surveys if they were very worried about the current safety of the respective facilities (figure 4.4).[12] In every pairing of locales, the Russian sample was more likely than its American counterpart to see a grave risk of future harm. For instance, fully 70 percent of the Kyshtym residents and 95 percent of Muslyumovo residents were "very worried" about the safety of the Mayak facility. Many residents of the rural communities located on the Yakama Indian Reservation (18 percent) also were "very worried" about the safety of the Hanford facility. For instance, a resident of the Yakama Indian Reservation echoed these concerns: "Hanford is our main worry affecting our food and rivers, and whether it will contaminate us to death. Day to day we do not know what the politicians have in mind about our livelihood." This level of concern seems modest in comparison to the

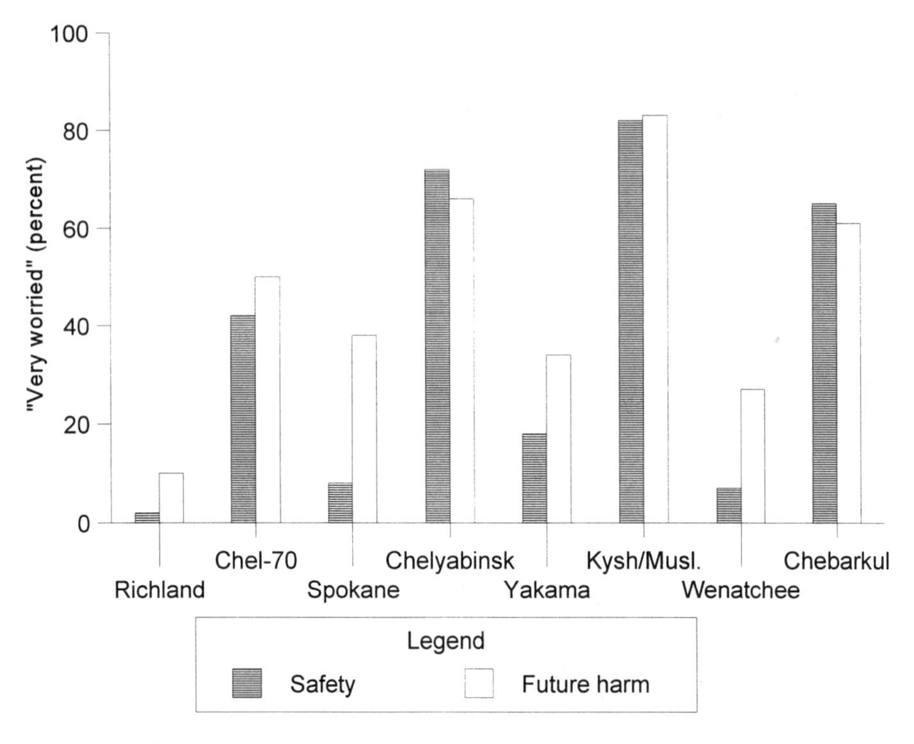

Figure 4.4
Perceptions of future problems at the facilities

Mayak results, but the comparison only dulls our realization that both statistics are alarmingly high if you think of the life conditions of these individuals.

The Russian survey contained an additional battery of items inquiring about the specific future risks citizens associated with Mayak.[13] Based on the weighted sample, we found disturbingly high percentages of Russians who felt that problems with the storage of nuclear wastes (75 percent), radioactive releases into the air (82 percent), and radioactive wastes migrating into water supplies (85 percent) were "very great" risks over the next ten years.

The depths of these concerns, and their manifest reality, can be seen in one other question from our survey. We asked respondents whether the activities at Hanford or Mayak ever affected their health personally or that of a family member. Roughly 5 percent of the Americans who answered this question responded in the affirmative; among residents of the communities on the Yakama Indian Reservation this statistic rose to 7 percent. These are contentious statistics to be certain, but the evidence of Hanford's health effects has steadily been growing (Pacific Northwest Laboratories 1991). In spontaneous comments, one Spokane respondent voiced fears that were shared by others, "I believe Hanford is the cause of concentrated MS disease in Northeast Washington and that releases of radioactive gas in the 1940s are the cause of a high rate of cancer in downwinders. I personally know of six to ten people who have died of cancer from that area and I believe their deaths are attributable to Hanford." Such references to health problems—whether real or mistaken—were common in the downwind communities, but negligible in the Richland surveys.[14]

Although the Hanford statistics are disturbing, they pale in comparison to Mayak's destructive legacy. Over half of the residents in the Chelyabinsk region said their health or the health of a family member had been affected by Mayak. Even in the closed city of Chelyabinsk-70, a plurality (49 percent) answered yes to this question. Mayak's harmful effects were believed to have touched nearly every family in Kyshtym (82 percent) and Muslyumovo (96 percent). Our in-depth interviews with residents of the region echoed these feelings.[15] A resident of Kyshtym saw the impact of Mayak around her: ". . . young people die much more often than before, when we were young. Children are weak. My friend's daughter is always sick, always in the hospital. Her father was working on the contaminated site and was irradiated. Then he hung himself when he found out that he was impotent." A resident of Muslyumovo echoed these sentiments: ". . . all the children are sick here. . . . There are no healthy children, they are born sick." Another resident of Kyshtym linked the early death of his son to cancer caused by radiation exposure. Such sentiments were voiced by most of the people we interviewed. Even if these concerns were somewhat exaggerated, the damage done to the area and the population around the Mayak facility has been too real for these people to entertain much doubt about the public health dangers associated with Mayak.[16]

Skeptics may question the accuracy of the environmental concerns expressed by respondents in our survey, arguing that people tend to exaggerate environmental risks as compared to more mundane and known risks to human safety (Wildavsky 1995; Gerrard 1994). For instance, residents of Richland, many of whom are employed in the nuclear industry and thus were presumably more knowledgeable about nuclear issues than the average person, do not share many of the environmental concerns of their somewhat distant neighbors in eastern Washington. The same contrasts existed between residents of Chelyabinsk-70 and the other Russian communities we surveyed. We admit that it is relatively easy to express environmental concerns in a public opinion survey, and it may have been relatively fashionable to do so at the time of the survey. Yet, as John Whiteley documents in other chapters (chapters 2 and 3), these concerns do have a firm basis in fact. Moreover, in seeking to explain citizen behavior, these perceptions constitute the reality that people take as established fact. If individuals indeed have come to believe the Hanford and Mayak facilities represent a threat to their well-being, their behavior will likely reflect these perceptions to a considerable extent.

The conclusion is thus rather inescapable. Russians were very worried about the quality of their immediate and more general environment, and specifically the health risks associated with Mayak. We uncovered an extremely high level of environmental concern that created a sense of risk or grievance that is seldom found in general population surveys. If strongly felt and widely shared grievances lead to action, then the potential for environmental action among residents in the Chelyabinsk region is substantial. Americans were also concerned about Hanford, although their perceptions of risks associated with Hanford seem modest in comparison to our Russian findings. However, it is only in comparison to the horrors of Mayak that the problems of Hanford appear modest. We should not take solace that only one-third of those surveyed in eastern Washington felt future generations will be harmed by Hanford's legacy of radioactive contamination of the environment. Substantial grievances exist there, too.

Correlates of Environmental Perceptions

Overall perceptions of environmental quality are important in defining the political climate surrounding Hanford and Mayak, but it is equally important to understand the social distribution of these perceptions.

In part, we wish to know if opinion differences across communities arose from the attributes of their populations. In addition, we want to determine how environmental perceptions were conditioned by social position, values, and other individual characteristics. Such analyses can reveal how those perceptions are filtered by—or perhaps contribute to—the cognitive tools individuals bring to their evaluations of those conditions (Rohrschneider 1988). Similarly, these analyses will describe the potential social basis for criticism about the facilities. For instance, knowing the degree to which highly educated individuals differed from others in their environmental perceptions provides evidence on whether the individuals who were concerned about the environment possessed the resources leading to political action.

The correlates of environmental concerns also may affect the nature of the conflict over environmental policy and determine the vehicles through which mobilization occurs. If, for example, there were major differences between the environmental perceptions of younger and older generations, or among those of higher and lower status, environmental conflicts may become intertwined with other dimensions of political division between these groups (Lowe and Rudig 1986). It is also possible, in some exceptional circumstances of extreme environmental conditions, that citizen perceptions are overwhelmed by broadly shared contextual knowledge, and thus there is little explainable variation in opinions or behavior (Walsh 1988).

Social Characteristics

The potential social class basis of environmentalism is a common interest in the analysis of Western publics (Hamilton 1985; Jones and Dunlap 1992; Rohrschneider 1991). To the degree that class or social status differences affect environmental attitudes, researchers usually find that higher-status individuals are more sensitive to environmental problems and more supportive of environmental goals. This class bias arises from several factors. Better educated and higher-status individuals tend to be more knowledgeable about public affairs generally; they can sort out the political and economic complexities of environmental issues and assess the likely consequences of different public policy options (Pierce et al. 1989). Higher-status individuals also have greater personal liberty and financial freedom to choose environmental policies at the expense of jobs, real estate development, or general economic growth that incur environmental costs (Kassiola 1990).

Table 4.2 presents the correlations between social status (education and family income) and two measures of environmental perceptions: the number of local problems listed as very serious and the number of environmental problems associated with the nuclear weapons facility.[17] The correlations are based on the pooled samples in each nation, weighted to reflect the population size of the sampled communities.

In both nations social status had a fairly weak impact on environmental perceptions. For example, higher-income respondents in the American survey were less likely than lower-income families to express concern about local environmental conditions or the problems associated with Hanford. We initially thought that this might arise because of the unusual demographic characteristics of our study: high-status respondents were concentrated in Richland and the downwind communities had lower av-

Table 4.2
Social characteristics and environmental perceptions

	Hanford		Chelyabinsk	
	Local problems	Hanford problems	Local problems	Mayak problems
Social characteristics				
Education	−.07*	.02	−.10*	−.05
Income	−.14*	−.04*	−.03	−.06*
Age	−.12*	.02	−.09*	−.14*
Values and ideology				
NEP	.23*	.27*	.29*	.10*
Postmaterial index	.07*	.18*	.12*	.16*
Liberal orientation	.14*	.28*	.11*	.12*
Party preference	.10*	.26*	—	—
Knowledge				
Knowledge of nuclear terms	−.02	−.02	−.06	.05

Note: For the construction of the local problems and facility problems measures, see footnotes 7 and 9. Table entries are Pearson r correlations, except for party preference in Chelyabinsk, which is an eta correlation; coefficients significant at .05 level are denoted by an asterisk. Results are based on the weighted samples in both nations.

erage levels of education and income. However, there were small negative correlations between income and perceptions of local problems in Richland, Spokane, and the Yakama Reservation communities.[18]

Essentially similar results were obtained from our Russian respondents. There was a weak tendency for higher-status individuals to display less concern about local environmental problems and the risks associated with Mayak. In all cases, however, these relationships were weak and often statistically insignificant.

Age (or generation) is another social characteristic that commonly structures environmental perceptions. A number of studies have observed significant generational differences in environmental attitudes, with younger cohorts exhibiting greater concern for environmental issues, worries about environmental risks, and support for the environmental movement (Dalton 1994; Jones and Dunlap 1992).

To test for these differences, we correlated age with our two measures of perceived environmental problems (table 4.2). Both measures of environmental concern displayed significant age group differences in the Hanford survey—younger respondents were more concerned than their elders about these environmental issues. For example, those under age thirty listed an average of 1.79 local environmental problems as being very serious, while those over age sixty listed an average of 1.20 problems as very serious. Among the four communities we surveyed, the generational gap was greatest in Richland—possibly representing the differences between parents who are employed at Hanford and their skeptical children who may have been more sympathetic to the environmental movement.

Similar age differences appeared in the Russian study. Younger respondents were more concerned with both local environmental problems and Mayak's environmental legacy.[19] These findings suggest the potential for distinct generational dynamics in the mobilization of political support for the environmental movements protesting Hanford and Mayak.

Political Values

Judgments about local environmental conditions are inevitably colored by the perceiver's own personal values. Where a liberal environmentalist might see a development project as destroying nature, a conservative might see it as an attempt to improve the quality of life by providing parks, employment, social and cultural amenities, and pleasant surroundings for the new residents. Such core political values are more central to the belief system than are judgments about specific environmental issues or other policy concerns. Indeed, core political values such as liberalism/conservatism or party attachments have been widely shown to frame and influence more specific political beliefs (Conover and Feldman 1989; Rohrschneider 1991).

One of the major breakthroughs in our understanding of the public's perception of contemporary environmental phenomena occurred with the conceptualization of the "New Environmental Paradigm" (Dunlap and Van Liere 1978; Milbrath 1984; and chapter 7 in this volume). The New Environmental Paradigm (NEP) provides a systematic framework for citizens to organize and interpret political issues, including environmentalism. The NEP incorporates a view of the world that is holistic, integrated, and nonanthropocentric. The world is seen as a system in which all living things—humans, plants, and animals—are equally important and interdependent. Individuals adhering to the NEP should be more sensitive to environmental problems and more attuned to environmental risks.

Another body of research suggests that environmentalism is linked to a rise of postmaterial values in advanced industrial societies (Inglehart 1990, ch. 11; Dalton 1994). Ronald Inglehart, the main proponent of this theory, argues that Westerners who matured in the relatively prosperous and peaceful times following World War II have shifted their value priorities from the materialist and security interests of prior generations to a new set of postmaterial values. Postmaterialists pursue "higher order" needs, such as the enjoyment of full participation in political decisions affecting them and the pursuit of policy priorities emphasizing the quality of life and equity issues pertaining to race and gender. The convergence of these two orientations—greater emphasis on participation and greater concern for the environment—may be important in understanding the dynamics of political mobilization in a case such as Hanford (Steger et al. 1989). Postmaterial values orientations should generate the policy incentive (environmental concerns) and the motivation to achieve those goals through citizen action.

Theoretical expectations for the Russian public are less certain. Both of these conceptualizations—the NEP and the advent of postmaterial values—link the rise of the environmental movement to a process of value change in advanced industrial societies. Can these theories of sociological change in advanced industrial societies apply to contemporary Russia? We address this theoretical and empirical question in the course of our study. Research on the Russian environmental movement has frequently drawn parallels between the values of Russian environmentalists and the NEP/postmaterial values espoused in the Western movement (see chapters 7 and 8; DeBardeleben 1992). In addition, empirical research has found some evidence that those western value orientations exist among the Russian public, even if their genesis may be different. Consequently, we believe that we should treat the impact of such values as an empirical question to study.

The middle portion of table 4.2 displays the correlations of NEP values and postmaterial values with two measures of environmental conditions.[20] Among respondents in our Hanford survey, support for the NEP had a strong effect on perceptions of local environmental conditions, and especially on concerns about the Hanford facility. There was a clear parallel in the Russian survey; individuals with biocentric values were more likely to express concern about the local environmental problems facing their community, and felt there are serious problems resulting from Mayak.

The postmaterial values scale in table 4.2 was a subset of the items that Inglehart has employed in his many studies (1990). The scale contrasts those who emphasize material goals of a stable economy and the fight against crime versus postmaterial goals of a society that is more humane and emphasizes ideas over money (see appendix B). Postmaterial values produced a strong effect among the American survey respondents. The influence of postmaterial values was strongest for perceptions of Hanford's environmental risks, which suggests that these views were most colored by these underlying values. These values exerted their strongest impact in the host city of Richland and the surveyed communities on the Yakama Reservation. For instance, postmaterialists in the Yakama communities listed a number of local environmental problems (mean = 3.56) that was more than twice as high as for materialists in these same communities (mean = 1.59); the ratio among Richland residents was even wider. Environmental judgments seemed strongly conditioned by individual values.

Despite our ambiguous expectations for the Russian study, table 4.2 indicates that individuals who expressed postmaterial priorities were more likely than materialists to be concerned about the local environmental problems facing Mayak and the consequences of these problems. Significantly, value priorities exerted a similar impact in the closed city of Chelyabinsk-70 as in the downwind city of Chelyabinsk. The one notable exception to this pattern is in the communities of Kyshtym and Muslyumovo, where values made little difference (either the NEP or postmaterialism). We suspect that the overwhelmingly grim reality of the environment in these communities made the value orientations of citizens relatively unimportant as an influence on perceptions.

In addition to political values with specific environmental connections, such as the NEP and postmaterialism, broader political orientations might also influence environmental perceptions. For instance, liberal/conservative labels have consistently proven to be important political cues in framing the political beliefs of the American public across a wide range of policy areas (Conover and Feldman 1989). Even if specific policy-relevant information is lacking, the description of a policy as liberal or conservative helps citizens to evaluate the policy and determine their own position.

Such political positions were certainly more fluid in Russia in the early 1990s, as the political and social systems were undergoing fundamental and rapid changes. Yet one could argue that in such a turbulent environment the public would find even more value in developing broad ideological categories to help organize (and simplify) the diversity of political interests at play. Moreover, the environmental movement in Russia established a clear reformist identity during the waning years of the Soviet Union, and may continue to be perceived and evaluated in those terms.

Table 4.2 presents the correlation between the ideological identities of the public and the listing of environmental concerns (see appendix B). Liberal/conservative positions were strongly related to environmental perceptions among the Americans we surveyed, especially for evaluations of Hanford's problems. Despite the fluidity of Russian politics, reformist/

conservative identifications were also linked to concern over the environmental conditions at Mayak. Taken together with the previous findings, these results suggest that environmental concerns were partly a reaction to local environmental conditions, but they were also to a significant extent a projection of an individual's political values.

Policy-Relevant Knowledge

There is extensive controversy about the appropriate role of mass publics in decision-making processes involving highly technical or scientifically complex public policy issues—such as the radiation pollution controversies surrounding Hanford and Mayak (Kuklinski, Metlay, and Kay 1982; Mitchell 1984; Pierce and Lovrich 1986). The focus of the controversy is whether individuals can reasonably come to possess the knowledge required to make sensible choices in these complex policy areas.

Environmental policy is one area where this question of policy knowledge arises with considerable frequency. On the one hand, many environmental interest groups are inclined to push for maximum openness of official records and decision-making processes on the grounds that the public is entitled to make "informed choices" on matters of natural resources and environmental protection (Gundersen 1995). Environmentalists frequently suspect government agencies and industry interests of hiding information from the public, which citizens could use to develop reasonably informed "elite-challenging" positions on public policy questions related to nuclear weapons plants (Steele 1990). On the other hand, many natural scientists, policymakers and policy experts argue that the environmental movement frequently misuses scientific evidence to mislead and manipulate public opinion. They claim that environmentalists often engage in scare tactics to gain support for their positions among the unsuspecting and scientifically illiterate public (Rothman and Lichter 1982; Schneider 1991). Indeed, one of our Richland respondents strongly expressed this position: "The public is, in general, not qualified to be involved in decision making about problem resolution (treatment) of nuclear waste." These are somewhat oversimplified characterizations of much more sophisticated positions, but the point is clear: the level of the public's policy knowledge is thought to have important consequences for public perceptions of environmental matters.

The effect of citizen knowledge on environmental perceptions is especially important in the twin cases of Hanford and Mayak. Secrecy and official deception based on national security needs have been the norm for decades. As information became available about the abuses of public trust committed at both facilities, or as public statements were made about the dangers of intentional or accidental releases of radiation, how did the public formulate autonomous conclusions about the appropriate course of collective societal action? Does the level of technical or scientific knowledge about nuclear energy magnify or dampen environmental concerns?

In order to estimate the knowledge levels of respondents, we asked for self-assessed familiarity with six specialized items associated with nuclear weapons production and toxic waste storage processes: radioactive waste, spent fuel, plutonium separation, fuel fractionalization, strontium, and the PUREX process. Respondents indicated whether they knew the meaning of the term, had heard of the term but did not know the meaning, or did not recognize the term. We created a simple count of the number of terms that the respondent claimed to know. Knowledge is obviously a difficult trait to measure, and self-assessments in a general population survey are not the best means of determining an individual's true level of knowledge. However, earlier environmental studies using this type of measure indicate that there is a high correlation between citizen self-assessment and actual knowledge as determined by follow-up questioning (Pierce et al. 1988). In any case, this index will tell us what people believe they know, which may be only an approximation of what they actually do know.

Both surveys found that self-assessed knowledge was essentially unrelated to environmental perceptions (table 4.2). In the Hanford study, there was virtually no relationship between knowledge level and perceptions of local environmental problems or concerns about the risks of Hanford. In Richland, however, where the potential knowledge gradient was the sharpest because the city contains many residents who work at Hanford, there was only a weak negative correlation between knowledge levels and concern about local environmental conditions (−.11) or Hanford's environmental problems (−.14). In the other communities in our Hanford survey, knowledge and environmental concerns were essentially unrelated.

Knowledge levels were also relatively independent of environmental perceptions in our Chelyabinsk survey. There was a slight tendency for the more knowledgeable to see fewer local environmental problems, but greater problems with Mayak's pollution. In both cases, however, these relationships were not statistically significant. As with the Richland findings, within the community of Chelyabinsk-70, where nuclear expertise abounds, there was a strong negative relationship between knowledge and perceptions of local environmental problems. However, concern about Mayak's problems was essentially unrelated to knowledge. Furthermore, among the entire Russian sample, worries about the future risks of Mayak were greatest among those who are most knowledgeable about nuclear technology.

In summary, there is no conclusive evidence that knowledge (and presumably informed judgments) substantially affects public evaluations of the environmental problems connected with Hanford and Mayak.

Cumulative Impact

The final step of our analysis considers the cumulative capacity of the social characteristics and political beliefs to explain perceptions of the environmental conditions at Hanford and Mayak. We have examined each of these factors separately, but their influences are potentially overlapping. Thus we are also interested in the independent impact of each variable in explaining environmental views.

We combined the data from our various communities into a pooled analysis. For these pooled samples we collected the variables we have previously analyzed into a single multiple regression analysis, adding "dummy variables" to represent the effects of local context on environmental perceptions. We calculated separate regression analyses to explain perceptions of local environmental problems and perceptions of the environmental problems specifically associated with Hanford and Mayak.

The first two columns of table 4.3 present the results from the Hanford survey, the next two present the results from the Chelyabinsk survey. In large part, these results reflect the basic patterns arising from the preceding analyses. The independent impact of most variables is lessened, as is normally the case in multivariate regression where the overlapping effects of variables are held constant. Political attitudes have a more direct and stronger impact on environmental views. The New Environmental Para-

Table 4.3
A multivariate analysis of environmental perceptions

	Hanford		Chelyabinsk	
	Local problems	Hanford problems	Local problems	Mayak problems
Education	−.06	.05	−.10	−.06
Income	−.14*	−.01	−.09*	−.09
Age	−.12*	.00	.01	−.10*
NEP	.17*	.15*	.24*	.10*
Postmaterial index	.03	.11*	.03	.10*
Liberal orientation	.05*	.15*	.10*	.05
Policy knowledge	.02	.02	−.05	.08
Spokane/Chelyabinsk	.13*	.15*	.11*	−.13*
Richland/Chelyabinsk-70	.04	−.10	−.14*	−.28*
Yakama/Kyshtym-Musl.	.09*	.05	.05	−.05
R	.34	.40	.37	.36

Note: Table entries are standardized regression coefficients; coefficients significant at .05 level are denoted by an asterisk. Results are based on the unweighted samples in both nations.

digm is strongly related to perceptions of environmental problems in the Hanford and Mayak surveys. Similarly, postmaterial values (which partially measure similar attitudes) also display a significant relationship to environmental worries. These results suggest that environmental concerns are magnified by value priorities that accentuate such quality of life issues.

The lower rows of the figure represent the effects of local conditions in our three comparative sites, relative to the control community. These effects can be interpreted as the impact of local conditions once the social and attitudinal differences between communities are statistically controlled. In the American study, residents of Spokane were significantly more concerned than residents of Wenatchee in terms of local environmental conditions (.13) and the effects of Hanford (.13). In contrast, Richland residents differed only slightly in terms of local environmental conditions, and were less concerned (−.10) with Hanford's past problems. In short, local context was still very important in structuring opinions, even when social and attitudinal factors were taken into account.

The Russian findings reinforce this conclusion. Residents of Chelyabinsk City (.11) displayed greater concerns about their local environmental problems than did residents of Chebarkul, while residents of Chelyabinsk-70 display significantly less concern (−.14). The last equation indicates that the residents of Chebarkul held relatively high concern about Mayak's problems even though this was our supposed control site. This pattern can also be seen in figure 4.4, and was even more pronounced once social and attitudinal factors were controlled. The important comparisons are the relative rankings of our three other sites. Residents of Chelyabinsk-70 were least concerned about Mayak's problems (−.28), residents of Chelyabinsk City were slightly more concerned (−.13), and residents of Kyshtym-Muslyumovo were even more concerned (−.05). Geographic context is an independent influence on opinions.

Environmental Perceptions and their Implications

This chapter examined how residents of eastern Washington and the Chelyabinsk region perceived their local environment and the environmental impact of Hanford and Mayak. The respondents in the Russian survey were uniformly negative about their local environment and Mayak's environmental consequences. In comparison, Americans in the Hanford area were more positive about local conditions, and less worried about Hanford's environmental legacy. Yet even in eastern Washington state, the levels of concern with the potential health and safety problems associated with Hanford were troublingly high. Neither nation managed to escape the long-lasting negative environmental consequences of the race to produce nuclear weapons.

The one overreaching similarity between the two countries involves the common pattern of perceptions across the four study cities. In both the Hanford and Chelyabinsk regions, citizens in communities that are closely connected with the nuclear weapons facilities were relatively more positive about their local environment and the environmental impact of the facilities. Residents in communities that are somewhat removed from the facilities—either in spatial or employment terms—were much less positive about the region's environment.

The two nations differ in the correlates of environmental perceptions. The overwhelmingly negative views expressed by Russians apparently have washed away the capacity for individual level variables to generate substantial differences in environmental perceptions. The feelings of "victimization" by Russian respondents focused on the far-reaching and imminent dangers they perceived in their environment. The victim orientation may also be an important consideration among American downwinder populations, but to a more limited extent. In comparison, American respondents differed in their environmental perceptions based on their political orientations, their position in society, when they were socialized, and their general value orientations.

These findings hold two broad implications for the questions underlying our research. First, both publics harbored substantial doubts about the environmental impact of both facilities. Extensive public relations campaigns by both facilities over the past several years, after decades of rigid secrecy, have failed to reassure Hanford's or Mayak's neighbors. Even if chemists and physicists debate the scientific evidence of the facilities' environmental problems, these problems were real in the minds of much of the publics in these two regions.[21] When one-third of the Hanford area residents and two-thirds of those in the Mayak area felt there was a "very great danger" of future harm from the facilities, this represents a level of environmental concern that was large and real to these individuals.

These findings also have implications for the prospects of political mobilization in both nations. Chapter 1 noted several conditions for mobilization, including perceptions that some wrong must be corrected. In both countries this stimulus to mobilization was present, especially in the Chelyabinsk region. In addition, we stated that there must be some social, ideological, or political structure through which the elite-challenging perceptions of the environment can be converted into collective action. This potential seems much greater in the American context than in the Russian one. The much greater statistical predictability of environmental perceptions in the Hanford survey defines the social framework through which individuals could find a context and cognitive framework for action. Those variations imply shared interests beyond the environment itself around which people can be mobilized, and which can constitute additional salient stimuli for action. For example, interpreting environmental conditions within the broader cognitive framework of the New Environmental Paradigm provides a much more comprehensive view of the need for action and a more sophisticated view of the political and social implications of inaction. With regard to these findings, the potential for political mobilization was greater in Russia based on the first-order views of the negative quality of the environment, but greater in the United States based on the more powerful grounding of these views in other relevant social and political divisions.

Having made these observations on the basis of survey findings, it is proper to note that political mobilization obviously depends on much more than perceptions of the environment. Other attitudes, such as evaluations of political actors and feelings of political efficacy, will also influence the likelihood of political action. The political skills and resources that an individual possesses will also influence his or her behavior. The opportunities for participation and the framework of action will also affect mobilization patterns. We must remember, too, that through the 1990s Russia was suffering from severe economic problems and other policy challenges associated with the transformation of the political and economic systems. Thus environmental problems competed with a very full political agenda. Subsequent chapters will examine these elements in more detail. However, we begin this inquiry knowing that residents of the two regions harbor serious doubts and concerns about the past and present activities of both Hanford and Mayak.

Notes

1. Richland is one of the Tri-Cities that are adjacent to Hanford. Richland is most closely identified with Hanford and is home to many of the technical and scientific personnel employed at the facility. Pasco and Kennewick are adjacent communities located across the Columbia River.

2. Chelyabinsk-65 is the closed city that actually contains the nuclear production facilities. We attempted to include this in our study, but were excluded by Mayak officials despite the prior approval of the Russian Ministry of Atomic Power.

3. Federally recognized Native American tribes also hold a unique status as a sovereign government. In theory, this status provides tribal governments with a privileged legal and political position for dealing with policy issues and U.S. government officials that is distinct from other stakeholders.

4. The question reads, "Overall, how would you rate the quality of the environment here in our nation, in your local community, and the world as a whole? Would you say very good, fairly good, fairly bad, or very bad?"

5. The Gallup national surveys found the following percentages rating the environment as "very good" or "fairly good":

	United States	Russia
Local environment	71	28
National	52	7
Global	26	11

6. Chelyabinsk-70 status as a closed city creates economic advantages and other resources that may explain the more positive views of these residents. In addition, the Russian secret atomic cities were often established in the most beautiful natural areas; this applies to Mayak.

7. We also asked citizens in both surveys to describe their level of overall concern regarding environmental problems. The Russian samples were much more likely than their respective American counterparts to express a "great deal" of concern for contemporary environmental problems. In this instance, too, residents of the Chelyabinsk region were somewhat more concerned about the environment than were Russians overall, and residents of the Hanford region slightly less concerned about the state of the environment than were Americans overall (Dunlap, Gallup, and Gallup 1993). Similarly, Andrei Zaitsev (1993) found that environmental concerns are greater in the Chelyabinsk area than in other areas in Central Russia.

8. The numerical entries below each community are the number of problems perceived as "very serious."

9. The question read: "There has been a lot of discussion about whether Hanford represents a threat to the environment. In the past, do you think there have been problems with the leakage of radioactive gases? Would you say this has been: a major problem, a slight problem, or no problem at all? The leakage of radioactive waste into the groundwater? The storage of nuclear wastes from the plant?"

10. The problem of leaking storage tanks is a very significant official concern at Hanford, and those closest to the site could be expected to be highly sensitive to it (Dunlap et al. 1993). This may explain why Richland residents were more concerned about past waste storage problems than either residents of the Yakama Indian Reservation communities or Wenatchee.

11. In addition to the environmental problems chronicled in chapter 3, there was a significant holding tank explosion at the Mayak plant in 1986. Just before our survey went into the field, technicians from the facilities identified a significant amount of radioactive material in Muslyumovo and had to remove it from the community.

12. The two questions read: 1) "Do you worry about your safety or that of your family because you live near the facility?" and 2) "How likely is it that future generations will be harmed by the facility?"

13. The question read: "How likely is it that harmful amounts of radioactivity will leak out and contaminate the air in the next ten years? Is there a very great risk, a slight risk, or no risk at all? How likely is it that radioactive waste will leak into the water supply in the next ten years? How likely is it that the long-term storage of radioactive waste will be a problem?"

14. Through our surveys and informal interviews it became apparent that downwind residents frequently drew a link between Hanford's activities and incidences of cancer and other medical problems among their friends and family. Even if oncological studies are not able to separate radiation effects from "natural" cancer sources, the linkage was seen among downwinder groups. In contrast, residents in Richland seemed predisposed to equate cancer and other illnesses to non-Hanford sources.

15. A separate study by the Kaluga Institute similarly finds that health complaints are higher in Chelyabinsk than in "control" environmental zones in Central Russia (Zaitsev 1993).

16. The Chelyabinsk region is dotted with examples of the health effects of Mayak. For instance, the Chelyabinsk Children's Hospital contained a ward overflowing with children suffering from leukemia that doctors largely attributed to Mayak; most of the children in the ward at the time of our fieldwork died before this book came into print. Similarly, Garb (1997) cites the depressed life expectancies of Muslyumovo residents. Also, see Whiteley (chapter 11).

17. The measure of local problems is a count of the number of items in table 4.1 that are rated as very serious local problems (range from 0 to 10); the other measure is a count of the number of concerns associated with the facilities in figures 4.3 and 4.4.

18. Dunlap et al. (1993: 155) similarly found that support for a nuclear waste repository at Hanford was more common among the better educated and higher-income respondents in the Tri-Cities area.

19. Surprisingly, the young were not more concerned about future risks associated with Mayak ($r = .04$), a separate battery of questions that was only available in the Russian survey (see above). This may reflect a fatalistic view of youth, or the simple uncertainty of predicting the future.

20. For more information on the construction of these indices see chapter 7. In the American study we created a simple additive index of the ten items constituting the NEP battery. In the Russian survey we used the biocentric index, which displays the closest empirical approximation to the full NEP measure in the Hanford study.

21. At the research conference where these results were first presented, they stimulated a debate between a nuclear chemist and a physicist over whether these concerns had a scientific basis. This is an important consideration, but both individuals missed the primary point. To our respondents, these were real concerns. Thus these concerns would influence their attitudes and behaviors toward the facilities.

III

The Environmental Movement

5

The Development of Interest Group Activism at Hanford

William Schreckhise

Hanford is a mismanaged, unsafe taxpayer rathole . . . [It] is a microcosm of the national security state, inefficient, lacking the usual checks and balances . . . they got away with it because of secrecy.

—Tom Carpenter,
Director, Government Accountability Project
August 31, 1992

We asked [former Energy Secretary] O'Leary for a drink of water. She not only gave us a drink of water, but the whole treatment plant. We give her an "A," or even an "A+" for her efforts.

—Tom Carpenter,
June 16, 1994

Over fifty years of nuclear weapons production have left a troublesome legacy for the post–Cold War world. In both the former Soviet Union and the United States, the production of weapons of mass destruction left the areas in which they were produced dangerously contaminated. Furthermore, the rushed production of these weapons led to radioactive emissions that may have affected the health of those living near the areas where nuclear weapons were developed, manufactured, and tested (see chapters 2 and 4). This chapter concentrates on how some of the residents near one of these sites—the Hanford Nuclear Reservation—have mobilized into interest groups in response to the environmental contamination caused by the production of plutonium, and by the related radioactive releases made into the region's natural environment. The chapter describes how citizens have come to terms with a past that was kept secret from them for decades as new information became available about the nature of Hanford's past operations (Gerber 1992a; D'Antonio 1993).

It describes the process by which these groups moved from the role of highly critical out-groups on the fringe of the policy process to wielding a noteworthy degree of influence in the contemporary operations of the Hanford site.

Systematic study of the type of environmental interest groups presented here also can provide us with important information about the contemporary governmental process. These groups typically evolve at the grassroots level with relatively little elite support, but nonetheless they have proven able—within the space of a few short years—to influence public policy substantially. Our research contributes to the identification of the major dynamics of interest group formation and development by those individuals seeking to influence public policy in post–Cold War America. Of course, we also assess the extent to which the American political system has adapted to the demands of these new claimants for government to be responsive to public health and environmental safety concerns.

Unlike the civil rights and women's movements, environmental interest groups seek changes in public policy and social attitudes that do not require an increase in social standing for their members. Instead of arguing for social progress to benefit an identifiable group in society, they advocate collective goods, such as setting aside wilderness areas, providing for the protection of endangered species, reducing risks to public health, or the preservation of the natural environment in general (Ingram and Mann 1989). In the case of strictly "preservationist" groups, these advances for nature may come at considerable expense to humans. In the case of groups advocating air, water, and food system safety, their advocacy on behalf of the protections of nature come hand in hand with benefits to human health and societal well-being.

In addition, the work of environmental groups represents a key advocacy mechanism for the citizen activism vital to the functions of a democratic system. Interest groups provide an important source of information for their membership to remain knowledgeable about their particular interests (Berry 1984; Pierce et al. 1992), and they provide independent information for the media, the government, and the general public. Most important, of course, environmental groups serve the functions of interest articulation and interest aggregation in an American political system where a pluralistic struggle among contending groups is a hallmark of the policy process (Schattschneider 1960; Lowi 1979, 1995). The environmental groups protesting the management and operation of Hanford were key actors in keeping the issue on the political agenda, in mobilizing public support for their positions, in developing rival expertise on the facility's environmental consequences, and ultimately in pressuring the government to act (or react). These activities provide a trail of evidence as to how the Hanford advocacy groups originated and developed over time, and how their contemporary role as interest representatives within the democratic process came into being. What follows is an abbreviated history of the activism of these environmental groups at the Hanford Nuclear Reservation.

The Particular Case of Hanford Area Environmental Groups

The study of Hanford environmental interest groups is especially instructive because these groups have struggled to extract information from a public agency shielded by "national security" protection, and they were located in a region that was hostile toward those harboring antinuclear views and other forms of environmentalist advocacy. As compared to preservationist environmental groups (Nash 1967), the Hanford area groups have been somewhat less concerned with the general welfare of the environment and more concerned with the health and safety of humans. These groups have evolved out of a perceived direct threat to the health and safety in their immediate region stemming from the plutonium production activities of the Hanford nuclear site.

Perhaps most instructive about these groups is the key role information played in their creation. Significant feelings of mistrust toward Hanford were generated among the general public of the region within the space of only a few short years. This mistrust was not so much a result of a lack of information, but rather the consequence of an abundance of hitherto secret information revealing that area residents had been deceived by their own government (see chapter 2). As information was released from the site, attitudes toward Hanford dramatically worsened in the region (Gerber 1992a). Our central question concerns the role that environmental interest groups played in causing the Hanford Nuclear Reservation to be transformed from a source of great pride among those living in eastern Washington to a social and political pariah institution. The role of information gathering and strategic dissemination was crucial in this transformation, and the part played by environmental interest groups in the area was substantial by all accounts.

The Climate of Public Opinion

Beginning with its construction in 1943, and going on until the end of World War II (which was occasioned by the dropping of a bomb produced with Hanford plutonium), only a few people knew the nature of the work being done at the site. In 1946, the Army Corps of Engineers turned over authority of the Hanford site to the newly formed Atomic Energy Commission (AEC). Although the overall nature of the work done at Hanford had been revealed to the public, it still involved highly sensitive national security information and the AEC continued to operate in official secrecy. Written into the Atomic Energy Act of 1946 was a key provision that explicitly allowed the AEC to produce nuclear weapons in secret (Rosa and Freudenburg 1993; Grainey and Dunning 1995/1996). This "canopy of secrecy" continued to exist during the postwar years and allowed the AEC (and its successor agencies) to operate without independent oversight for the health and safety of the workers at the site or the surrounding population. Because there was no independent oversight, the AEC pledged itself to conform to the radiation safety levels as outlined by the National Committee on Radiation Protection (established to ensure the safety of commercial nuclear power production facilities). However, on numerous occasions these levels were either surpassed or even raised to permit greater productivity during the heyday of the Cold War (Gerber 1992a). Even when significant radiation releases occurred, as during the Green Run of 1949 and in several incidents occurring during the 1960s, few people in and around the site felt they were in any danger from plant operations and no noteworthy area-based political opposition to the nuclear site was in evidence.

Because there was no independent oversight or citizen scrutiny of Hanford's operations, area residents were inclined to be highly supportive (or at least indifferent) to the operations there. In fact, the first active interest group to focus on the workings of Hanford was a decidedly pronuclear organization—the Tri-City Industrial Council (TCNIC). The group was formed in 1963, in part to forestall (or at least to minimize) the regional economic impact of what they anticipated to be pending budgetary cutbacks for Hanford operations. With the aid of Washington's two influential senators, Henry J. "Scoop" Jackson and Warren G. Magnuson, the organization was able to ensure not only that Hanford's budget was not cut, but also that the contract for its operations was "segmented," meaning that a larger number of contractors were brought in, a decision that supported the group's economic diversification goals. The local advocacy group influenced the Atomic Energy Commission to hire a contractor at the Hanford site specifically to conduct research into nuclear power production. The group was also instrumental in slowing down the scheduled closings of Hanford's older production reactors, and in convincing the AEC to transfer a parcel of land to the state of Washington. The state, in turn, leased that land to a private corporation for the creation of a low-level radiation waste repository (Fleischer 1974). Although such "pronuclear" feelings would remain well into the present day with many residents within the Tri-Cities themselves, such supportive feelings toward the Hanford site would shift dramatically around the region in the years ahead.

Oppositional activities at Hanford followed a slow evolution over the next several years, as did a similar evolution in public concern over nuclear power and nuclear weapons production at the national level. At the national level, Ebbin and Kasper (1974) point to the late 1950s and early 1960s as the period in which nascent opposition developed against nuclear power. In 1956, the International Union of Electric, Radio, and Machine Workers brought suit against a nuclear power company over concerns about the dangers to public health and workplace safety involved in nuclear weapons production. In 1961, a small group of citizen activists protested the construction of a Pacific Gas & Electric (public utility) nuclear power plant in Northern California (Mazur 1981). However, substantial opposition to nuclear power did not develop until the late 1960s, when a substantial amount of citizen advocacy occurred and a smattering of academic literature began focusing on the environmental and the health effects of nuclear power plant construction (Mazur 1981; Mazur 1990; Rosenbaum 1993; see also Ebbin and Kasper 1974, 11–13 for a discussion). A large portion of this concern for the environment was directed at the safety and health effects of nuclear power production stimulated by academic researchers working in the area of nuclear power plant construction (Tamplin 1971; Forbes et al. 1972; Ebbin and Kasper 1974; Nealy, Melber, and Rankin 1983). The first "Earth Day" in 1971 also helped focus attention toward the potentially negative effects of nuclear power.

With any dramatic shift in public opinion toward nuclear power still several years away, the oil embargo of the early 1970s refocused policy debates on nuclear power production. The oil embargo in 1973 forced attention on nuclear power as an alternative energy that the United States could substitute for foreign oil (Mazur 1981). Support for nuclear power remained strong through the early 1970s despite some early discussion of its possible dangers (Rosa and Freudenburg 1984). By the mid-1970s, public support for nuclear power began to move noticeably lower; in 1976, citizens groups in eight separate states (including Washington) placed statewide initiatives on their ballots calling for the limitation of nuclear power construction. All of these initiatives were defeated, but it was clear that opposition to nuclear power was of growing significance in American politics. Group-based opposition to nuclear power manifested itself as environmental groups around the country pursued "direct action" methods to protest plant construction (Downey 1986).

The most dramatic shift in public attitudes toward nuclear power occurred in 1979 with the accident at Three Mile Island (Gordon and Knapp 1989; Rogovin 1980). Coincidentally, the TMI incident occurred when the movie *The China Syndrome*[1] was playing countrywide and as President Carter was planning a major program for the production of energy (Mazur 1990). Perhaps because of these particular circumstances, the TMI fiasco received extremely heavy media attention. This media focus in turn changed public opinion on nuclear energy dramatically. Several scholars have argued that no other event has had such an impact on our feelings toward nuclear power (Nealy, Melber, and Rankin 1983; Rosa and Freudenburg 1984; Morone and Woodhouse 1989). Prior to 1979, either a clear majority or a substantial plurality of Americans favored nuclear power. In the months following TMI, however, this situation was starkly reversed; most Americans were critical toward nuclear power plant construction, and the number of critics of nuclear power continued to grow long after the accident (Rankin, Nealey, and Melber 1984; Rosa and Freudenburg 1984). By the early 1980s, electorates or legislatures in six states enacted legislation either prohibiting or greatly restricting the construction of nuclear power plants within their borders.

During this same time, the Tri-Cities area (Richland, Pasco, and Kennewick), where the Hanford site is located, experienced a substantial boom period because of the construction of new commercial reactors by a consortium of regional utilities (the Washington Public Power Supply System). Because of this prosperity, local voices of dissent were minimal and not warmly received (Loeb 1986). In fact, residents of the cities closest to the Hanford site were the most supportive of nuclear power, at least far more supportive than the rest of the nation. In a 1987 survey, the percentage of Americans opposed to constructing more nuclear power plants had reached 65%, while in the Tri-City area only 28% were opposed (Dunlap et al. 1993). If one were to carefully examine the sentiments of the residents most proximate to Hanford during this time, one would not predict that any regional opposition to Hanford operations would arise anytime soon.

Many scholars contend that the serious accident at Chernobyl further solidified national opposition toward nuclear power (e.g., Schneider 1986; Marone and Woodhouse 1989). The dramatic fire at the Soviet site, and the apparent helplessness of the Soviet authorities to limit damage to their citizens and their environment, led greater numbers of people to feel that such an accident could happen in the United States. Other scholars argue that Chernobyl only increased opposition to nuclear power slightly (e.g., Rosa and Freudenburg 1984). Regardless of the ultimate impact attributed to Chernobyl, opposition toward nuclear power undeniably grew during this period. The Reagan administration's arms buildup also played a critical role in changing public opinion toward both nuclear reactors and nuclear weapons (Rosa and Freudenburg 1993; Mazur 1990). As Reagan maligned the USSR as "the Evil Empire," deployed Pershing II cruise missiles in Europe, and argued that it was possible for one side to win a nuclear war, many voices in the public became increasingly leery of nuclear power and nuclear deterrence as a means to insure peace. Citizen protests began to shift from opposing nuclear power plant construction to challenging nuclear weapons production (Mazur 1990).

Such misgivings concerning nuclear technology were not limited to citizen activists; more and more of the general public and politicians began to call for controls on the nuclear arms race. Senators Mark Hatfield and Edward Kennedy sponsored a "freeze" resolution in the U.S. Senate. On the West Coast, over one-half million citizens signed a petition to place a nuclear "freeze" referendum on the California ballot. Similar referendum petitions were circulated in Michigan, New York, and Delaware. The threat of nuclear war cut into the national psyche in an unprecedented manner in November 1983, as millions of Americans watched the powerful film, *The Day After,* which depicted the pitiful plight of the survivors of a nuclear war (Mazur 1990).

As national public opinion began to shift toward opposition to nuclear arms, Hanford's operations were actually stepped up as part of the Reagan-era rearmament effort, and several of the site's facilities were brought back into production after several years of sitting in "standby" status. In 1983, chemical separations were started up again at the PUREX facility after being maintained for eleven years on standby status, and in 1984 the Plutonium Finishing plant was also restarted after several years of standby status and upgrades (Department of Energy, n.d.). Not only were Hanford's facilities restarted because of the Reagan rearmament efforts, but the area was also being considered by the Department of Energy for a nationwide nuclear waste repository. The repository proposed was opposed by growing numbers of Washington and Oregon residents as site selection activities unfolded (Dunlap et al. 1993).[2] As the nation's attention began to include concern for nuclear weapons, and as regional concern began to focus on the possibility of a high-level nuclear waste site nearby, the early voices of dissent toward Hanford were beginning to be heard—mostly in the regional urban centers of Spokane and Seattle.[3]

A Brief History of Environmental Action in the Hanford Area

On May 20, 1984, Rev. William Houff of the Spokane Unitarian Church gave a sermon entitled "Silent Holocaust," in which he described how the nuclear industry, in both the public and private sectors, had pursued its own interests with scant regard to the "public health and safety" concerns of American citizens. In the sermon, Rev. Houff likened the American military and commercial nuclear establishments to the Nazi regime during World War II, and noted that during the war there were a few courageous Germans willing to sacrifice their safety to oppose the regime. Too many otherwise good people stayed "silent." Houff challenged those in the audience to not remain silent during this newer holocaust (Ratliff 1994). Although the sermon concerned the nuclear power and nuclear weapons production industry in general, Rev. Houff's remarks focused on the Hanford site specifically. After the sermon, thirty people came forward and began meeting every two weeks during the following summer. The purpose of these meetings was to discuss the Hanford site and to share research that the members had conducted on the facility (Kaplan 1992).

For the first time, individuals organized in opposition to the operations at the Hanford site and provided some public monitoring of Hanford's daily operations. In September 1984, this group officially adopted the name "Hanford Education Action League" (or simply, HEAL). One of HEAL's first actions was to sponsor two "educational" forums in October 1984. The first, cosponsored with the Portland Chapter of the Physicians for Social Responsibility, was a slide presentation titled "Plutonium: Everything You Want to Know, but Were Afraid to Ask," presented by an Oregon health officer and physician. The second was presented by Dr. Alice Stewart, a British expert on the dangers of radiation exposure on children. The following month, HEAL invited Hanford's manager, Michael Lawrence, to speak to their group in Spokane about the potential health hazards that Hanford posed on the area residents. The meeting was contentious. Lawrence tried to convince HEAL's members that the Hanford site was indeed safe, while his audience was predisposed to discount his assurances (Ratliff 1994; D'Antonio 1993). Later that year, Rev. Houff and several other HEAL members expressed concern to the Spokane City Council about the dangers Hanford posed to the city. The HEAL members requested that the council investigate the January 1984 mishap at PUREX where plutonium was released through the facility's main stack. They also requested that the council ask the state of Washington to refrain from entering into any agreement with DOE to allow Hanford to become a nationwide waste repository, and to pass an ordinance banning the shipment of nuclear waste through Spokane city limits on Interstate-90 (Steele 1984).

In October 1985, HEAL sponsored another symposium with the Physicians for Social Responsibility focusing on the health effects of Hanford, and on the operations of PUREX. Over 250 people attended the event. The keynote speaker was Dr. Benjamin Spock, a name that conferred a great deal of legitimacy to the fledgling group (Ratliff 1994). The symposium also featured Dr. Carl Johnson, the Denver-area health officer who found elevated levels of cancer in the Denver area. His study, in turn, prompted the U.S. Geological Survey study that found plutonium levels in the soil forty-four times higher than previous AEC studies (Denver is home to the Rocky Flats facility) (Steele 1985a). At the symposium, the speakers called for the release of official documents that would trace the history of plutonium production at Hanford. Three days later, the DOE stated that the agency would comply because of the interest expressed by area residents (Steele 1985b).

In February 1986, with HEAL's membership reaching 500, the Department of Energy released the previously secret documents outlining Hanford's history (chapter 2). In total, the Department released 19,000 pages of documents outlining the amount of radioactive particles released during the forty-year history of the Hanford Nuclear Site (Gerber 1992a; 1993). A few weeks prior to the documents' release, HEAL submitted a Freedom of Information Act request to evaluate the information released and to acquire any further information that DOE might have otherwise withheld (Kaplan 1992). Over the course of the next five years, the DOE released another 50,000 pages of documents about Hanford's past activities. It became increasingly apparent that the secrecy involved in weapons production during the Cold War, justified by national security reasons attendant to the Cold War, also extended to information about the true levels of radioactive particles emitted since Hanford's inception. The newly released documents described much greater radiation releases than was previously thought. With the Chernobyl disaster, antinuclear opposition at Hanford became even more solidified—primarily because of one of HEAL's researchers, Tim Conner. He announced that he had found design similarities between one of Hanford's weapons production reactors (the N-reactor) and the Soviet reactor at Chernobyl. After being shut down temporarily for safety improvements, the N-reactor was finally closed indefinitely in 1989, along with the PUREX facility.

The release of DOE documents resulted in the development of additional downwinder groups. The DOE documents provided a wealth of information on the radiation problems encountered during the first thirty years of Hanford's existence. Because the amounts of airborne releases and environmental contamination were greater than previously thought, there was considerable concern about the health effects of the emissions. In May 1989, the Hanford Education Action League sponsored a conference on such health effects on downwinders. Residents who lived in the area spoke about the health problems that they attributed to the Hanford releases. One of those that spoke was Judith Jurji, the daughter of a former Hanford worker who had been raised in the Tri-Cities area. She had suffered from various symptoms that were a result of hyperthyroid, a condition that can be caused by high amounts of radioiodine concentrating in the thyroid gland. After this conference the downwinders eventually coalesced into a group known as the Hanford Downwinders' Coalition, with Jurji as president.

Around this time other groups also began to form; groups such as the Columbia River United and the Hanford Downwinders' Health Concerns came into being in the late 1980s. Also during this time, downwinders' concerns were reinforced by the journalistic work of Karen Dorn Steele, a reporter for eastern Washington's principle newspaper, The *Spokesman-Review*. Along with HEAL, Steele submitted a Freedom of Information Act request to ensure the full and timely disclosure of the Hanford documents in 1986. She has published dozens of articles focusing on the operations of Hanford and the health effects of its emissions over time (Dunlap et al. 1993). In turn, concern about the health effects of the emissions prompted DOE to establish a large-scale study into the probable doses area residents might have received during this time (the Hanford Environmental Dose Reconstruction Project).

With the demise of the Soviet Union and the end of the Cold War, Hanford's official mission within DOE was transformed from weapons production to the daunting task of "environmental cleanup," and the major production facilities, PUREX, and N-reactor were placed on standby by 1990. In May 1989, the state of Washington's Department of Ecology, the U.S. Department of Energy, and the U.S. Environmental Protection Agency signed the Tri-Party Agreement (formally known as the Hanford Federal Facilities and Consent Order), which outlined the cleanup of Hanford by the year 2018 (chapter 2). Also, during visits to the site in 1989 and 1990, DOE Secretary Admiral James Watkins announced that Hanford would cease producing plutonium for weapons. As the mission of Hanford was officially changed, the groups' focus also switched to some degree. No longer stressing opposition to the facilities' production operations, they now emphasize oversight issues in cleanup and environmental remediation.

The Era of Cooperation

Whereas in the late 1980s the Hanford area groups were struggling to gain legitimacy and to influence the operations of the DOE at Hanford, the early 1990s could be characterized as a period when the groups received that legitimacy and began to cooperate with the DOE in a number of significant ways. In fact, when one such cooperative effort between the groups and the DOE was completed (the Tank Waste Remediation Task Force), one ever-suspicious reporter covering the presentation of the final report started his article thus:

> For anyone familiar with the more than two decades of acrimony over the nation's nuclear weapons manufacturing facilities, it was a most unlikely sight: Two dozen people, from pony-tailed activists to grey suited Department of Energy officials faced each other around a ring of tables, outdoing one another with compliments. (Wilhelm 1993)

In large part, the Richland offices of DOE had little choice but to work with the groups. As mentioned above, the DOE became, for the first time ever, subject to oversight from outside agencies under the Tri-Party Agreement. The overall purpose of the TPA was to "ensure that the environmental impacts associated with past and present activities at Hanford are thoroughly investigated and appropriate response actions taken as necessary to protect the public health, welfare and the environment" (TPA, Article III paragraph A). This was a somewhat radical departure from the past activities conducted by the DOE at the Hanford site. The document also holds the Department of Energy to a timetable for cleaning up the past areas of contamination and pollution on the site. The "Action Plan" of the TPA holds the DOE to specific dates on the closure of the single-shell waste tanks (see TPA, Action Plan). The department was also committed to remediating (and often restoring) the areas on the site that were contaminated through past activities and to the construction of a "vitrification plant" on the site, where nuclear waste and contaminated materials are to be turned into glass (TPA, Action Plan Executive Summary).

The TPA is a legal document that not only binds DOE to the milestones of cleanup, but also forces the department to comply with existing federal and Washington State environmental statutes. Specifically, the agreement forces DOE to comply with the Washington State Hazardous Waste Management Act (RCW 70.105; TPA Article 1). This was a large change for DOE because it made the Washington State Department of Ecology, in conjunction with the U.S. Environmental Protection Agency, regulators and enforcers of the statutes and the TPA. In the past, the DOE had been exempted from review by other agencies in the federal government and the state.

The TPA imposed on the Hanford DOE administrators extensive mechanisms for public involvement. Although the DOE was already legally committed to some forms of public involvement in its decision-making processes (Steinhardt 1995), such as through hearing processes of environmental impact statements, the TPA broadened the scope of public involvement even further. First, the TPA requires that the EPA and Washington State Department of Ecology, "with the assistance of the DOE when requested," conduct quarterly information meetings where interested participants can gain information regarding the progress in the cleanup effort at Hanford, and where they can offer their own input into the cleanup process (TPA, section 10.5). Moreover, the TPA also provides for the issuance of grants to nonprofit citizen groups to facilitate "the active participation of persons and organizations in the investigation and remedying of releases or threatened releases of hazardous substances" (TPA, section 10.9). In effect, the Tri-Party Agreement gave to the environmental interest groups monetary assistance to provide an independent watchdog role over DOE at Hanford.

Other mechanisms also enabled the groups to oversee, if not influence, operations at Hanford. In 1993, members of HEAL (and other groups discussed later) were given seats by DOE on the Tank Waste Remediation Task Force. DOE created the Task Force to discuss the remediation efforts beginning on the site, and to oversee the DOE's handling of those efforts. The group's final report was used by negotiators and regulators in TPA renegotiations (O'Connor et al. 1995). Also in 1992, DOE convened the Hanford Future Site Uses Working Group, also composed of members from Hanford area groups, to discuss the future of the site and its possible uses after it had been decontaminated (Hanford Future Site Uses Working Group, 1992). In 1994, DOE established the Hanford Advisory Board composed of thirty-three members, many of them Hanford area interest groups. The board advises DOE on major cleanup issues, risk management assessments, economic development actions, future land use possibilities, and budget allocations.

Current Status of Hanford Interest Groups

Hanford's activities have raised environmental concerns of an unprecedented nature in American politics, and they in turn have spawned a diversity of environmental groups that address various elements of this legacy. In addition, several national public interest groups or environmental groups are active on issues related to Hanford. This section describes some of the major groups involved in Hanford-related environmental questions. The observations are based on personal interviews with group representatives and documentary research that was initially conducted in 1993–94, and updated in 1996–98.

Hanford Education Action League

The Hanford Education Action League (HEAL) is probably the central actor in the popular movement against Hanford. HEAL is the most visible of the groups, it garners significant attention from the regional media, and it may very well surpass any other interest group in eastern Washington in its overall name recognition. It is difficult to characterize the group's overall ideology because HEAL is only wary of the environmental impact of nuclear weapons production and not opposed to nuclear weapons production. It could be argued that HEAL is more conservative than pure antinuclear groups. However, HEAL's membership is drawn from individuals who tend to be somewhat more liberal than their eastern Washington counterparts.[4]

HEAL currently has a dues paying membership of 400, and possesses a mailing list of 3,000–4,000. It is one of the most professionalized of all the groups that deal directly with Hanford, having three full-time staff, which includes an office manager, an executive director and a researcher. At any time there are also additional researchers, journalists, college interns, and others who are working with the HEAL staff and their substantial files of information.

Today HEAL remains active in the study of issues surrounding the Hanford site, especially on matters regarding future land use planning for the site once the waste and contamination are remediated. Moreover, the group is also concerned with the storage of liquid wastes and with the DOE's efforts to remedy this problem. HEAL has attempted to ensure that sufficient cleanup funds are coming to Hanford from the U.S. government. Finally, HEAL also participates in a formal way as a member of the Hanford Advisory Board.

Heart of America Northwest

Heart of America Northwest deals with issues somewhat outside of the environmental activists' normal realm, focusing on issues of governmental oversight and consumer protection (Healy 1989). It was organized in 1987 in reaction to the reports released the year before Hanford's radiation releases over the prior forty years. The group solidified as a result of its opposition to the nuclear waste depository plan (Basalt Waste Isolation Project). According to the group's administrative director, Heart of America Northwest is neither pronuclear nor antinuclear (power and weapons production), but is concerned that "whatever is done, is done right."

The group consists of 1,600 members in Washington, Oregon, and Idaho, with 8 members on the board of directors, an executive director, an administrative director, and 12 additional staff members (even surpassing HEAL in its professionalization). Past actions include the stopping of waste shipments, lobbying for whistleblower protection, and keeping the public informed on developments at Hanford.[5]

Hanford Downwinders' Coalition

The Hanford Downwinders' Coalition (HDC) was "informally founded" in 1988. In 1989, they conducted an exhibit commemorating the fortieth anniversary of Operation Green Run "in a deliberate way."[6] In 1990, the group filed with the state of Washington as a nonprofit organization after the HEAL conference dealing with the downwinders. Today the group primarily acts as a communication network for individuals who lived in the area during the period of emissions, and serves as a "support group" for those who feel their health may have been affected as a result of Hanford. HDC also maintains a database for downwinders who wish to be interviewed by the media. The group currently has a membership of "1,000 receiving newsletters."[7]

In 1990, the Hanford Dose Reconstruction Project began to determine if individuals who had lived in the Hanford area suffered from health effects from radioactive particle emissions. This project stimulated extensive regional media coverage on the plight of the downwinders. The HDC and the state health departments were inundated with calls from former residents who were concerned about their health. In response to the deluge of inquiries, the federal government allocated $5 million from the Department of Energy's budget to the public health agencies of Washington, Oregon, and Idaho to assist in information dissemination and medical history gathering among the affected public. This allocation was a direct result of lobbying by the Hanford Downwinders' Coalition for such a publicly funded and officially legitimated project. The three states created the Hanford Health Information Network (HHIN), headquartered in Seattle. The HHIN maintains a toll-free number for citizens who are concerned about Hanford-related exposures that might have affected their health. The HHIN conducts mailings on the health concerns of downwinders and collects information on them.[8]

At the time the HHIN was created, a group of members from the Hanford Downwinders' Coalition split from that group because they were opposed to receiving any money from the federal government for such a program. These individuals formed a new group called the Hanford Downwinders' Health Concerns. The HDHC currently consists of a loose association of roughly 300 members, with a local board of directors of 7 members and a regional board of 10 members. Unlike the other groups mentioned, the HDHC is not considered a nonprofit organization and requires an application for membership that consists of questions on health problems and on where the person lived from 1940 to the present.

Three members of the group, along with Columbia River United (discussed next), have filed lawsuits against the DOE and its past and present contractors, seeking to establish a viable tort claim against those parties if the Dose Reconstruction Project establishes an empirical link between exposure levels and the incidence of disease. However, the groups have faced problems with their own legal teams. The attorney representing the HDHC, Nancy Oreskovich, was removed from the case by U.S. District Court Judge Alan MacDonald when he learned that Oreskovich had leaked to the media a sealed report by an independent court master that was critical of the Hanford Environmental Dose Reconstruction Project. MacDonald further stated that Oreskovich had run a substandard practice and had overbilled her clients (Geranois 1996). The case has been put on hold, indefinitely (Steele 1996).

Columbia River United

Columbia River United was formed in 1988 in opposition to the transport of the Shippingport reactor up the Columbia River for disposal at the Hanford site in the spring of 1989. It has a membership of 20 to 400 active members (depending on the issue or activity), and it produces a quarterly newsletter with a mailing list of 1,300 addresses. The primary focus of this group is water quality issues on the Columbia, and it is concerned with "all life dependent on the Columbia." Its mission statement commits its members to ensure the river remains healthy for plant, animal, and human life.[9] The group's president, Greg DeBruler, served on the EPA's Columbia River Water Quality Study. The group helped to bring about the study, which was financed by a $5 million grant awarded from the EPA's National Estuary Program, which provides funds for studies of critical water quality issues.[10]

The study group held hearings in communities along the Columbia River to discuss water quality issues, but stirred up considerable opposition due to what DeBruler called "opposition funded by the pulp and paper companies." The group conducted one activity that came close to an outright protest during the "Complex-91" process. In July 1991, the

DOE began to hold public comment hearings on possible future sites for a reconfigured nuclear weapons complex (Complex-91). DOE selected five potential sites across the country, including Hanford. The hearings on each site were not held in a major city nearby, but at largely remote localities. The Columbia River United group "brought the cities to them." Over 350 members from Washington and Oregon attended the meeting held at Hanford to voice their stern opposition to making Hanford a site for the new weapons complex. Members of the group also frequently sit on advisory committees that deal directly with the Department of Energy in the role of stakeholders, such as the Site Specific Advisory Board and the Tank Waste Remediation Task Force. The group continues to be active in public information campaigns as well.

Hanford Watch

Hanford Watch was founded in August 1992. The primary cause of the group's formation was a series of articles that appeared in the *Oregonian* on the possibility of Hanford's waste tanks exploding and contaminating the region. The group now has 10 "very active" members with a mailing list of roughly 500 people to whom they disseminate information. The group's chairperson, Paige Knight, hosts an occasional radio show on a publicly owned Portland radio station on the topic of current activities at Hanford. Past activities include sponsoring a petition to the governor of Oregon urging him to become involved in the Tri-Party Agreement. The group's overarching goal is to educate the public of Oregon through the regional print and broadcast media and through its own newsletter. In the past the group has brought in experts to discuss Hanford and its operations in conference settings.[11]

Government Accountability Project

The public policy controversies surrounding Hanford have also involved a diverse set of other public interest groups or national environmental organizations—the Government Accountability Project, for example. GAP was formed in 1976 as a result of a conference sponsored by Ralph Nader. Its primary mission is to encourage and help represent whistleblowers from the Hanford area who can document environmental violations in federal or state court. GAP is a persistent critic of DOE on management issues brought forward by these revelations, and stimulates governmental oversight where it is lacking (through the Governmental Accounting Office or congressional hearings).

Currently GAP is involved in nine lawsuits initiated by Hanford-related whistleblowers, and there have been several other employee complaints (but no formal lawsuit) about which GAP has advised Hanford workers. Within the last year they have become involved in four new lawsuits. The GAP leadership has also met with the current Hanford contractors, with Washington State, and with the DOE to set up the Hanford Joint Council. The purpose of the council is to provide "case management mechanisms" to deal with the whistleblowers' complaints before they proceed to litigation. The board is an independent conflict-resolution-oriented entity, funded by the DOE with "a heavy emphasis on independence."[12]

The group enjoys nonprofit status and tax tretment; it claims a membership of 10,000 nationwide. GAP employs 20 staff members and 9 attorneys. The hierarchy of the organization consists of an 8-member executive committee and a board of directors that meets four times a year.

Physicians for Social Responsibility

The Physicians for Social Responsibility (PSR) is another advocacy organization with broader political interests, which involve them in the Hanford controversy. PSR is primarily concerned with human health issues rather than the quality of the environment per se. They see the environment as a pathway through which pollutants may travel and eventually harm human populations. The PSR is primarily concerned with reducing pollution and building an appropriate set of barriers for the protection of people's health.

PSR was formed in Boston in 1963, when local physicians in the New England area examined what would happen to a city if a 1-kiloton bomb were dropped on a metropolitan area. They found that there could be no locally organized and coordinated response to a nuclear strike. A local chapter of the national organization subsequently was formed in Portland in 1977.

Although the Portland chapter concerns itself with other issues of human health (such as youth gang-related violence), the health and safety issues at Hanford remain one of its more important concerns. PSR was active in the formation of the Hanford Health Information Network and continues to work with the other downwinder groups to study health effects attributable to Hanford. In addition to its original concern about the effects of nuclear weapons production on public health, the organization is interested in the public health effects of toxic waste storage, and about worker health and safety at Hanford as the facility's cleanup activities expand.

PSR's national membership initially grew due to several charismatic leaders. It expanded to 50,000 members in 1990. Since then, many of the original leaders have left and membership has dropped to 35,000. The local chapter has undergone the same expansion/contraction due to the arrival and departure of leaders differing in ability and broader appeal; leadership instability notwithstanding, the group remains active. Unlike other PSR organizations around the country, the Portland chapter allows anyone who is interested to join. Elsewhere, only physicians are permitted membership.[13]

Sierra Club

The Sierra Club has not been involved in a big number of area interest activities around Hanford. However, it does boast a large membership of 25,000 in Washington and in Oregon in its Cascade chapter. It should be considered a significant actor because of its sheer size and unquestioned influence at the state government level and in the Seattle regional office of the EPA.

The Sierra Club's past activities have included some monitoring around Hanford, and a considerable amount of lobbying with other groups for the Federal Facilities Compliance Act of 1992.[14] This legislation forces federal facilities engaged in nuclear weapons production, such as Hanford, to submit to state environmental regulation. The Sierra Club also participated in the negotiation of the Tri-Party Agreement.

Overview of Hanford-Area Environmental Interest Groups

Having described the interest groups concerned with past and present activities at Hanford, I now consider some of the larger implications of the patterns of environmental action. By identifying differences in how these groups are organized, in their orientations to politics, and in their styles of political action, I hope to expand our understanding of this contemporary movement, and I anticipate being able to link our findings to other research on the nature of contemporary environmental action in American politics.

Resources/Structure

Much of the literature on social movements suggests that there is an organizational and political incentive for citizen groups to develop a strong organizational structure that will enable the group to mobilize resources efficiently, and to focus its political activities (Jenkins 1983; McCarthy and Zald 1977). Our description of Hanford-related environmental groups provides striking evidence of the diverse forms and divergent foci that environmental action has taken in the Hanford setting. These groups differ widely in organizational styles and resource bases, as well as in their professed aims and methods of influence building.

For example, Paige Knight of Hanford Watch described decision making in her group as being highly informal, with decisions being made by those individuals who can attend the group's meetings at any one time. The Hanford Downwinders' Coalition, in contrast, relies upon a survey of the group's membership to determine a course of action. Both of these groups rely primarily upon fairly informal means to decide the group's actions. At the other end of the spectrum stand Heart of America Northwest and the Sierra Club. Both of these groups rely upon a formal board of directors to determine their goals and set their policy directions. Not only do the Hanford area groups differ in their decision-making structure, but also they differ widely in their organizational capacities. Some of the groups such as the Physicians for Social Responsibility and Hanford Watch employ no full-time office staff, whereas both HEAL and Heart of America Northwest employ a full-time staff; both organizations employ a professional office manager to coordinate the activities of the staff members and maintain the administration of organizational business.

The criteria for membership also vary considerably across these groups. The Physicians for Social Responsibility considers "anybody interested who wants to deal with peace and justice issues" eligible to become a member. Columbia River United, Hanford Watch, and the Hanford Downwinders' Coalition consider anybody on their mailing list to be a member. The Sierra Club, GAP, HEAL, and Heart of America Northwest have a formal dues or fee-paying membership. In some cases membership is equivalent to receipt of a free newsletter, while in other groups membership implies a significant level of group or political activity. Reflecting these differences, the size of group membership ranges from twenty to several thousand.

Similarly, these groups depend on a diverse set of funding sources. The Sierra Club and the Portland chapter of the Physicians for Social Responsibility rely primarily on membership dues. The Hanford Downwinders' Health Concerns receives no funds from its membership or from external sources, due to its loose configuration. Hanford Watch and the Hanford Downwinders' Coalition rely primarily on donations. Columbia River United, Heart of America Northwest, and the Spokane-based HEAL receive Public Participation Grants from the state of Washington (via the Department of Ecology). Heart of America Northwest also receives other grants from sources such as the Peace Development Fund. The Government Accountability Project receives money from numerous grants, such as the Rockefeller Family Fund and the Fund for Constitutional Government, and it solicits money through direct mailings.

What implications do these different structures have for the vitality of the environmental movement? Dalton (1994) found that the diversity of organizational structures, range of activities, and ideological orientations across European environmental groups represented a distinct asset to the movement in realizing its overall goals. Countering the conventional application of resource mobilization theory, Dalton asserted that the diversity evident among environmental groups can actually benefit the groups in allowing them to address a broad range of citizen concerns. Dalton's observations can be readily applied to the case of the Hanford groups. They differ in their relative intensity in dealing with the problems at Hanford—some groups enjoy a large following, yet rely upon key members for the bulk of the work to be conducted. This allows the groups to enjoy a large numerical base of support, while leaving free those more active members (and the group's leadership) to pursue goals they themselves deem worthy of pursuing. At the same time, the smaller, more unified groups can concentrate on more specific (sometimes rather technical) issues that the larger groups may not address. Moreover, the different goals pursued by each of the groups allows for a greater number of actors to participate in the process—the larger number of groups pursuing their own goals will ensure a greater chance that nearly all concerns will be addressed by one or more of the groups. A more centralized structure unifying the groups' memberships may very well have limited greatly the diversity in the goals the groups pursued, and most likely would have limited their success.

Furthermore, the regional diversity of the groups aids in ensuring the diversity of their goals. Groups in eastern Washington (HEAL and the HDC) are concerned with the effects of airborne radioactive releases, while groups located lower on the Columbia River from Hanford (CRU, PSR, Hanford Watch) have focused on the effects of water-borne releases. Finally, Seattle-based groups (GAP and Heart of America Northwest)—which are relatively free from the threat of Hanford-borne pollutants—are more concerned with the legal and financial concerns surrounding Hanford. It could be argued that the latter groups are primarily concerned with "good government" at Hanford in a more general sense. They want to see a "governmental response" at Hanford that can restore public confidence in a federal agency that the public perceives as having violated its trust through deception, cover-up, and callous disregard for human health and safety.

Political Positions

Another important comparison across environmental groups involves their respective perceptions of the most pressing public policy issues that Hanford represents. The political goals the groups project to prospective members and the policy goals they pursue through their activities are important to document and assess for patterns of shared and disparate occurrence.

One perspective shared across groups is negativity toward the officials running Hanford, and a negative opinion of the government agencies responsible for the facilities (although there is a slight variation across groups in the amount of negativism). Julie Reitan of the Sierra Club observed that the Department of Energy has done an "unconditionally terrible job," and that the milestones of the Tri-Party Agreement that the Sierra Club had worked on during the late 1980s have all been missed, except minor parts that were (in her opinion) "easy and convenient." Greg DeBruler of Columbia River United called the Hanford site "run-down and shoddy," and asserted that the DOE and its contractors "don't have the foggiest idea of what has happened or is happening out there. Authority is still living under a veil of secrecy." He also stated that the DOE should be removed from the site and a cabinet-level organization should replace it, which concentrates solely on waste cleanup. Judith Jurji of the Hanford Downwinders' Coalition complained that Hanford officials "squander money, putting workers at risk, putting the off-site population at risk. . . . I am pessimistic about Hanford in general." In perhaps the most frank assessment of all, Tom Carpenter of the Government Accountability Project called the site ". . . a mismanaged, unsafe taxpayer rathole," and he said the site "is a microcosm of the national security state, inefficient, lacking the usual checks and balances . . . they got away with it because of secrecy."

Paige Knight of Hanford Watch stated that her group was making a concerted effort to have honest relations with Hanford officials, but noted that the "reception is far from ideal." She observed that one problem lies in the fact that when confidence is established and a clear communication channel is put into place with a DOE official, he or she typically is transferred elsewhere within the DOE system within short order. Lynn Stembridge of HEAL claimed that there were indeed a fair number of "good people who were 'getting it,'" but maintains that the primary goal of Hanford's contractors "is to maximize its bottom line, and everything else be damned."

Alliance Patterns

The methods employed and the degree of success achieved by public interest groups are often conditioned significantly by the pattern of political and resource alliances that support their individual efforts (Klandermans 1990). This aspect of interest group behavior is also apparent across the groups we surveyed in the Hanford nuclear facility study.

When we asked about potential allies, almost all group representatives cited most or all of the other groups in this study as political allies. Although each of the groups is a separate and independent entity, and often has quite different reasons for pursuing its particular goals, there has been a large degree of contact and cooperation between and among the groups. As mentioned earlier, the Hanford Education Action League was instrumental in the formation of the Hanford Downwinders' Coalition. After the HDC's formation, HEAL continued to work with them closely, to the extent of actively mentoring the group's leadership and providng training to its organizers. Furthermore, leaders and key members of one group will often publish articles in the newsletters of other groups. The common view seems to be like the opinion of Paige Knight of Hanford Watch: "the more united, the more power they have."

Many group representatives also mentioned national groups that deal with other weapons production and testing sites across the United States as significant allies—such as the Military Production Network and the National Association of Radiation Survivors. HEAL was instrumental in the formation of the national-based Military Production Network. When the MPN was first forming, HEAL sent representatives to assist in its creation, and a HEAL member drafted the MPN's "Community Bill of Rights" which calls for, among other things, community right-to-know provisions. Both groups still assist each other in pursuing their respective goals. In summary, there is ample evidence of a strong network of communication and mutual support among the groups that reinforces Charles Tilly's (1978) observation that other challengers are potential, big allies for groups in many situations of political advocacy against status quo practices and policies.

Although only one group (HEAL) specifically mentioned an alliance with the media, the significance of the media coverage attained by the Hanford area groups cannot be overlooked. The work of Spokane's *Spokesman-Review* reporter Karen Dorn Steele was crucial in generating support for the groups' causes early in their formation, and Steele's work continues to keep Hanford on the minds of area residents. She has written dozens of articles chronicling the course of major events at Hanford—from covering the initial release of DOE documents that displayed evidence of past human radiation experiments, to covering the long-term fate of the downwinders at Hanford. Because the *Spokesman-Review* has a circulation covering a large portion of eastern Washington, her reports enjoyed one of the widest audiences in the region. News of Steele's early local media coverage soon spread, making Hanford a national issue of how the U.S. Department of Energy was going to restore public faith in its stewardship over all of the nation's nuclear weapons complexes, located across the breadth of the country. Through the 1980s and early 1990s, new reports about Hanford became a regular feature of other newspapers in the Pacific Northwest. Hanford's legacy was featured on ABC's *Nightline,* the *MacNeil-Lehrer News Hour,* and the evening news programs of the major television networks. The discussion of Hanford in the media was important in keeping this issue before the public, and in helping environmental groups inform the public of their concerns.

Group representatives were hesitant to name any specific elected official as an ally. Overall the Oregon congressional delegation was seen as more sympathetic to their causes than the Washington counterpart. Oregon state legislators were also mentioned as somewhat sympathetic. Some elected officials from Washington have been critical of Hanford—in particular Governor Mike Lowry (when he was running for U.S. Senate in 1988) and current U.S. Senator Patty Murray. However, at the time of our interviews these politicians did not appear to be a prominent source of support to groups protesting Hanford.

One group did mention organized labor as a source of support on some key areas. HEAL has worked with labor organizations at Hanford in the specific areas of the impact of privatization on the workforce there. Through their mutual work on the Hanford Advisory Board, HEAL and labor organizations have examined the impact of the privatization of tank waste management on worker safety. HEAL and labor groups both criticized the means by which the DOE was pursuing privatization.[15]

When pressed to name specific opponents the groups might encounter, leaders were reluctant to name specific individuals or groups that they felt opposed their interests. A later interview with HEAL researcher Todd Martin revealed an insight that sheds further light on the political dynamics surrounding Hanford. Because virtually all actors now involved with Hanford favor some form of cleanup, there is relatively little disagreement over the overall goals for the management of the site. The source of disagreement lies in narrower, specific areas about which the actors may find themselves on opposite sides. However, actors who disagree on one issue may find themselves on the same side on another, so they consider each other allies as a general rule. These shifting alliance patterns make it more difficult to assess who is opposed to any particular group's overall goals.

One could argue that the dynamic structure of the Hanford groups' alliance patterns works to the advantage of the groups in pursuing their goals. Because curently many individuals inside and outside the federal government have some influence in formulating DOE policy, the groups' interalliances help ensure that their concerns will be heard by some people who can influence the policymaking process. The fragmented nature of DOE policymaking may prevent groups from reaching total victory on their preferred goals, but they will be heard and their concerns addressed to some extent. This situation was unheard of at Hanford only a few short years ago when the DOE was immune to outside influence.

Political Strategies and Tactics

While the Hanford area advocacy groups share a substantial range of common environmental goals, they do differ in their specific interests in relation to Hanford, as described earlier. The Hanford Downwinders' Coalition and the Physicians for Social Responsibility have focused on the health effects of Hanford's radioactive releases on individuals living nearby or downwind from the facilities. Columbia River United concerns itself with Hanford in the broader context of the health of the Columbia River. The Government Accountability Project focuses on whistleblower protection for Hanford site workers. Heart of America Northwest is concerned with the costs and delays associated with cleanup, and with ensuring the fiscal accountability of DOE officials.

The groups also differ in the ways they manifest their concerns. The Hanford Downwinders' Coalition is primarily concerned with information—receiving citizen health and Hanford operations information and packaging it for the media and for citizens who have lived near Hanford. HEAL mixes information dissemination activities with direct action. The Government Accountability Project also mixes their methods between litigation for whistleblower compensation and lobbying the DOE and

Congress for further whistleblower protection. The Hanford Downwinders' Health Concerns group is primarily concerned with litigation aimed at compensation for past residents of the area for their health problems.

Not only do the groups differ in regards of the source of their concerns relating to Hanford but they also differ to some extent in the scope of their Hanford concerns in relation to their overall organizational goals. Hanford is the sole concern for the Hanford Education Action League, the Hanford Downwinders' Coalition, and Hanford Watch. The Physicians for Social Responsibility are concerned about Hanford because they believe it more generally adversely affects the health and safety of citizens residing in the Pacific Northwest region. The Government Accountability Project is concerned with all whistleblowers in government, and has represented whistleblowers in the U.S. Forest Service (Government Accountability Project, 1994), the military (Government Accountability Project, 1995), and DOE employees. Examination of these groups displays a protest movement that is diverse in its goals, methods, and overall group orientation.

Conclusion

It is apparent that the groups protesting Hanford formed, at least in good part, as a result of an intense mistrust of the officials at Hanford—a feeling that all of them shared at the outset of their policy advocacy efforts. This unfavorable and untrusting attitude helps to explain the mobilization of some of the area citizens to watch over the operations of the nuclear reservation, and to monitor continuously facility activities that they fear could constitute a hazard to the general public. In particular, this was the case with their opposition to restarting the N-reactor, the PUREX facility, and the siting of long-term nuclear waste storage at Hanford.

This widening mistrust can be accounted for by the release of the formerly classified information about the site. It was not until the Department of Energy opened its files to public inspection in 1986 that the general public learned that they had been exposed to health risks both as a result of intentional decisions and as a consequence of accidents. Citizens would probably have found the risks unacceptable if they had known about them. This observation reinforces the claim made at the beginning of this chapter, that the collection and strategic dissemination of information is a key factor in understanding the mobilization of protest against the Hanford facility.

Most of the representatives of the Hanford environmental interest groups feel they were victims of governmental processes over which they had little or no control. It was not until the massive release of documents in 1986 (and later) that they gained access to information on Hanford's past. Before the release of the documents, opposition to Hanford was far less evident and received far less attention in the public affairs discussions of people in the Pacific Northwest. The current level of mistrust evident across the region arose largely as a consequence of the coverage given to the public admission of serious previously denied emissions, covered in the regional press and broadcast media.

As John Pierce and his colleagues (1988) observed, a key function of interest groups can be their ability to disseminate credible policy-relevant information to their members and to the mass public—a fact that has been true for the Hanford environmental interest groups. They helped disseminate the information collected during the 1986 release of documents, and they continue to analyze and disseminate information through their contacts with the media and through their membership newsletters. Furthermore, group membership increased as a result of a strong public desire for more information about the site. Some of these groups collected and translated the often highly technical information associated with the environmental and public health issues into more readable messages. In addition, Hanford environmental groups acted as a collective voice through which group members and cobelievers could get their message to DOE decision makers and the oversight authorities in the citizen participant forums called for in the Tri-Party Agreement.

A study of the Hanford area groups provides an opportunity to examine the conception, birth, and maturation of a strong network of interest groups within a relatively short period of time. In less than a decade most of these groups were conceived and went on to wield strong influence on the U.S. Department of Energy, an agency that had been relatively free of political opposition or influence outside of the major institutions of the federal government. As the DOE continues to release information regarding the agency's past activity—including newly released documents pertaining to human radiation experiments—interest in Hanford across the region, and even nationally, will likely continue at a high degree. As a consequence, the groups studied here will maintain their respective memberships and perhaps even add new members, in turn keeping Hanford in the forefront of public awareness.

On perhaps a more optimistic note, it is interesting to observe that the groups were able to move from hostile critics of the site, fearful and suspicious of the DOE, to legitimate influences within Hanford's decision-making structure. The groups were allowed to participate in the negotiations over the Tri-Party Agreement as stakeholder groups, and they continue to wield significant influence with their seats on the Hanford Advisory Board. Less optimistically, however, for such influence to continue the groups still need the goodwill and cooperation of the Department of Energy and the greater openness of agency personnel that the post–Cold War political climate has fostered. Events in recent years have signaled a regression in DOE and federal government support for the Tri-Party Agreement and its plans for Hanford (see chapter 11). Thus the "era of good feelings" and cooperation might be coming to an end. At this writing, the future of these relationships is far from clear.

Notes

1. *The China Syndrome* featured an accident, and a cover-up of that accident, at a nuclear power plant.

2. Dunlap et al. (1993) noted that significant opposition to the nuclear waste repository was growing even within the Tri-Cities.

3. See Dunlap and Olsen (1984) for a discussion of the divergent views toward nuclear power held by residents living in the Tri-Cities area versus those living outside. The former were substantially more supportive of nuclear technology than the latter.

4. Interview with Todd Martin, Hanford Education Action League, 6/4/96.

5. Interview with Stebbow Hill, administrative director, Heart of America Northwest, 8/31/93.

6. The HDC's chairperson, Judith Jurji, when asked if the group had ever held a protest, stated the group had not done so. However, the public exhibit created to commemorate the anniversary of Operation Green Run was an important, symbolic way of commemorating the critical event "in a deliberate way."

7. Interview with Judith Jurji, president, Hanford Downwinders' Coalition, 8/30/93.

8. Interview with Judith Jurji, president, Hanford Downwinders' Coalition, 8/30/93.

9. Interview with Greg DeBruler, president, Columbia River United, 8/26/93.

10. Interview with Greg DeBruler, president, Columbia River United, 8/26/93.

11. Interview with Paige Knight, chairperson, Hanford Watch, 8/26/93.

12. Interview with Tom Carpenter, Government Accountability Project, 8/31/92.

13. Interview with Dick Belsi, Physicians for Social Responsibility, 8/30/93.

14. The Federal Facilities Compliance Act of 1992 (Pub, L. 102-386) constitutes an amendment to the Resource Conservation and Recovery Act of 1982 (42 U.S.C. Sec. 6901). The FFCA places federal facilities under state regulation by waiving the federal government's sovereign immunity for RCRA violations; this waiver applies specifically to violations conducted by the U.S. Department of Energy at its nuclear production facilities. This statute allows states to sue the federal government for compensation for violations of EPA standards and environmental reporting requirements (Oge 1994).

15. Interview with Todd Martin, Hanford Education Action League, 6/4/96.

6

A History of Environmental Activism in Chelyabinsk

Paula Garb and Galina Komarova

When the people began to fight against pollution, new factories in Chelyabinsk and Magnitogorsk were designed taking the environment into account. It shows that people can do a lot.
—Vasily Mizev
Resident of Kyshtym

This examination of the evolution of environmental activism in Chelyabinsk traces the growth of citizen protest in this formerly secret corner of Russia. It looks at the movement's origins, first steps, organizations, issues, tactics, friends, and foes. The story is told in the context of Soviet and later Russian environmental protest in order to identify how local environmental politics compare with national trends.

This study provides rare information about local and national environmental action in a post-Communist state, and the nongovernmental groups in Russia that are still ill-formed or underdeveloped. Democratic reform in Russia partly depends upon the growth of an independent nongovernmental sector that can monitor and publicize the conformance of government agencies to the laws, and lobby for additional laws to protect citizens from government abuse. The findings can have fundamental implications for our understanding of democratic theory, the processes of resource mobilization used by citizen groups, and the nature of citizen action.

Results of the research are based on six months of field work over ten visits to Russia from June 1991 to March 1998, including seven trips to Chelyabinsk from December 1992 to August 1994. Information was obtained by working with, observing, and interviewing environmental activists in various Moscow-based and Chelyabinsk environmental organizations.

Pre-Gorbachev Environmental Action

In the two and one-half decades that preceded the Gorbachev era, Soviet environmentalists were primarily involved in the All-Union Society for Nature Protection [SNP] (*Vsesoiuznoie obshchestvo okhrany prirody*) or in student nature protection patrols (known as *druzhiny,* the term used for the armed forces of ancient Russian princes).

The Society for Nature Protection was founded soon after the Soviet government was established, on November 24, 1924, by conservationists who were mainly researchers in the natural sciences. Douglas R. Weiner concluded that through the early 1930s the Soviet Union was on the cutting edge of conservation theory and practice (Weiner 1988). By 1933, with the advance of Trofim Lysenko,[1] the Society was no longer a progressive force for conservation. Thereafter, the SNP carefully avoided locking horns with the establishment. The organization limited its functions to holding educational forums about conservation, keeping track of endangered species, and assisting government agencies responsible for maintaining nature preserves.

The student nature protection patrols, established in the mid-1960s largely by biology students (after the demise of Lysenko), originally guarded forests and reserves against poachers. Inevitably they confronted corrupt authorities who had been bribed to close their eyes to the poaching. Yet the students held their ground. Later they expanded their activities to identifying natural wildlife habitats, studying them in detail, and lobbying for them to be turned into preserves. Throughout their history the student activists in these groups pressured the establishment to its limits, yet they managed to remain a legitimate and viable force (Schwartz 1991; Yanitsky 1993).

As conservationist groups, the primary emphasis of the Society for Nature Protection and the student patrols was to defend nature. In this respect they shared common goals, so they sometimes worked together. However, they had significant differences. Members of the Society tended to be middle aged and older, professionals, and politically pro-establishment, who rarely, if ever, crossed their Communist Party sponsors. Members of student patrols were usually in their twenties and early thirties; they were primarily students of biology, geography, and engineering. The student patrols tended to be critics of the Soviet political system, although they learned to work with the establishment. These organizations did not constitute an environmental movement, but were essentially the efforts of a relatively small group of people who were able to exert some political influence through back channels (French 1991). Neither the Society for Nature Protection nor the *druzhiny* were funded by the government; both were member supported.

During the Soviet period, conservationists had some notable triumphs in influencing the state to abandon environmentally destructive projects. One of the most important examples was the effort to stop the diversion of water from several Siberian rivers for irrigation in southern Russia and Central Asia. Another important activity focused public attention on the pollution of Lake Baikal. Furthermore, by the end of the Brezhnev era environmentally oriented legislation provided a certain degree of legitimacy to environmental advocacy. Individuals used the print media to draw public attention to these problems, prompting many of these activities and legislation (Green 1991; see also chapter 12).

There was no equivalent pre-Gorbachev environmental action in Chelyabinsk. A local chapter of the Society for Nature Protection held its regular membership meetings and collected dues in customary tranquility. Students in the Chelyabinsk region were not organized into nature protection patrols.

The Greens of Perestroika (1987–1991)

In the late 1980s, when the grassroots environmental movement in the former Soviet Union had emerged as the first mass movement of the early reform period, Chelyabinsk was just giving birth to its local environmental movement. The Chelyabinsk movement was started from scratch, its members had no previous interest in environmental issues nor relationships with established conservation organizations. This was in contrast to the situation at the national level, where the movement was led by environmentalists who had been highly critical of the old system. Many activists in the new national groups had been the leaders and members of the student nature protection patrols, and were among the most vocal and energetic organizers of the burgeoning political movement that became the vanguard of democratization in the early Gorbachev period. This new environmental activism was largely an urban movement that focused on the defense of people, whereas the earlier conservationist movement of intellectuals had been entirely focused on the defense of nature.[2] Instead of calls for saving flora and fauna for their own sake, the new environmentalists emphasized restoring the damaged ecology for the sake of human health. The movement was essentially challenging the political and economic system that was jeopardizing the health of the nation.

During that period, several significant national environmental organizations came to the forefront (Yanitsky 1993; Green 1991). The largest and most influential was the Socio-Ecological Union (SEU), an umbrella organization founded on December 25, 1988. The SEU claimed that in 1990 its mass actions involved around 1 million citizens. In 1991, it united as many as 160 local, nongovernmental environmental groups and individuals in 260 towns in nearly every republic (Zabelin 1991). The SEU became the Soviet Union's largest and most visible national environmental organization.

Another influential environmental organization was the Ecological Union. Its members parted ways with the Socio-Ecological Union at their founding convention in 1988. They formed a separate organization a day before the SEU did, on December 24, 1988. These two groups apparently differed over a strategy of activism versus education. The Socio-Ecological Union advocated political action that would lead to tangible changes in environmental policy and practice. The Ecological Union was oriented to studying environmental problems and educating the public in environmentally sound principles. The leadership wanted to develop environmentalism as a science, and promote small businesses for environmental protection and restoration.

Another important environmental actor was the Green Party, which formed on March 24, 1991. It stressed political reform as the only way to bring about changes in environmental policies. Its leaders hoped to unite all environmentalists to constitute a major new political force in Soviet politics. Its mass actions focused on influencing the general political course of the nation, rather than specific environmental issues. The party believed that such political changes were prerequisites to improving environmental policies.[3] Leaders of the Socio-Ecological Union opposed establishing a party; they believed the idea was premature and that the organization would not hold together because the movement was too weak and politically divided.[4]

To counterbalance the emergence of the autonomous environmental groups, the government-sanctioned Soviet Peace Fund and Soviet Peace Committee initiated a new group, Ecology and Peace. Its founding members were scientists who had worked against the river diversion project, and it also involved prominent environmental scientists. Unlike the other environmental organizations, it had the advantage of a well-supported infrastructure (building space, office equipment, and funding). Its activities have consisted mainly of publishing studies, sponsoring conferences, and supporting local activists (Green 1991).

The final component of the environmental movement was local groups in diverse regions of the Russian republic that particularly attracted Russian nationalists who viewed the restoration of the environment as part of their agenda for Russia's cultural revival. Nationalism in the non-Russian republics tended to reinforce environmentalism because people identified with the degradation of their ethnic territories. This was generally not the case in Russia. During this period the main difference was that Russian nationalist environmentalists were unable to attract large followings, perhaps because of their antidemocratic and chauvinistic thrust. Some antidemocratic Russian nationalist groups used environmentalism to cloak their anti-Semitism or attacks on other non-Russians, which was contrary to grassroots trends among Russian activists (Green 1991).

Chelyabinsk Greens Are Born

The Chelyabinsk environmental movement traces its origins to the government's official acknowledgment of the radiation contamination caused by plutonium production at Mayak.

The newspaper article in the *Chelyabinsk rabochy* (Chelyabinsk Worker) on July 6, 1989, which gave impetus to the Chelyabinsk greens was the transcript of statements made in the Soviet parliament by Deputy Chair of the USSR Council of Ministries, L.D. Ryabov. The article outlined the environmental and health consequences of the Mayak accidents. After it was published, the local democratic reform movement, the Chelyabinsk Popular Front, which had been active for nearly a year and one-half, decided to create the Ecological Group to focus on environmental issues.

This environmental group's first action was to call a public hearing to debate the continued construction of the Southern Urals Nuclear Power Plant in light of the new environmental information released about earlier accidents at Mayak. The hearing, held in September, was essentially the first public meeting on an environmental issue that electrified the public. It attracted about 600 people. The Popular Front was the official sponsor, but the Ecological Group initiated and organized the event.

The public confronted the local government officials at the hearing with hard questions about the region's history of radiation contamination, the advisability of building another nuclear power plant, and the impact of Chernobyl contamination. One of the most dramatic moments was when the region's official in charge of sanitation inspection, Dr. Eleonora Kravtsova, was asked why she had permitted meat from Chernobyl to be brought to Chelyabinsk for processing, knowing the degree of radiation exposure that already existed in the region. Kravtsova's only defense was that she was a government official first, and a citizen second.

Mayak was not a focus of this public protest, but it was the first local institution to respond to the hearing. The Mayak authorities initiated a conference to discuss the Southern Urals Nuclear Power Plant. They invited 250 participants, including 50 representatives of the public. The others were nuclear power scientists, many from Mayak and from a weapons facility in Obninsk (near Moscow). Mayak officials invited the Popular Front's Ecological Group to select the 50 representatives from the public, the Ecological Group ensured that a number of antinuclear scientists attended from *Academgorodok* (Academy City)[5] in Novosibirsk. This was the first time scientists in a public forum opposed one another openly on such sensitive issues as nuclear power and other weapons related issues.

The next important stage in the formation of the Chelyabinsk environmental movement was the 1990 election campaign in Russia. By that time the national environmental movement was at its peak. It was playing a key role in reform politics, supporting or opposing candidates in the contested elections across the nation (Feshbach and Friendly 1992). The Chelyabinsk greens, however, were only getting started. The Popular Front ran its own candidates for the regional legislature and the Chelyabinsk City Council. They won 5 to 10 percent of the seats in the regional legislature, about 30 percent in the city council, and around 20 percent in city districts.[6] Only a small percentage of these successful candidates ran on strong environmental platforms.

The Popular Front and its environmental candidates were not strong enough to have a dramatic impact on the 1990 election outcomes, but the Chelyabinsk greens gained valuable organizational experience from these campaigns. During the first year of their activism they learned how to organize rallies, make posters, and write newspaper articles and letters for their campaign. It was relatively easy to mobilize hundreds (or even thousands) of people to rallies in the prevalent atmosphere of general political activism. The environmentalists still had no contact with foreigners or with environmentalists from other parts of the Soviet Union. Everything they did they created by themselves. It was a case of self-taught, grassroots democracy.

It was also during the 1990 election campaign that candidates chose either political or environmental priorities in their platforms. After the elections, those who opted for environmental platforms formed the core of a movement independent of the Popular Front.

Chelyabinsk Environmental Organizations

After the 1990 elections, four distinct environmental organizations evolved in Chelyabinsk: the Movement for Nuclear Safety (*Dvizhenie za yadernuyu bezopasnost*), the Democratic Green Party (*Demokraticheskaya partiya zelyonykh*), the Association of Greens (*Assotsiatsiya zelyonykh*), and the Kyshtym-57 Foundation (*Fond Kyshtym-57*).

All four focused on nuclear issues related to the weapons complex. This stress distinguished Chelyabinsk environmentalists from most other local environmental groups around the country, which emphasized conventional pollution. However, environmental groups in Tomsk and Krasnoyarsk also targeted their local weapons complexes. While the focus on nuclear issues was the same, the four Chelyabinsk organizations represented different tendencies on the political scene—from the far left to the far right. However, their political differences were not clearly enough defined to prevent them from working together in a loose, united front for the next year and one-half.

Soon after the 1990 election, Natalia Mironova founded the Movement for Nuclear Safety (MNS). She was an engineer who had never taken an interest in politics until the advent of the Popular Front and the exposé of the region's ecological damage. She was a family-oriented mother of two children, passionately concerned about her youngsters' education in a school system starving for funds, which she felt were being consumed by the weapons complex. From this starting point, she catapulted to a candidate to elected office and the leader of a militant social movement. Mironova took the leadership role in the Popular Front's environmental group, and was one of the candidates running for oblast deputy[7] who chose to focus on environmental issues in her successful campaign.

Mironova worked regularly with ten to fifteen activists in the MNS, and could count on another forty to fifty individuals to help her organize large mobilizations. By early 1991, the Movement for Nuclear Safety affiliated with the SEU. This association connected the MNS with the SEU's valuable intellectual resources. This relationship also brought Mironova out of Chelyabinsk, where no foreigners had ever been permitted, to attend an international environmental conference sponsored by the SEU in Moscow in February 1991. There she made a presentation about the MNS, and was subsequently invited to attend another international conference in April at the University of California, Irvine. This conference focused on the environmental consequences of nuclear weapons development (Whiteley 1991). When Mironova returned to Chelyabinsk, journalists from the United States, Western Europe, and Japan began contacting her for interviews and assistance in obtaining permission to visit Chelyabinsk.

Around this same time, the Movement for Nuclear Safety launched the Democratic Green Party (DGP), primarily as a vehicle to run candidates in elections. Most of the party's core organizers were male members of the MNS who wanted to tackle general political issues as a means of resolving environmental problems. During the August 1991 coup, the young men in the party formed defense teams to protect the newly elected democratic representatives in local government. Mironova became co-chair of the party together with Nikolai Kalachev.

The organization eventually signed up approximately 2,000 dues-paying members throughout the region, and could count on a steady 10 to 20 members to mobilize others for specific actions. Mironova felt that in the early 1990s the ideas of the Democratic Green Party and the MNS were endorsed by at least the 300,000 people in the region who signed a petition demanding an end to nuclear and other industrial development in the area, and protesting continuation of the construction of the Southern Urals Nuclear Power Plant.[8]

Among the original members of the Popular Front's Ecological Group were individuals who affiliated themselves with a Russian nationalist organization known as *Rodina* (Homeland). When political differences became more defined within the Ecological Group, these individuals shifted their attentions to an organization called the Association of Greens, led by Vitaly Kniaginichev, who later became the chair of the Russian nationalistic Great Russia (*Velikaya Rossiya)* Ethnic-State Bloc.

Nikolai Kalachev, cochair of the DGP, claimed that Kniaginichev was instructed by the Oblast Communist Party Committee to found the Association of Greens as a means of controlling the environmental movement.[9] Others in the democratic reform movement maintained that Kniaginichev was linked to the KGB, which wanted to control the environmental movement through him. It was claimed that the KGB connection was indirect, through the Communist Party, before it was banned, and subsequently through members of thc old party elite who still held influential positions in local government. In a December 1992 interview with Kalachev, right after the appointment of Prime Minister Victor Chernomyrdin,

Kniaginichev expressed his unequivocal support for Chernomyrdin solely on the basis that, unlike his predecessor, "he is a pure Russian, and therefore will do what's best for Russia."[10]

During the early period of environmental activism in Chelyabinsk (mid-1989 to mid-1991), which was marked by mass rallies and petition drives against the Southern Urals Power Plant, the MNS, the Democratic Green Party, and the Association of Greens found enough common ground to work together on various issues. Gradually, however, by the fall of 1991, the organizations went separate ways, following a nationwide trend. As Eric Green (1991) pointed out, politics in the Russian republic were generally far more complex than the "us versus them" attitude that dominated politics in the non-Russian republics. This complexity resulted in greater ideological heterogeneity, which eventually undermined the environmental movement's unity.

The fourth organization in Chelyabinsk that played a significant role in addressing local environmental issues related to Mayak was the Kyshtym-57 Foundation. It was set up by Louisa Korzhova, a retired nuclear physicist who spent her career working on nuclear power plants, and Alexander Penyagin, who once worked on a nuclear submarine. Neither Korzhova nor Penyagin had been involved in politics before perestroika. Both presented themselves as professionals who understood the complex nuclear issues better than other Chelyabinsk environmentalists.

Kyshtym-57 provided diverse assistance to the region's estimated 1 million victims of radiation exposure from Mayak. For instance, Korzhova, the only full-time, paid staff member, worked with an elaborate network of volunteers throughout the region to distribute humanitarian aid packages (food and medicine) to those who were exposed to radiation during the accidents and to members of their families. The Kyshtym-57 Foundation also promoted legislation that gave Chelyabinsk status as an environmental disaster zone, the same as Chernobyl. This enabled the region to receive additional allocations from the national budget, which were intended to provide better medical and other social services to the stricken population. Korzhova did not categorize her organization as a movement, but many of the activists viewed themselves as environmentalists contributing to the movement through the Foundation.

Mironova, Kalachev, and Korzhova originally worked together, but by late 1991 they began operating independently of one another, although some rank and file members were volunteers for both organizations. The national referendum in March of 1991 seems to have been the final turning point, after which the Chelyabinsk organizations went their separate ways, except for the MNS and the DGP, which remained aligned for another few years.

The debate over preserving the Soviet Union brought to the forefront general political differences between the group leaders that previously had not prevented joint action. One complication was that the leaders themselves were deliberately unclear about their positions. Korzhova, Penyagin, and even Kniaginichev (who was more clearly aligned with right-wing forces) did not express general political views that were radically different from those of Mironova. Democracy and economic reform were goals expressed across the political spectrum. The tensions arose in the movement among the leaders who responded to each other's subtle statements, subtle action, or lack of action that indicated the real positions behind the words.

The political differences related to how fundamental the reforms should be and how fast they should proceed. In addition, the environmentalists differed over the Southern Urals Nuclear Power Plant and government plans to import spent fuel from other parts of the Soviet Union and from Western countries for reprocessing and storage at Mayak. The Movement for Nuclear Safety and the Democratic Green Party were unequivocally against finishing the plant and importing spent fuel. If necessary, they advocated civil disobedience to stop these plans. Neither the Association of Greens nor the Kyshtym-57 Foundation took such firm positions. Their leaders were open to negotiating with the local authorities on how much might be charged to foreign agencies wishing to reprocess their fuel in Chelyabinsk, and what percentage of that money might be spent on social benefits for the local population. Depending on the terms of the agreements, the Association of Greens and the Kyshtym-57 Foundation would take a final position. These two groups also did not take a clear stand against the Southern Urals Power Plant.

All these four organizations were based in the city of Chelyabinsk and had no branches in the outlying towns and villages. This was primarily because reform politics were still concentrated in large urban centers, and the Chelyabinsk region was no exception. Individuals in some outlying areas (most notably, Muslyumovo and Kyshtym) were influenced by citizen environmental action in Chelyabinsk, but still were not involved in any systematic way. The main obstacles to involvement were the typical infrastructure inadequacies throughout Russia—the absence of phone service and the poor roads—making communication within a region more difficult than between large cities.

To varying degrees the Chelyabinsk environmentalists had notable success. In mid-1990, for instance, in response to the movement's newspaper and letter-writing campaign and a petition drive against the Southern Urals Power Plant, Gorbachev set up a commission that within months published a detailed assessment of the region's environmental conditions. As a result, the government declared the Chelyabinsk region a national environmental disaster area, enabling it theoretically to receive national allocations similar to the territories affected by the Chernobyl accident. Environmentalists in Chelyabinsk also felt that their public protest forced the government to suspend construction of the Southern Urals Power Plant, although the opposition claimed that the project was halted only because of lack of funds. Mironova, representing the Chelyabinsk Movement for Nuclear Safety and the Democratic Green Party, contributed to the draft of the Radioactive Waste Law that is still awaiting passage in the Russian Parliament. No progress, however, has been made toward banning the import of spent fuel for reprocessing and storage in Chelyabinsk.

Perhaps the MNS's most large-scale and effective mobilization concerned the March 1991 nationwide referendum on the fate of the Soviet Union. The Chelyabinsk greens (primarily the MNS and the DGP) wanted to add two more questions to the referendum:

1. Do you agree with building the Southern Urals Nuclear Power Plant—yes or no?
2. Do you agree with importing radioactive material from other republics and countries—yes or no?

It was not possible to put the questions on the ballot throughout the whole region, but the local environmentalists included them in city elections in Chelyabinsk. Just before the referendum the first Russian Congress of People's Deputies had passed a resolution forbidding the import of radioactive materials; this lent solid support to the referendum. The movement used this resolution as the main argument to vote no on both questions. These points were put forth for two months during the election in weekly newspaper columns in the *Vecherny Chelyabinsk* (Evening Chelyabinsk); this stimulated a discussion on both the pros and cons of the two questions. Support for the referendum and opposition to the power plant and import of spent fuel came from the Democratic Front (formerly the Popular Front), the Social Democratic Party, and the Popular Labor Alliance. The most vociferous opposition came from Mayak officials and the local Communist Party elite, who opposed adding the questions to the referendum. When they were added anyway, these groups represented the loudest voices in favor of the power plant and reprocessing imported spent fuel. The outcome of the referendum was dramatic. Seventy-six percent of the city's population voted against the Southern Urals Power Plant, and eighty-four percent against importing radioactive materials to Mayak.

In December 1991, the Regional Administration (formerly the Regional Executive Committee) offered Mironova a job to organize and head its new Committee on Radiation Safety. As in many other regions of the country where pro-Yeltsin politicians were put in charge of Regional Administrations, there was a partial merging of government bodies and nongovernmental organizations. This appeared to be a deliberate policy on the part of local officials who had come into office through the activism of these nongovernmental organizations.

During this period of 1991 and 1992, Mironova's committee employed many of the leading activists in the MNS and the DGP. They launched a major campaign to register everyone in the region who had been exposed to excessive radiation through the 1957 Kyshtym disaster and thus was eligible for government compensation. In the process they expanded their movement's base of potential activists. They collected and publicized valuable information about Mayak that previously had been classified, and hired Socio-Ecological Union specialists to conduct an independent monitoring project to identify plutonium contamination in the village of Muslyumovo.[11]

The high point of the movement's activities under the sponsorship of the Radiation Safety Committee was the organization of an international conference in Chelyabinsk on the environmental consequences of nuclear weapons production in the Urals. In the process, Chelyabinsk environmental activists forged contacts and exchanges with counterpart environmental organizations, independent scientists, medical professionals, and journalists in Russia and elsewhere in the world. The conference, open to anyone, was intended to be a major educational forum for the public. In May of 1992, after overcoming countless logistical and political hurdles, the hosts of the landmark Chelyabinsk conference greeted 550 delegates from Russia, Ukraine, Belarus, Kazakhstan, Germany, France, England, Sweden, Italy, Japan, Canada, and the United States. Among them were scientists, physicians, government officials, and environmentalists.

The week-long conference comprised a busy schedule of workshops on a wide range of issues: reprocessing, medical and epidemiological concerns, environmental law, groundwater contamination, radioactivity in agricultural crops, and grassroots organizing. People with divergent views on these controversial matters had daily opportunities to hear one another out. Participants noted that in the same month of October 1957, not only did Mayak suffer a major accident, but so did similar plants at Rocky Flats, Colorado, and Windscale (now Sellafield) in the United Kingdom. In her final remarks at the conference's closing, Mironova called those nearly simultaneous disasters a tragic warning of the dangers of nuclear power which, had they not been hidden from the world, might have prevented Chernobyl.

These activities brought the Nuclear Safety Committee and the Democratic Green Party generous television and newspaper coverage. Local officials began to see their connection with the independent-minded nongovernmental organizations as a liability. Almost immediately after the May conference Mironova felt strong pressure from within the administration to discontinue her grassroots work. Local officials attempted to limit her contact with foreigners and her ability to spend time away from

Chelyabinsk in order to travel inside or outside the country. Pressure was also placed on the other environmental leaders whom she had involved on the committee, particularly her cochair of the Democratic Green Party, Nikolai Kalachev, who was not an elected official and therefore did not enjoy Mironova's immunity. The MNS and the DGP agreed that it was important for Mironova to continue heading the committee because the position gave her unique access to sensitive information that was crucial to the movement.

Throughout the country other nongovernmental organizations that had accepted the funding and hospitality in the offices of local governments were also beginning to lose their sense of independent action. They faced the dilemma of either staying on the inside without an independent voice or perishing on their own.

Meanwhile the national environmental movement had already slid into serious decline. The collapse of the Soviet Union, followed by the severe economic crisis, dealt the movement a heavy blow. The Yeltsin government did not have the resources to put environmental considerations ahead of economic development. By the winter of 1992–93, when the Soviet Union no longer existed and economic hardships completely consumed the population's time and energy, activism for environmental or political reform became a luxury that most people could not afford. Nikolai Reimers, head of the Russian Ecological Union, offered this explanation:

> In the first phase of the movement the focus was on tearing down this, preventing the building of that, not on restoration. That phase was passionate. Passivity set in when energy was sapped and gains were few, and when life's everyday hardships multiplied.[12]

With the demise of the Soviet Union, environmentalists in the various republics no longer looked for ways to unite with activists outside their republics. The exceptions were those organizations most closely affiliated with the SEU. The Moscow-based organizations (except for the SEU) that had been building all-Soviet movements had to shift gears and focus on shaping solid networks within Russia. Only now they had numerically smaller and less optimistic memberships, and fewer resources to run their operations.

As Yanitsky (1996, 72–73) noted, between 1992 and 1994, the link between the population and the environmental movement weakened considerably:

The government abolished one by one all of the past achievements of the green movement. . . . In fact, from 1991 to 1994, local initiative groups as civil society units had been extinguished almost everywhere. The ecological movement did not notice the quick decline in the population's living standards; growth of the crime rate, unemployment, and refugee problem; and infringement on the rights of the Russian-speaking population in the former Soviet republics. In these four years, the movement leaders did not organize any mass campaign for defending the social or economic rights of any Russian people or a social stratum.

Yanitsky also stressed that the SEU and other environmental organizations declined because they did not cooperate with groups outside the environmental movement: "The leaders of the Social Ecological Union and other ecological organizations always preferred to involve new people and groups in their own actions rather than to cooperate with other movements."

By 1992 and 1993, the Russian government started openly pursuing policies dictated by the nuclear lobby (chapter 12). The MNS tried to combat the enemy image of environmentalists that officials at the weapons facilities promoted; Mayak officials charged that environmentalists were incompetent to evaluate complex nuclear issues. The MNS responded by urging that the closed cities be opened. "The scientists in the closed cities," said Mironova, "are acting like a rejected lover no longer adored for their patriotic contribution to defense."[13]

After 1992, the Chelyabinsk environmental movement could no longer boast the mass following that it had enjoyed from its outset. It could rely on some core militants who had launched the movement. Unlike at the national level, however, most of these local activists had not taken an interest in environmental issues before the democratic reform movement. Environmental activists were still motivated more by a sense of victimization and a concern for environmentally related health risks than by broader environmental values (see chapter 10).

The Movement for Nuclear Safety was the only environmental organization in Chelyabinsk that survived the fluctuating political scene of post-Soviet Russia. This was, in part, due to its close relationship with the

Socio-Ecological Union and its ability to receive funding from foreign foundations interested in promoting Russia's floundering nongovernmental, democratic sector. The MNS began sponsoring small affiliate groups that formed in some towns and villages. In 1998, it claimed to have fifteen affiliated organizations in the Chelyabinsk Oblast and Ekaterinburg (formerly Sverdlovsk), and 150 activists.

The MNS has sponsored independent monitoring and assessment of local environmental conditions, promoted public education programs intended to alert local residents to possible health and safety risks, helped citizens use the court system to protect their legal rights to environmental health protection (chapter 10), and lobbied for nuclear legislation.

Over time, however, the political enthusiasm in support for the movement, and the political impact of the movement has declined. For instance, in the early spring of 1998, the focus of environmental protest, both in Chelyabinsk and in the national antinuclear movement, was against plans by the United States and Russia to dispose of surplus plutonium by converting it into mixed oxide (MOX) fuel and burning the fuel in nuclear reactors. Mironova said this about these plans:

> I am very worried and very sad about this kind of cooperation. In 1991, I was very enthused about Russian-U.S. cooperation because the cold war was over and this was a time to think about a new kind of friendship, a new kind of relationship. But step by step I began meeting with the American authorities and saw that they are much like the Russian weapons authorities. I am very sad that the American nuclear elite has developed cooperation very quickly with the Russian nuclear elite and together they developed very quickly a new plan for the use of MOX.[14]

The MNS today has little faith in its ability to stop MOX in Russia. It relies on the U.S. movement to pressure the United States into abandoning the project, believing that will stop Russia.

Analyzing the Movement

This section discusses factors that may help explain the behavior and social implications of the environmental movement: the organizational structure, resource base, alliance patterns, political strategies, and tactics. From these analyses I hope to draw larger implications about the Chelyabinsk experience for environmentalism and citizen movements.

Research on Western environmental groups indicates that these movements often do not follow the model of centralization and bureaucratization that is common among other citizen interest groups (Kriesi and van Praag 1987; Kitschelt 1989). One explanation is that the participatory and democratic norms embodied in the New Environmental Paradigm led environmental groups to avoid centralized and bureaucratic organizational structures (see chapter 7). Green activists prefer a fluid organizational framework based on a horizontal coordination of grassroots groups in loose networks. In addition, the small size of locally based groups makes an extensive organizational structure less necessary and less desirable.

In Chelyabinsk, the environmental organizations evolved from amorphous structures with loose decision-making processes into distinct organizations with charters outlining specific structures, rules, and regulations about membership, leadership, and decision making. The organizations were not entirely participatory. The charters of the Chelyabinsk organizations provided for membership-elected coordinating councils and a chair or cochairs that met regularly, weekly, biweekly, or monthly to make decisions between general membership meetings. They also placed a high value on democratic decision making in principle. In practice, most day-to-day decisions were made by phone among the leading activists, who were usually amenable to whatever the chair or cochair advised. This style of leadership was more in keeping with old political practices than with the consensus-oriented process characteristic of the SEU at the national level. This traditional Russian style of leadership was further reinforced in the political chaos that prompted a prevalent public opinion that the ideal leader was a strong personality prepared to provide political direction. The only local Chelyabinsk organization to survive the dramatic decline of the environmental movement throughout Russia, especially after 1994, the MNS, reflected this style of leadership.

Resources

To survive and become politically active, environmental groups must mobilize finances and other resources to sustain the organization. This was an especially difficult challenge for the environmental groups in Chelyabinsk.

Another important factor is the resource base of the movement. None of the Chelyabinsk organizations ever had formal dues, and except for the MNS, no funding source other than contributions by its own activists and supporters. Some donations have been free, such as xeroxing (probably at the expense of a state enterprise or private business, where some sympathetic employee may work), materials for making posters and announcements for rallies, and, of course, free labor. For instance, the MNS, during the whole year of 1990, when the average monthly salary was 220 rubles, spent between 3,000 and 4,000 rubles donated from its activists.

For a transitional period, direct support from the government was an important factor in sustaining environmental groups. When Mironova held a formal position with the Committee on Radiation Safety, the MNS and DGP benefited from the government's support. The conservative Association of Greens also appeared to be receiving indirect support from government agencies. But with Russia's deepening economic problems, and a conservative turn in the local and regional governments, this support ended. Indeed, the difficulty in marshaling sufficient resources to support organizational activity is a factor that contributed to the fading away of the local environmental groups. Unlike the other local environmental organizations, since 1993, the MNS has received funding from international nongovernmental sources, chiefly from the United States, enabling it to endure.

Alliance Patterns

Identifying potential friends and allies is a crucial step in planning strategies of action and building coalitions that can produce environmental reforms. Resource mobilization theorists stress the importance of ties linking environmental organizations to other social groups and political organizations (Tilly 1988; Klandermans 1990; Diani 1995). For instance, local citizen groups formed after the Three Mile Island incident had numerous (and diverse) ties to other political groups (Walsh 1988). Antinuclear protest groups in the West also have developed complex patterns of alliances (Rudig 1991).

Alliance networks are important because they can provide funding sources and influence strategies of political mobilization and political action. For instance, groups that have allies within the political establishment may be less willing to pursue tactics of confrontation and direct action. Conversely, a group that emphasizes its conflicts with the political establishment may be less willing to work with government agencies on possible reforms.

Western environmental groups, such as Friends of the Earth or Greenpeace, are often estranged from established social and political groups. Some scholars explain the autonomy of Western environmental groups in terms of their commitment to environmentalism, which places them at odds with established interests. Thus to work with established political actors implies a betrayal of the movement's call for a fundamental change in economic structures, political processes, and humankind's relationship to nature. Other scholars link the supposedly autonomous tendencies of environmental groups to their lack of resources and hence their inability to engage in the bargaining exchange of normal politics.

In our examination of alliance patterns in Chelyabinsk we saw a fragile process of coalition building within a quasi-pluralist setting among Russian environmentalists. In the midst of revolutionary changes in society and the political system it was not always clear who were one's potential friends and foes. In addition, there was no ethos like that expressed by Hanford organizations: "the more united, the more power they have" (chapter 5).

The MNS's main institutional support originally came from the Regional Executive Committee, and especially from the head of environmental affairs, who responded positively to any request for his department's approval of the movement's various initiatives. This official provided the MNS, in a timely manner, with important information about the operational plans of the weapons facilities.[15] Both sides shared an interest in such an alliance. The Movement for Nuclear Safety and its affiliated Democratic Green Party were the region's most prominent, mass-based environmental organizations, the ones most closely associated with the democratic reform movement, and with the national SEU. Perhaps by showing their support for popular movements, local officials wanted to increase their popular legitimacy. They may also have wanted to control Mironova and her followers, who had no other funding sources or office space, and therefore agreed to the patronage. The MNS felt that a primary advantage to this alliance, other than the funding, was the quick access to information about the weapons complex that they could not obtain from the outside.

The major base of popular support for the movement was the individual activists in the MNS. They were primarily professionals in the fields of engineering, medicine, and education. Another base of support was people concerned about the alarmingly high incidence of illnesses, especially allergies, among children, and the high mortality rate, particularly among men. The movement's emergence in 1989 was accompanied by intensive industrial output and correspondingly severe pollution in Chelyabinsk, so environmental issues were particularly salient. In 1989, the residents of Chelyabinsk also learned of the extent of the radiation contamination from Mayak, and began realizing that their ailments and early deaths might be related to the weapons complex.

In the very early stages of the movement the various organizations saw one another as political allies, but by 1992–93 they were political competitors for public attention and international funding. As Fomichev (1997, 61) has pointed out, "The shortage of resources leads to internal competition, which can't help but weaken the movement." Around this time, all the organizations disappeared from the scene, except for Mironova's Movement for Nuclear Safety. Indeed, the failure of other groups to survive may be traced to the lack of political allies and resources like those that sustained the MNS during difficult periods: a combination of support from the SEU and other Moscow-based sources, and international support. Furthermore, like the trend at the national level, the Chelyabinsk groups generally avoided any cooperation with other social movements that were not targeted at environmental issues.

Even these patterns have changed, however. In the early 1990s, the MNS and the SEU communicated regularly and supported each other's efforts. By the late 1990s, there was almost no communication between the SEU and the MNS or its affiliated Chelyabinsk organizations. Today there is no national organization helping local organizations with Chelyabinsk issues, unlike at Hanford where several national public interest groups or environmental groups are active on Hanford-related issues (chapter 5).

The main opposition to the MNS originally came from the Regional Communist Party's department in charge of defense. Party officials voiced a common charge from the Communist establishment, claiming that environmentalists were guided by Chernobyl phobia and were not focusing on the real environmental culprits—civilian industry in the city of Chelyabinsk. The KGB never took visible action against environmentalists. However, the defense department in the Regional Communist Party may have acted on behalf of the KGB (chapter 12).

The management of Mayak voiced repeated opposition to the environmental movement. In public statements plant officials accused environmentalists of not having the interests of the population at heart by not wanting Chelyabinsk to have the much needed electrical power from the Southern Urals Power Plant. They also criticized environmentalists' indifference to the valuable hard currency that could be earned from reprocessing imported nuclear fuel. Mayak officials presented environmental leaders as individuals who exploited the popularity of environmental issues to gain political capital in the pursuit of personal political careers. This enemy image of environmentalists was projected particularly vividly in the Chelyabinsk-65 local newspaper in recent years. In this campaign plant officials blamed the environmentalists for wanting to deprive Mayak of its privileged status and related benefits.

The media are normally an important group for social movements. This was true in Russia in the early years of the movement when the media were in the forefront of the democratic reform movement, generously providing space to the most militant voices for radical transformation and against the entrenched interests of the party elite (chapter 12). In Chelyabinsk, however, it has been somewhat more difficult for local environmentalists to have access to the local media due to the strong control the military complex wields throughout the region's power structure. Nevertheless, newspapers, such as the *Vecherny Chelyabinsk* (Evening Chelyabinsk), the *Ekologichesky vestnik* (Ecological Bulletin), and especially the local television station, have been the key forums through which local environmental organizations have spoken to the public at large. A few local journalists have readily assisted environmentalists in getting their articles and letters published, and getting air time on television and radio. Editors, however, are pressed by the weapons complex to allow them equal access to their critics.

Without doing a special study of the media it is not possible to say precisely how "equal" the access has been. However, conversations with environmentalists throughout the region and with officials in the closed facilities, point to some tentative conclusions. In the city of Chelyabinsk environmentalists had some access to the media, whereas Mayak officials claimed that they had difficulties being heard in Chelyabinsk. In the outlying towns and villages local activists maintained that they had almost no access to local papers that preferred to publish the point of view of the facilities.

Political Strategies and Tactics

Although environmental interest groups are active participants in the political process of most advanced industrial democracies, scholars remain divided on the overall quantity and quality of their participation. The early literature on environmental organizations focused on the unconventional political tactics and antiestablishment aspects of these groups. This one-dimensional image of unconventional action has been challenged by research that documents the diverse range of tactics actually used by these groups (e.g., Dalton 1994). Social movement theorists predict that the diverse goals and opportunity structures that environmental groups face lead them to pursue political reform through many routes, ranging from the conventional to the unconventional (Klandermans 1990; Kitschelt 1986).

In the case of environmentalism in Russia, and particularly in Chelyabinsk, our research team wanted to understand how groups make decisions on which tactics to use. Throughout its existence the Chelyabinsk environmental movement organized rallies, newspaper and letter-writing campaigns, petition drives, and referendums to promote its causes. The primary goal of these tactics was to compel national agencies to investigate the environmental consequences of the complex's nuclear disasters, to pry information from the complex directly, to promote legislation on atomic power and the handling of radioactive waste, and to ban construction of the Southern Urals Power Plant and the import of spent fuel.

Their decisions were pragmatic based on the opportunities, rather than a reflection of their identity as opposition groups. They were willing to work with national and local government agencies. For instance, the

MNS and DGP leaders were involved with and employed by the Regional Administration's Nuclear Safety Committee in 1992–93. The MNS and DGP, however, pulled back from this relationship when it began to threaten the goals and autonomy of the movement. They were eventually dropped from the administration for advocating their movement's anti-nuclear and anti-weapons complex thrust while the administration was moving in the opposite political direction.

With the loss of status within the Administration the MNS and DGP found themselves without local funding or office space. The MNS, however, found funding from the United States through its association with the Socio-Ecological Union. This shaped their strategies accordingly. Since the funding was coming from private organizations interested in promoting the education and implementation of democratic citizen action to improve environmental conditions, as well as coalition-building, the MNS was pressured to engage in actions to move them in these directions.

Within that framework Mironova and the members of her organization attempted to forge a stronger intraregional network. Part of creating this network was to build on the increased political activism of the Turkic-speaking, Muslim Tatar and Bashkir minorities in the region, who had been most affected by environmental damage by the weapons complex because they reside in the villages closest to the site. They represented a key social base for local activism because of the upsurge in ethnic identity and cultural cohesion among them over general political issues as well as nuclear environmental issues.

Part of the strategy of building an intraregional network was to involve individuals in the multiethnic outlying towns and villages who had expressed concern about local environmental issues and frustrations over inadequate contact, impeded by poor phone service and transportation. Even within the same city, such as Kyshtym (a few miles from the weapons complex), potential activists did not know of one another's existence. Other potential participants in this network live in the town of Voroshilovsk, inhabited mainly by employees of Mayak who are bussed in and out of the closed city daily, and by employees of the complex's environmental monitoring station. They have circulated a petition underlining the long-neglected environmental and health problems they face. However, local activism has been sporadic and limited.

Conclusion

Nongovernmental organizations have an especially significant role in a society attempting to change from an authoritarian to a democratic political system. This is particularly true of nongovernmental environmental organizations. In Chelyabinsk, where the heart of the military industrial complex is backing the resurgent right wing in impeding reform and the grassroots agents of reform, such as the Movement for Nuclear Safety, the challenge to these organizations is all the more critical. This chapter has assessed the strengths and weaknesses of this fragile movement, and suggests the following conclusions.

The Chelyabinsk environmental movement is characterized by a leadership that has no links with pre-Gorbachev politics or environmentalism. This may be its strength, in that it is not hampered by the old habits of constrained activism, but it is also a weakness because its experience with environmental issues is only very recent. This is, perhaps, one reason why local environmental action did not make or break political candidates in the 1990 election campaign, whereas the national movement based in major industrial centers actually shaped the campaign.

The political disunity that is characteristic of the movement throughout Russia is also evident in Chelyabinsk. There is no "us versus them" attitude that dominated politics in the non-Russian republics and united diverse nongovernmental organizations around the issue of reasserting ethnic identity against Moscow. Another related problem is that there is no strong intraregional network, which is also a common phenomenon throughout the country.

Across Russia, and Chelyabinsk is no exception, nongovernmental organizations have not developed internal funding sources, perhaps because ordinary people are too poor and too unaccustomed to donate their funds to political and social causes, and new entrepreneurs who are potential big donors do not see how their self-interests can be met in such philanthropy.

The authoritarian political culture of the Communist regime, which has survived the reshuffling of official personnel at all levels of government structure, is also alive and well in the nongovernmental sector. While it is less evident in the national environmental movement, particularly the SEU with its consensus-oriented democratic process, the amorphous democratic structure and decision-making process of the Chelyabinsk Movement for Nuclear Safety borders on the autocratic.

Therefore the Chelyabinsk movement entered this period of political chaos and economic devastation with only a brief and modest history of a mass following and political victories, a core leadership with much determination, but inadequate organizational and environmental expertise, political disunity among fellow environmental activists, no strong intraregional network, no internal funding base, and a feeble democratic structure.

The much stronger SEU has not been able to compensate for the weaknesses of its local affiliates by providing political guidance and environmental expertise, and furthering its inclusive approach to promote greater unity among divergent forces nationwide and at the local levels. The poignant environmental problems in Chelyabinsk that have resulted from the U.S. and Soviet arms race, the global relevance of these issues, and the admirable dedication to solving the problems on the part of Chelyabinsk environmentalists have compelled U.S. and other international foundations to provide financial support to further these efforts. It seems unlikely that local organizations, like the MNS, would survive without international funding.

Notes

1. See Weiner (1988). Trofim Lysenko was the Soviet biologist who during the Stalin Period rose to dominate biology, discrediting and ruining the study of genetics, as well as the early Soviet conservation movement and the discipline of ecology.

Weiner documents the golden age of early Russian conservation activity, locating it in the decade before the 1917 revolution until the advent of Trofim Lysenko in the early 1930s. Weiner points out that due to these early conservation efforts, the Soviets were first to propose and set aside protected territories for the study of ecological communities, and to suggest that regional land use could be planned and degraded landscapes rehabilitated on the basis of those ecological studies.

He attributes Russian and early Soviet ecologists with pioneering phytosociology, the individualistic theory of plant distribution, and the trophic-dynamics, or ecological energetics, paradigm, so critical to the field of community ecology.

2. Conversation with Sviatoslav Zabelin, chairman, Council of the Socio-Ecological Union, Moscow, September 1991.

3. Conversation with Maria Ivanian, secretary of the Moscow Green Party, Moscow, March 1992.

4. Conversation with Sviatoslav Zabelin, Moscow, September 1991.

5. An academic community established in the 1960s whose major industry was scientific research.

6. Conversation with Natalia Mironova, Irvine, June 1993.

7. An oblast deputy has essentially the same function as a state legislator in the United States.

8. Conversation with Natalia Mironova, Irvine, California, June 1993.

9. Conversation with Nikolai Kalachev, cochair of the Chelyabinsk Democratic Green Party, Chelyabinsk, May 1992.

10. Conversation with Vitaly Kniaginichev, Chelyabinsk, December 1992.

11. At that time the monitoring laboratory at Mayak denied any contamination at Muslyumovo, other than in the nearby Techa River.

12. Conversation with Nikolai Reimers, head of the Russian Ecological Union (formerly Ecological Union), professor of ecology, March 1992.

13. Conversation with Natalia Mironova, Moscow, March 1993.

14. Conversation with Don Moniak, program director, Serious Texans Against Nuclear Dumping, by phone from Amarillo, Texas, and printed on e-mail, March 13, 1998.

15. Under the Soviet regime a region had two administrative bodies: the Regional Party Committee and the Regional Executive Committee. As reforms proceeded and the Communist party declined in prestige and power, the authority of the Regional Executive Committee rose accordingly.

IV

Citizens and the Environment

7

Environmental Attitudes and the New Environmental Paradigm

Russell J. Dalton, Yevgeny Gontmacher, Nicholas P. Lovrich, and John C. Pierce

It is possible and necessary, comrades, to improve nature, to help nature reveal her living forces completely.
—Leonid Brezhnev

In both eastern Washington and the Chelyabinsk Oblast the primary motivation for environmental action among the region's citizens was a concern with the specific environmental problems resulting from the nuclear weapons production and research activities of Hanford and Mayak. Once tapped, however, the wellsprings of environmental action can run much deeper. Environmentalists concerned about pollution of the Russian Techa River or the American Columbia River can draw parallels to other environmental problems affecting their respective nations. Alliance building with national environmental organizations, or with other local groups, can similarly broaden conceptions of the goals and methods of environmental action. These interactions can stimulate new ways of thinking about the environment, which in turn may influence future patterns of collective political action. While environmental protest might start immediately beyond the fences of Chelyabinsk-65 or the Hanford Reservation, these initially narrowly focused events can stimulate a broader public to develop an interest in a wider range of environmental (and broadly political) questions facing the United States and Russia.

Not only does the environmental action typically encourage a broadening of the scope of environmental interests, but also, we would argue, a political movement must develop an articulated conceptual framework to function politically (Dalton 1994). Even if only implicitly, political action presumes that activists possess a framework for thinking about politics that helps them understand, interpret, and integrate events into a logical chain of thought. This framework would include beliefs about the ultimate goals of the movement, the origins of contemporary problems, orientations toward the existing sociopolitical order, and beliefs about the possible means of achieving their desired goals. For example, a belief system provides individuals with a framework for assessing the sources of environmental problems and their ultimate resolution. Whereas a downwind activist might consider the environmental problems of Hanford and inevitable results of the excesses of technology, an engineer at Hanford might see the same environmental problems as an engineering problem to be resolved by technology. Kempton and his colleagues (1995) describe these patterns of thinking as cultural models of environmentalism; others write of an environmental paradigm (Milbrath 1984; Dunlap and Van Liere 1978). Such a system of political beliefs would guide the orientations and behavior of individuals concerned about the facilities.

Simply stated, this chapter asks what it means to be "green" for citizens in Chelyabinsk, Russia, and eastern Washington State. Has a green philosophy or way of thinking developed from the political experiences of both movements? Is this conceptualization of environmentalism shared by citizens in both nations despite the basic differences in their experiences and political environments? Or is there instead a distinctly different green pattern of thinking for both cultures?

Past research on public opinion and environmentalism in Western democracies has described a commonly shared cluster of beliefs that constitute the core values of environmentalism in Western democracies, what researchers term the New Environmental Paradigm (NEP) (Dunlap and Van Liere 1978; Milbrath 1984, 1989; Cotgrove 1982; Pierce et al. 1986; Dobson 1995; Kempton, Boster, and Hartley 1995). The Western NEP includes elements such as the acceptance of natural resource limits, an orientation to collective values over individualism, a biocentric view of nature, distrust of technology, and other beliefs. It is not immediately clear whether the same environmental paradigm exists in the formerly Communist nations of the Soviet Union and Eastern Europe. Marxist-Leninist principles reinforced many elements of the NEP, while simultaneously undercutting other values identified with environmentalism in the West. On the surface, there are good reasons to stress the wide ideological and political gulf that separated Russian and American societies—and Russian scholars such as Yanitsky (1996, 1995) are highly skeptical of finding NEP equivalents in Russia. Yet there are similarities between both societies in their portrayal of nature and society's relationship to the environment. The collapse of the old order of Soviet society further obfuscates the nature of environmentalism in post-Soviet states. An examination of contemporary American and Russian environmental beliefs in the post-Communist era can tell us much about the basic similarities of the societal consequences of industrialization that transcend the East/West ideological divide, as well as provide an exceptional opportunity to explore what it means to be green in two such different nations.

This chapter begins by describing the components of the New Environmental Paradigm as found among Western environmentalists, and speculates about the possible contrasts between American and Russian beliefs comprising the NEP. Then we present the opinions of residents of the Hanford area and the Chelyabinsk region on the elements of the NEP. Finally, we examine the distribution in these opinions within the two populations. How both publics conceptualize environmental issues will to a considerable extent shape their policy views, their style of political expression, and their political goals. Our inquiry into the environmental thinking of American and Russians is directed toward providing this understanding.

The New Environmental Paradigm

Although we may sometimes speak of a single green philosophy, there is no single framework that unites all environmentalists. Environmentalism is a multi-hued movement that includes many different shades of green, potentially attracting support from a wide range of social groups and individuals.[1] Indeed, a significant strength of the movement is the breadth of its popular support from a diverse set of individuals interested in various environmental issues. The green movement can unite the nature preservationist, the wildlife habitat protector, and the public health professional.

Although environmentalism is often considered a new political development, its roots run deep in both Western and Eastern societies. The first major wave of environmentalism at the turn of the century involved a nature conservation initiative (Dunlap and Mertig 1992; Dalton 1994). The creation of parks and nature preserves in time became a global phenomenon. The conservation movement took on new energy in the late 1960s and early 1970s, reflecting a growing public interest in the perpetuation of species, the protection of habitats, the protection of public health, and the preservation of national cultural monuments and areas of distinctive natural beauty. The impulse for the protection of nature reflected a romantic's view of one's own nation's heritage, in part harking back to an earlier time when humans had a closer connection with their natural surroundings.

Similar tendencies in popular thought exist in Russian history. Russian culture, for example, retains a special value for the merits of rural life that reflect the peasant base of the country (Yanitsky 1995; see also chapter 8). There is an idealization of the simplicity and innocence of rural life that has retained a special appeal as Russia industrialized, urbanized, and experienced the fruits of scientific socialism (Ziegler 1987, 6–8). As Lev Tolstoi created images of a pastoral utopia in his writings, Henry Thoreau and John Muir stirred such feelings among Americans. The preservation of natural parks and the pride in the nation's natural beauty are distinct features of both cultures. The value Russians attached to a weekend retreat in the country, and the emotional symbolism of "Mother Russia" reflect these lingering sentiments. Thus Douglas Weiner (1988) speaks of a pastoralist tradition in Russian environmentalism. Though these views partially conflict with the scientific technical revolution of Marxist-Leninist ideology, the views did not directly challenge the goals of the Communist state. This valuation of nature often took symbolic forms, such as protection of certain species or areas of natural beauty, rather than being linked to contemporary environmental problems. The Soviet regime promoted the anthropocentric belief that nature could be protected while industrialization and expansion of the economy continued. This orientation was represented by groups such as the All-Russian Society for the Protection of Nature and Nature Protection Corps.

While conservationist environmentalism has existed for decades in both American and Russian societies, a new type of environmental thinking began to develop in several Western democracies in the 1970s. This new consciousness represented an alternative model of social and economic relations that challenged both the dominant social order of industrial societies and the traditional values of earlier conservation movements. Riley Dunlap and his colleagues have provided a useful framework for analyzing environmental beliefs in terms of the contrasting ideal-types of a Dominant Social Paradigm of industrial societies and an alternative New Environmental Paradigm (Dunlap and Van Liere 1978, 1984; Catton and Dunlap 1980).

Catton and Dunlap (1980, 17–18) summarize the ideal-type of the Dominant Social Paradigm (DSP) as four broad beliefs: 1) people are fundamentally different from all other creatures on earth over which they have dominion; 2) people are masters of their destiny; they can choose their goals and learn to do whatever is necessary to achieve them; 3) the world is vast, and thus provides unlimited opportunities for humans; and 4) the history of humanity is one of progress; for every problem there is a solution, and thus progress need never cease. The DSP provides a cultural basis for advanced industrial societies—of either the capitalist or socialist economic form. It places humankind, and its needs, above all other species or nature itself. This justifies a pattern of natural resource usage that underlies industrial development. Capitalists and socialists share a common belief in progress and in the growth of society's capacity to provide for human needs, and both see social gain as by-products of economic expansion. This belief system conditions people in these societies to accept pollution and other aspects of environmental degradation as by-products of progress, which future advances in technology and scientific knowledge would doubtless correct. Faith in science and technology is unquestioned. The DSP thus describes a set of values that underlie the basic structure of modern industrial societies, capitalist and socialist alike.

The New Environmental Paradigm is presented as the polar opposite of the DSP (see also Milbrath 1984, 1989; Cotgrove 1982; Dobson 1995; Eckersley 1992; and Kempton, Boster, and Hartley 1995). The NEP reflects quite different beliefs premised on different core values, and thus it presents a fundamental challenge to the prevailing order in contemporary industrial societies.

A core element of the NEP involves specific beliefs about the relationship between humans and nature. The anthropocentric emphasis of the DSP posits humankind as the sole basis of society; nature is a means to addressing human needs, with little value in itself. This view became the norm of industrial society in both the West and East, and is sometimes linked to the cultural and religious traditions of Western (and Russian) society (Pryde 1972). In contrast, the NEP reflects a biocentric approach to nature. Far from being superior to the rest of nature, humans are actually dependent on a healthy natural environment. From a deep ecology perspective, nature may even hold a superiority over humankind's needs. The primary value of nature calls into question acts that damage it in the name of human-directed goals, such as pursuit of economic progress (e.g., burning Brazilian rain forests) or conspicuous consumption born of affluence (e.g., building golf courses in the desert).

The DSP also includes a belief in the economic rationality that underlies both capitalist and state-planned economies. The DSP presumes that the efficient management of resources, whether through marketplace forces or state planning, can maximize living standards and guide future development. Industrial societies are committed to the creation of wealth, an orientation of rational and systematic action, the maximizing of economic values, and the use of economic criteria as the benchmark for evaluating collective action—under either the market system or a planned economy (Cotgrove 1982, 58). The NEP, in contrast, emphasizes qualitative values that are not easily accommodated by economic cost/benefit models. The adoption of NEP beliefs leads to a critique of both market-based economic activities and centralized planning that are premised primarily on economic criteria.

Another contrast between the DSP and the NEP involves assumptions about the finite character of natural resources. The DSP views economic growth as the engine of economic and social progress. Instead of stressing (or even accepting) economic growth as a societal goal, the NEP accepts the limits of nature's bounty and a need to adjust human expectations to reflect these limits. The NEP thus advocates an economy in harmony with nature, reflecting the genuine personal needs of its citizens as opposed to the artificial needs of economic systems based on the exploitation of nonrenewable resources. The NEP criticizes the materialism and excess consumerism of modern societies as causes of resource shortages and pollution. In its place, the NEP advocates the goal of establishing a "sustainable society," in which consumption patterns are balanced against the world's natural productive capacity (Kempton, Boster, and Hartley 1995; Milbrath 1989).

The Dominant Social Paradigm relies heavily upon technology as a driving force in social and economic progress. Technology provides solutions for limitations of energy, inadequacies in materials, and shortages of other resources, thereby enabling the process of modernization to continue (such as the development of nuclear power, petrochemical compounds, and synthetic fibers). In contrast, adherents of the NEP are often skeptical of technology, which they see as both postponing the consequences of past environmental mistakes and even worsening the damage humankind can inflict on the environment. For instance, in their *Green Manifesto,* Sandy Irvine and Alec Ponton (1988, 36) state: "Technological gadgets merely shift the problem around, often at the expense of more energy and material inputs and therefore more pollution." Advocates of the NEP do not abhor technology, and often are users of certain technological innovations to promote better communications or facilitate the operation of small-scale economic projects. There is, however, a basic skepticism toward the prospects of "technological fixes" for environmental problems, and a staunch belief in the need for more basic social change to protect nature from irreparable harm.

The NEP also emphasizes the value of social and collective needs over individual needs. Instead of individualistic behavior, many ecologists advocate a pattern of social relations that in some respects harks back to small-scale communal life. The NEP replaces the status differences implicit in the growth machine of the Dominant Social Paradigm with a preference for egalitarianism and attention to social needs. This pattern of social relations envisions a sense of collective identity coexisting with a tolerance of individual lifestyle choices. A concern with the collective and commonly shared resources is thus an important part of the NEP, and a major factor defining the political identity of many environmental activists (Cotgrove 1982).

Finally, the NEP advocates a structure of political relations in contemporary societies that differs from the Dominant Social Paradigm. The latter emphasizes the essentiality of hierarchical authority and bureaucratic decision making—the culmination of Weberian (or Leninist) models of societal efficiency. In contrast, drawing from the traditions of Thoreau and Kropotkin, the New Environmental Paradigm reflects the view that personal fulfillment and self-expression should be primary social goals (Cotgrove 1982). Even for Western democracies, this orientation leads environmentalists to advocate changes in the political structure leading to a more open political system, a greater opportunity for citizen input, and a more consensual style of decision making.

In short, the NEP does not consider current environmental problems as an aberration of the socioeconomic order, but rather sees them as a direct result of the social system. Consequently, the solution of environmental problems requires significant changes in the socioeconomic order. Because of their orientation, advocates of the NEP are inclined to question the economic and social structures that contribute to contemporary environmental problems. This encourages an expansion of the movement's political agenda beyond the specific issues that initially stimulated action. In addition, the elite-challenging orientation of the NEP leads to a broadening of the methods of political action (e.g., Dalton 1994, ch. 6–8). In overall terms, therefore, the NEP presents a culture-based criticism of the social and economic structure of contemporary industrial societies.

Theorizing about the New Environmental Paradigm has been based on the experience of Western democracies, on countries that share a common cultural and economic heritage. We might, therefore, expect the DSP/NEP framework to be useful in describing the beliefs of Americans in the Hanford area. It is less clear, obviously, whether this same pattern should apply to contemporary Russia. A central question of this research project is the extent to which we can see similar belief systems among American and Russian publics.

In many areas there is an uncanny similarity between the Western and Eastern versions of the Dominant Social Paradigm. Jonathon Porritt, former head of Friends of the Earth in Britain, described a commonality between the industrial orders of the West and East that is shared by many activists in the Western environmental movement:

Both (Western and Eastern) systems are dedicated to industrial growth, to the expansion of the means of production, to a materialist ethic as the best means of meeting people's needs, and to unimpeded technological development. Both rely on increasing centralization and large-scale bureaucratic control and coordination. From a viewpoint of narrow scientific rationalism, both insist that the planet is there to be conquered, that big is self-evidently beautiful, and that what cannot be measured is of no importance (Porritt 1984, 44).

The Russian sociologist, Oleg Yanitsky (1993, 91–92), discussed a dichotomy between "the System" and environmentalists in very similar terms:

For the System, nature was simply a means of guaranteeing its existence, while for the environmentalists nature was valuable in itself. . . . For the System, science was a bureaucratic instrument of self-defense, while for the environmental movement it was a source of strength in the struggle against ministerial expansion and a basis for the formation of a new system of values. The System and the environmentalists also had diametrically opposed views where the organization and development of the social sphere was concerned.

The DSP of Soviet society had many elements in common with the DSP of the West.[2] Both societies contain an anthropocentric emphasis in which humankind and economic progress take priority over environmental protection. Joan DeBardeleben's (1985, 80) review of the Soviet environmental literature highlights the anthropocentric orientation of Soviet Marxism. A common Marxist-Leninist slogan aptly illustrates this element of the DSP: "Everything for man, everything for the service of man." The DSP presumes that the efficient management of resources can increase living standards and guide future development—Russian and American systems differed in the methods of exercising this economic rationality effectively, not in their belief in this element of the DSP. In Marxist-Leninist theory, moreover, nature offered virtually inexhaustible resources for humankind to exploit. It has even been argued that Marx's labor theory of value provided an ideological justification for the unrestricted use of natural resources, thereby encouraging their overexploitation (DeBardeleben 1985). The Soviet belief in the power of technology to transform nature, whether through the development of virgin lands or the reversal of river flows, is legendary (Pryde 1991, ch. 13; Ziegler 1987;

DeBardeleben 1985, ch. 2). Indeed, both societies idolize technology to a considerable degree, as illustrated in the central role technology played in competition between the superpowers.

In summary, if environmental thinking in Russia represents a negative reaction to the DSP of the Soviet era, we might expect that environmentalists will display many of the beliefs identified with the NEP in the West. Indeed, Joan DeBardeleben concludes (1992, 77):

> In sum, the new 'green' phenomenon in the USSR is rooted in a rejection of the misguided materialism of the old structures. Like post-materialism in the West, Soviet environmentalism represents the ascendence of non-material, cultural, human values.

However, there are also significant differences between American and Russian political cultures and experiences. Despite the antielitist claims of Western environmentalists, the politically tolerant and open nature of democracy is a major source of strength for the American environmental movement. Democracy allows environmentalists access to information that was unobtainable in the East, and it allows groups to engage in challenging activities that were not tolerated in the pre-glasnost Soviet Union. Even after political reforms in Russia, this gap in political structures (and political cultures) is still wide. Consequently, orientations toward citizen participation and political involvement are expressed in a much different political context for Western and Eastern environmentalists.

The American and Russian societies also presumably differ in the value attached to individualism versus collective interests. Even beyond the structural differences between capitalism and a state-planned economy, American culture emphasizes individual initiative and individual responsibility while the Communist system stressed the value of the collective as a structure for social relations and decision making. Thus while Western adherents of the NEP may favor a more communal orientation to counterbalance the individualist emphasis of the DSP, it is unclear whether Russian environmental beliefs will follow this same pattern.

Even more ambiguous is the specific orientation toward market versus planned economies. Western adherents of the NEP are generally critical of capitalist markets because they ignore or discount the externalities of environmental damages. In addition, many Western environmentalists are skeptical of large-scale government, which they see as adherents of the DSP, and socialism is commonly rejected as a potential element of the green philosophy. The common environmental slogan "neither Left nor Right" reflects this ambiguity toward either economic orientation. Russian environmentalists might share these ambiguous views. On the one hand, they might see the Western economic system as preferable to the environmental policies produces by Soviet state socialism. On the other hand, the environmental weaknesses of Western market systems should be clear after a generation of Soviet environmental education stressing this point.

This chapter focuses on the environmental values of residents living near the two nuclear weapons facilities in order to determine whether their environmental orientations share common or dissimilar belief systems. We believe this provides a critical test of whether environmentalism and the NEP represents a distinct ideology that transcends the social and political differences between these two nations.

Support for the New Environmental Paradigm

To determine whether the NEP exists in a similar fashion among Russian and American publics we assembled a battery of survey items to tap the elements of this belief system (see appendix A on survey methodology). These questions draw upon the research of Dunlap and Van Liere (1978), Milbrath (1984), and Pierce et al. (1986). Ten questions in our Chelyabinsk and Washington regional surveys measure opinions toward five specific areas: the value placed on nature, the relationship between humankind and nature, the acceptance of limits to growth, the nature of attitudes toward technology, and the importance attached to collective values over individual self-interest. The expectation of the New Environmental Paradigm is that these elements are interrelated into a new framework of environmental thinking.

To focus on the broad comparisons of respondents in the Hanford area and those near Mayak, we pooled and weighted data from the four communities in both regions (community patterns will be described below). These cross-national comparisons find that both publics voiced strong approval for many of the core elements of the New Environmental Paradigm (table 7.1). For instance, our respondents almost universally valued the natural environment: 90 percent of the Hanford sample and 97 percent of the Chelyabinsk sample agreed with the view that humans should live in harmony with nature. However, this naturalist orientation did not reach as far as a biocentric view of nature. Only one-third of all Russian respondents held biocentric opinions on two questions: animals exist for human use (36 percent disagree), and humans were created to rule the rest of the world (41 percent disagree). Roughly comparable results were obtained from Americans (45 and 48 percent, respectively).

There is an important distinction between these two orientations toward nature.[3] Most residents of the Chelyabinsk and Hanford regions positively valued nature when discussed in abstract terms. As we noted above, such positive views of nature can draw upon the conservative agrarian heritage of both cultures without directly challenging the DSP. At the same time, for most individuals this appreciation of the natural environment coexists with an anthropocentric view of nature. For example, one of our Richland residents voiced these mixed feelings: "I do feel that wildlife, plants, and humans are equally important to the environment, and each depends on the other to survive. But when human beings are starving and suffering, every effort must be taken to alter this, even at the expense of the environment." Few people are willing to attribute an equal value to man's needs and those of nature, or accept the equal interdependence of each.

Continuing with table 7.1, residents of both nations accepted the NEP premise that there are inherent limits to humankind's use of nature. Virtually all of the respondents from the Chelyabinsk region agreed that "humans are using up natural resources too quickly," and three-quarters of the Americans shared these opinions. A majority of respondents in eastern Washington, as well as a plurality of Chelyabinsk residents, further believed that we are approaching the limit of the number of people the earth can support.[4] An insight into this thinking comes from the comments of one Wenatchee resident: "The Northwest will experience a serious setback in environmental quality in the future as we switch from hydroelectric to fossil-fuel-based energy consumption. . . . Our lack of good energy policy will contribute to global warming, acid rain, and air

Table 7.1
Support for elements of the new environmental paradigm

	American	Russian
	(percentages)	
Value nature		
Humans must live in harmony with nature in order to survive	90	97
Wildlife, plants and humans have equal rights to live and develop	62	92
Human dominance of nature		
Plants and animals exist primarily for human use (Disagree)	45	36
Humankind was created to rule the rest of the world (Disagree)	48	41
Limits to growth		
Mankind is using up the world's natural resources too quickly	75	88
We are approaching the limit of the number of people the earth can support	62	39
Technology		
We are in danger of losing control over the consequences of technological development	45	72
The good effects of technology outweigh its bad effects (Disagree)	26	29
Collective		
Society should use its resources to benefit future generations even at the detriment of today's living standards	60	87
The interests of society should take precedence over individual needs	57	60

Note: Results are based on the total weighted samples (Russian N = 1,187; American N = 1,532). Table entries are the percentage who give pro-environment opinions; for most items it is the percentage who "strongly agree" or "agree" with each statement, for items marked "(Disagree)" it is the percentage who disagree with the statement. Missing data are included in the calculation of percentages.

quality problems. I believe the only rational energy policy is one based on conservation."

Respondents in both nations also displayed an ambivalent attitude toward technology. On the one hand, many Russians and Washingtonians felt a sense of "future shock" as technological innovations seem to change the world at an ever faster pace. A full 72 percent of Russians believed that we are in danger of losing control over the consequences of technological development, and nearly half of the surveyed Americans agreed (45 percent). An engineer in Chelyabinsk expressed worries that were shared by many: "People have to slow down and think in developing technology. . . . All of this takes a toll on nature." However, only a minority of either nationality believed that the disadvantages of technology outweigh the advantages. To many, it seems, technology was still predominately a benefit.

Finally, citizens in both regions expressed support for collective needs over individual needs. Clear majorities of residents in both regions stated that the needs of society should take precedence over individual needs. Even higher percentages of Russians and Americans agreed that society should use its resources to benefit future generations even at the detriment of today's living standards. Taken alone, the Russian responses might be attributed to the collectivist political culture in Russia and the socialist norm of delaying short-term personal gratification in order to build a better future for everyone. For instance, a later question in the survey found a plurality of Russians favored society holding the government responsible for ensuring citizens' welfare, rather than the individuals themselves. The extensive research on the individualism of American political culture (Lipset 1990, 27–30) would suggest that this might be an area of contrast between the two societies, but we actually find a substantial similarity in these opinions.

In overall terms, the breadth of support for the basic elements of the NEP might lead us to conclude that the Dominant Social Paradigm is no longer so dominant. This echoes the findings of Kempton, Boster, and Hartley (1995) for American environmental values, but it is striking that the results from Hanford also mirror this pattern. Despite the vastly different historical experience and sociopolitical context of the two nations, the commonality in beliefs between residents in eastern Washington and the Chelyabinsk region is the overwhelming pattern that emerges from these data. There is a surprising cultural similarity in the attitudes of residents in our American and Russian communities toward the environment, and it reflects a general predisposition toward the values of the NEP rather than the DSP.

Structure of Opinions

One of the significant aspects of Western attitudes toward the New Environmental Paradigm is the fusion of distinct opinions—about nature, technology, the limits to growth, and other areas—into a single framework of environmental thinking. Dunlap and Van Liere (1978, 14), for instance, find that items spanning the same concerns as our constitute a single dimension of opinions. Similarly, Lester Milbrath (1984) speaks of a dimension of environmentalism that includes each of the opinions we have analyzed (see also Kempton, Boster, and Hartley 1995; Pierce et al. 1986). Indeed, it is the union of these discrete beliefs that creates a new paradigm for environmental thinking.

It was much less clear, however, whether a single NEP dimension would exist in the political thinking of Chelyabinsk residents. The environmental debate was a relatively new phenomenon in Russia in the early 1990s, and the development of environmental thinking lacked the relatively long experience of Western environmentalism. Furthermore, Russia was in the midst of tremendous political change wherein basic political values of all types were in flux. And finally, even similar levels of environmental support among Russian and American publics might result from different causal processes, or have quite different meanings to Russians. Thus we wanted to assess whether environmental attitudes among our Chelyabinsk respondents formed a single belief system as it did in the West.

To determine the structuring of environmental opinions we entered each of the items from table 7.1 into a principal components analysis for both samples. These analyses find basic differences in the structure of beliefs between the two nations (table 7.2). The American results on the top of table 7.2 display what we expected from prior research. The first principal component is linked to most of the separate items that comprise the NEP. As should be the case, the three pro-DSP items all load

Table 7.2
Factor analysis of environmental attitudes

	Unrotated dimensions			Rotated dimensions		
American Results						
Plants/animals exist for man	−.43	.63	.09	−.06	.77	.03
Humans created to rule world	−.36	.64	.01	−.06	.72	.12
Technology advantages outweigh disadvantages	−.35	.51	.21	.02	.65	−.08
Using up resources too quickly	.72	.17	.09	.63	−.22	.33
Must live in harmony with nature	.67	.14	−.12	.45	−.26	.46
Wildlife and humans have equal rights	.66	−.18	.01	.43	−.49	.20
Resources should benefit future generations	.46	.37	−.44	.19	−.01	.71
Society takes precedence over individual	.28	.44	−.58	−.01	.11	.77
Approaching population limits	.52	.29	.49	.77	.07	−.01
Losing control over technology	.58	.25	.41	.75	−.01	.06
Eigenvalue	2.74	1.64	1.01			
Variance explained (%)	27.4	16.4	10.1			
Russian Results						
Plants/animals exist for man	.47	.62	.13	.77	.04	.14
Humans created to rule world	.44	.69	−.07	.76	−.14	.26
Technology advantages outweigh disadvantages	.26	.38	.50	.58	.25	−.25
Using up resources too quickly	.44	−.32	.50	.08	.74	−.06
Must live in harmony with nature	.49	−.38	.34	.01	.70	.10
Wildlife and humans have equal rights	.49	−.50	−.05	−.21	.54	.41
Resources should benefit future generations	.48	−.15	−.47	−.03	.10	.68
Society takes precedence over individual	.46	.11	−.49	.16	−.07	.65
Approaching population limits	.41	−.03	.04	.18	.29	.24
Losing control over technology	.56	−.13	−.17	.10	.54	.41
Eigenvalue	2.10	1.55	1.14			
Variance explained (%)	21.0	15.5	11.4			

Note: Table entries are factor loadings from a principal components analysis and a varimax rotated factor analysis. Results are based on the total weighted samples; pairwise exclusion of missing data was used.

negatively on this primary factor.[5] This is evidence that a single DSP/NEP framework structures the environmental attitudes of residents in the Hanford region. The principal components analysis also yields weak second and third dimensions, which reflect a pro-DSP anthropocentric orientation and an individualistic sentiment; but these dimensions are clearly secondary to the DSP/NEP framework.[6]

Much different results emerge for the Chelyabinsk sample. Even though some questions were phrased in negative terms (that is, agreement meant a NEP response for some questions and a DSP response for others) all the questions load positively on this dimension. Furthermore, opinions are more weakly structured than in the American survey. While the first component in the American case shows three items with factor loadings of .60 and higher, none of the items reaches this magnitude in the Russian survey. Additionally, while the first principal component in the United States explains a significant portion of the variation in responses (27.4 percent), the first dimension in Chelyabinsk explains a smaller proportion (21.0 percent).[7] There is not a theoretically acceptable explanation of this solution.[8]

The rotated factor results for the Chelyabinsk survey yield a more understandable patterning of opinions. This solution indicates that Russian environmental opinions are structured around three dimensions. The first taps biocentric/anthropocentric attitudes, two items represent a rejection of humankind's dominance over nature, and a third is distrust of technology. These three items were endorsed by a small proportion of Russians; thus support for a biocentric view represented a very strong environmental commitment. The second dimension emphasizes the consensual value of nature that was supported by 90 percent or more of our respondents. Finally, the third dimension includes the two items tapping collectivist/individualist orientations.

In summary, the findings from the Hanford sample approximated a structure in which support for the NEP contrasted with DSP values. In contrast, environmental opinions among the Chelyabinsk respondents were more weakly structured, and formed three separate dimensions. These multiple dimensions suggest that environmental thinking among Russians constituted a looser set of beliefs than is normally found among Western publics. This loose structuring of Russian environmental attitudes may reflect the transitional nature of Russian politics, and the novelty of environmental discourse, in the early 1990s. Alternatively, this pattern of environmental attitudes also may reflect enduring differences in how Russians think about the environment because of their culture and history. The second dimension, the valuation of nature, evokes virtually universal appreciation for nature and wildlife that might be a lasting legacy of the agrarian roots of Russian society and the cultural value placed on nature. What appears distinctive about the Russian findings is the separation of a positive valuation of nature from biocentric views. Russian reverence of nature as an abstract entity apparently can coexist with the DSP of industrial societies, but biocentric attitudes seem closer to the reformist, alternative thinking underlying the NEP orientation found in the American survey. Finally, in contrast to the West, Russians have apparently not integrated beliefs about collectivism and individualism into a broader framework of environmental thinking.

Correlates of Environmental Attitudes

The NEP should reflect larger views about politics and social change that should be related to other political beliefs. Western scholars, for example, have treated environmentalism as typifying the new "postmaterial" issues and political concerns of advanced industrial societies. Numerous studies thus have presented a strong relationship between NEP and measures of postmaterialism (Milbrath 1984, 39; Inglehart 1990). Joan DeBardeleben (1992, 76–7) also sees elements of postmaterialism in the development of green politics in the former Soviet Union.[9] Analysts also stress the populist, participatory orientation of the New Environmental Paradigm (Milbrath 1984, 35–36; Cotgrove 1982), and we expect NEP orientations in Russia to be strongly related to democratic, reformist attitudes. There was, in fact, a clear environmental aspect to Gorbachev's glasnost and perestroika reforms, and environmentalists were at the vanguard of the political reform movement that produced even greater political transformations (White 1991; DeBardeleben 1992). We also expect that NEP orientations in both the West and East are related to support for the peace movement and opposition to the military competition that produced the Cold War conflict (Milbrath 1984, 49–50).

It is less clear how opinions about economic structures should be related to environmental orientations. Green activists in the West are likely to criticize the excesses of private enterprise and market economics as a major cause of current environmental problems, leading them to lean toward government planning as a preferred alternative (Milbrath 1984, 36–37; Cotgrove 1982; Dalton 1994, ch. 5). Certainly the context is much different in Russia, where the evidence of the environmental failures of a planned economy is overwhelming. A simple reaction against the old regime's economic structure would suggest that environmentalism in Russia should be related to support for a private economy. One could also hypothesize, however, that the environmental failures of Western market systems have been widely discussed as part of the past Cold War competition. Environmentally oriented Russians might endorse a government-coordinated economy that adheres to its environmental responsibilities—rather than the consumerism, materialism, and inequalities of a Western market economy.

Because there are different belief structures in the two nations, we could not simply create the same attitudinal indexes to measure environmental orientations in both samples. The American public displayed a single dimension of environmental thinking consistent with the NEP model; thus we constructed a simple additive index of all ten items to measure these general orientations (for the details of index construction, see the appendix B). Russian environmental attitudes, in contrast, distinguished among three different elements of environmental thinking. Consequently, we have three separate environmental indices for the Russian survey: biocentric/anthropocentric attitudes, the valuation of nature for its own sake, and collectivist attitudes. In addition, as a methodological verification of the differences in belief structures, we computed a combined index that included all ten questions from the Russian survey.

Table 7.3 presents the attitudinal correlates of environmental attitudes. In the American survey, the summary NEP dimension was strongly related to other "reformist" attitudes. For instance, postmaterialists were much more likely to score highly on the NEP index than others (r = .24). Similarly, NEP supporters criticized the military and supported unpopular forms of dissent. As further validation of the American NEP measure,

Table 7.3
Attitudinal correlates of new environmental paradigm dimensions

		Russian			
	American NEP	Combined Index	Biocentric/ anthropocentric	Value nature	Collective
Postmaterial values	.24*	.05	.12*	−.02	.01
Democratic values[1]	.10*	−.09	.34*	−.09	−.21*
Private economy	—	−.01	.27*	.03	−.12*
Militarism	−.24*	−.06	−.18*	.12*	.08
Reformism/conservatism scale	−.06	−.03	−.09	.02	−.02
Rating of national environment	.31*	.07	.07	.07	−.02
Local environmental problems	.26*	.06	−.09*	.29*	.04

Note: Table entries are Pearson correlations between NEP indexes and other attitudes; coefficients significant at .001 level are denoted by an asterisk. For the construction of items see appendix B. Results are based on the total weighted samples.
1. Toleration for unpopular forms of dissent in the American survey.

these attitudes were strongly related to ratings of the national environmental condition and the specific environmental problems in the respondent's locale. Thus the findings from our surveys in eastern Washington follow the basic patterns of other research on environmental thinking in Western democracies.

In comparison, the Chelyabinsk survey underscores the distinctiveness between the three dimensions of environmental thinking. Biocentric attitudes were interrelated with a variety of reformist or modernist values, such as postmaterialism, democratic attitudes, and opposition to the military.[10] There was also a strong relationship between biocentric attitudes and support for a private economy, reinforcing the sense that these attitudes tapped a general orientation toward reform of the former Soviet system. Surprisingly, biocentric attitudes were only weakly related to two general measures of environmental concern: evaluations of magnitude of national environmental problems, and concern about local environmental problems (in the latter case, biocentric attitudes were actually related to perceptions of fewer local environmental problems). The following section will show that this mixed pattern at least partially reflects the unique environmental problems of the Chelyabinsk region; the rural populations of Muslyumovo and Kyshtym, who have been most affected by a negative environment, were less likely than other residents of the region to accept biocentric views of nature.

Table 7.3 also indicates that the valuation of nature—a highly consensual attitude—was virtually unrelated to political modernization and reform attitudes for the Russian sample. There was not, for instance, a significant relationship between valuing nature and postmaterialism, democratic attitudes, or support for a private economy. Only concern about local environmental problems was substantially related to the valuation of nature index. Thus this measure tapped a specific and relatively apolitical orientation of environmentalism. Finally, collectivist attitudes were only weakly tied to other political attitudes.[11]

The combined index of all ten environmental items displays weak and statistically insignificant correlations with all the other attitudinal measures in the Russian survey. These findings underscore the point that the structure of environmental thinking in Russia is multidimensional, and it is necessary to separate these dimensions to understand how Russians perceive the environment.

Predicting Environmental Orientations

Determining the distribution of support for NEP orientations across population subgroups can enrich our understanding of the meaning of these attitudes among Chelyabinsk and Hanford area residents. In addition, the distribution of opinions across social groups provides a framework for interpreting the political potential of each value dimension; opinions that are held by the better educated and politically active, for instance, are more likely to lead to policy influence. This section examines the distribution of environmental attitudes according to the social, geographic, and political location of our respondents.

Geographic Context and Ethnicity

The distribution of environmental attitudes across these American and Russian communities can describe the political context of each area. The communities in both nations were chosen to provide cross-national pairings of sites with comparable relationships to Hanford and Mayak (see appendix A). Richland and Chelyabinsk-70 are both "host" cities that are located adjacent to the facilities and whose residents are economically dependent upon the operation of the facility. Spokane and the city of Chelyabinsk are the major downwind urban communities, somewhat removed from the nuclear weapons plant but close enough to be affected by major radiation leaks and to serve as a mobilizing center for environmental activists. Spokane is a rather environmentally aware city through which the once Salmon-bearing Spokane River flows and where close-by mountains provide a picturesque backdrop landscape. Chelyabinsk is a large industrial center for the Southern Urals, with extensive pollution problems from its resident industries. The three small American towns of White Swan, Wapato, and Toppenish correspond to the rural communities of Muslyumovo and Kyshtym in the Chelyabinsk area. These are rural towns located adjacent to the facilities containing a concentration of ethnic minority groups (Native American and Hispanic in the American case, Tatar and Bashkir in the Russian survey). Finally, Wenatchee and the Russian city of Chebarkul are small cities located in the region, but far enough removed from the respective complex to serve as a control site.

The distribution of support for the NEP by geographical location in the Hanford area (expressed as deviations from the overall sample mean score for NEP) reveals where the proponents and opponents of the nuclear weapons plant were located (table 7.4). The generally conservative, pro-DSP orientation of the region was evident in the low NEP score of the control site of Wenatchee. The stronghold of the Dominant Social Paradigm, however, was clearly Richland—the home of the professional scientists, engineers, and technicians who operated the nuclear weapons facilities (and are now in the "environmental cleanup" business). The relatively pro-NEP locales were in the close-by Yakama Reservation towns by a small margin, and more clearly, in the downwind city of Spokane.

The unique nature of the Chelyabinsk region intermixes contradictory influences on environmental attitudes; thus our expectations about the political climate in each community is uncertain. Residents of the two areas that suffered most directly from the radiation pollution of the Chelyabinsk-65 facilities, Muslyumovo and Kyshtym, were very concerned about their immediate environmental problems and thus might

Table 7.4
Geographic differences in environmental values

Hanford survey		Mayak survey			
Location	NEP	Location	Biocentric	Value nature	Collectivist
Richland	−.36	Chelyabinsk-70	.32	.14	.04
Spokane	.40	Chelyabinsk	.08	.01	−.02
Yakama Reservation	.02	Kyshtym	−.04	−.33	−.09
Wenatchee	−.08	Muslyumovo	−.46	−.11	.09
		Chebarkul	−.09	.22	.03
Eta	.14		.22	.19	.09

Note: Table entries are scores on the environmental values indexes expressed as deviations from the overall grand mean score on the indexes. Index construction is described in the appendix.

have developed an environmental consciousness including elements of the NEP (see chapter 4). At the same time, these residents come from predominately rural settings with less education and thus are not normally expected to be NEP adherents. The other extreme is Chelyabinsk-70, a city with a more urbane and educated population where NEP orientations might normally be expected, except for the fact that these respondents often were engaged in the research and design of nuclear weapons and related technologies. Are the scientists and technicians that are at the core of the Russian military-industrial complex likely to question the values underlying the Russian DSP? The city of Chelyabinsk was probably more environmentally aware than most Russian cities, but it also includes a heterogeneous population.

Table 7.4 also presents environmental attitudes in the five Russian communities (expressed as deviations from the overall sample mean). Most striking is the strong relative support for biocentric views among residents of Chelyabinsk-70 and the equally low level of support in the rural areas of Kyshtym and Muslyumovo. A biocentric environmental orientation clearly did not develop from the exposure to environmental problems in Kyshtym and Muslyumovo, but apparently evolved as part of the modernization of Russian society in urban areas and among the better educated. Surprisingly, residents of Kyshtym and Muslyumovo also scored below the sample average on the valuation of nature, an orientation we have previously linked to the agrarian traditions of Russian society. This may reflect the environmental degradation that has occurred in these two areas. However, residents of Chelyabinsk-70 also scored significantly above the overall mean, which further obfuscates the relationship between these opinions and rural/urban life.

Ethnicity is another potentially important factor in structuring environmental thinking, and one that is related to our five geographic locations. The Native American population in eastern Washington State has a distinct relationship to the facility. Their proximity means they were potentially exposed to radiation releases from the facility, and the hunting and fishing activities of reservation inhabitants created a lifestyle that might increase their exposure beyond that of other area residents. Moreover, the traditions and values of Native American culture might be more closely in tune with the values of the NEP than with the norms of the Dominant

Social Paradigm. Over the past several decades a substantial influx of Hispanic farm workers has increased the number who have "settled out" and established permanent residence, sometimes intermarrying with the Native American population. Approximately one-quarter of the respondents from the three towns on the reservation are Hispanic; and another quarter are Native American. In contrast, the populations of both Richland and Spokane are over 90 percent Caucasian.

Ethnicity is also a factor in the Russian study. The development of the environmental movement during the Gorbachev period often was related to latent conflict between ethnic minorities and the dominant Russian state in Moscow. Paula Garb and Galena Komarova's research (see chapter 6) explores similar ethnic dimensions to the environmental movement in the Chelyabinsk region. Ethnic Russians comprise more than three-quarters of our total survey; however, in the village of Muslyumovo there is a large Tatar population (82 percent) and a sizable Bashkir minority (16 percent). Overall, these two ethnic groups account for approximately one-sixth of our total survey.

Table 7.5 presents the relative NEP score for each ethnic group in the Hanford region. Pro-NEP sentiment was least among the "Anglo" (white) group, and greater among the Hispanic subpopulations and especially among Native Americans.[12] Native Americans in the Hanford area displayed a strong pro-NEP orientation, presumably drawing upon the

Table 7.5
Ethnicity differences in environmental values

Hanford survey		Mayak survey			
Ethnicity	NEP	Ethnicity	Biocentric	Value nature	Collectivist
Caucasian	−.05	Russian	.09	.02	−.01
Hispanic	.15	Tatar	−.39	−.01	.04
Native American	.84	Bashkir	−.24	−.17	.08
Other	−.01	Other	.00	−.06	−.04
Eta	.08		.17	.05	.04

Note: Table entries are scores on the environmental values indexes expressed as deviations from the overall grand mean score on the indexes. Index construction is described in the appendix.

naturalist and holistic tradition in Native American thinking about the environment. In this sense, the New Environmental Paradigm also has ancient roots in non-Western, preindustrial values. These minorities thus represent possible allies for environmental activists seeking to make common cause in challenging the authority and policies of the Hanford facilities.

Given the geographic distribution of opinions in the Russian survey, it is not surprising to find that support for environmental thinking was significantly lower among the Tatar and Bashkir populations, especially for biocentric attitudes (table 7.5). Opinion differences were much smaller for the valuation of nature index, but Tatars and Bashkirs also rated lower than the overall sample on this measure. At least for the Chelyabinsk region, environmental action has not created an alternative environmental philosophy among the two largest ethnic minorities.

Social Location

Western research on the NEP provides a clear model of where the core adherents of this paradigm are to be found. The NEP is identified with younger generations who have been raised under the affluence and social security of contemporary advanced industrial democracies, and especially among the postmaterialist, better-educated and middle class sector of younger generations (Milbrath 1984; Cotgrove and Duff 1980; Dalton 1994; Dalton and Rohrschneider 1999). Rising affluence, the modernizing influences of higher levels of education, greater exposure to diverse information in a media-rich environment, and similar experiences generate interest in postmaterial issues. As we have observed above, several scholars have noted parallels in the Russian experience, where environmental action initially was organized by intellectuals and other upper-status individuals, and where the movement appeared to have special appeal to the young. Consequently, we expect that support for the NEP in the West and biocentric attitudes in Russia might be concentrated among these same social groups.

Our expectations for the two other environmental dimensions in the Russian survey—the valuation of nature and collectivist values—are more ambiguous. If the appreciation of nature reflects the environmental values embedded in the NEP, it may display the same social correlates.

If, however, the valuation of nature reflects an older, more traditional Russian attachment to nature and the environment, these attitudes may be concentrated among older and less cosmopolitan sectors of society. A similar ambiguity exists for collectivist attitudes. These sentiments might be more common among younger Russians who have been socialized into this way of thinking, or they may reflect the residual commitment to the values of the revolution by the older generations. Therefore, the social correlates of these attitudes can help in our interpretation of these indices and the processes of social change occurring among the Russians of the Chelyabinsk region.

Table 7.6 presents the relationship between age group, or generation, and NEP attitudes for the Hanford area survey.[13] The figure displays the "normal pattern" found in other American and Western European surveys of environmental attitudes. The pre–World War II and Cold War generations were less supportive of NEP values; the generations of the 1960s, 1970s, and 1980s were progressively more likely to subscribe to NEP values.

In the Russian survey, also on table 7.6, we find a clear juxtaposition in generational attitudes toward biocentrism and the other two environmental dimensions.[14] Young people expressed biocentric attitudes more

Table 7.6
Generational differences in environmental values

Hanford survey		Mayak survey			
Generation	NEP	Generation	Biocentric	Value nature	Collectivist
Pre-WWII	−.39	Stalinist	−.24	.09	.09
Postwar	−.74	Khruschev	.02	.03	.02
'60s Protest	.15	Breshnev	.11	−.04	−.07
Détente	.37	Gorbachev	.40	−.19	−.10
Reagan/Bush	1.02				
Eta	.17		.20	.10	.13

Note: Table entries are scores on the environmental values indexes expressed as deviations from the overall grand mean score on the indexes. Index construction is described in the appendix. The definition of generations is presented in footnotes 13 and 14.

often than older persons. In contrast, older Chelyabinsk residents were more likely to value nature and express collectivist values. If we accept these age relationships as representing generational experiences, these findings suggest that the process of modernization has been transforming the beliefs of the Russian public. This implies that traditional beliefs of valuing nature and collectivist norms have declined over time, and in their place have arisen the reformist attitudes of biocentrism and its several correlates.

Social status is another characteristic routinely related to environmental attitudes in the West. Table 7.7 displays findings regarding the relationship between several social status measures and the indexes of environmental attitudes. In contrast to the normal pattern linking the NEP to higher social status, we found that education was unrelated to NEP beliefs in the Hanford survey (r = −.01). This anomalous finding occurs because the Richland subsample—the most politically conservative, pronuclear and antienvironmentalist subgroup—is heavily laden with highly educated, well-paid scientists, engineers, and specialized technicians. Similarly, some groups espousing NEP values (Hispanics and Native Americans; see table 7.5 above) are far below average in their formal education. The normally observed connection between NEP values and education is not present in this unusual study population. The same anomalous finding observed for formal education also appears in the relationship between family income and the NEP values scale.

Table 7.7
Social status correlates of new environmental paradigm dimensions

		Russian		
	American NEP	Biocentric/ anthropocentric	Value nature	Collective
Education	−.01	.25*	−.03	−.12*
Family income	−.16	.07	.06	−.02
Household conveniences	—	.14*	.05	−.04

Note: Table entries are Pearson correlations between NEP indexes and social status measures; coefficients significant at .001 level are denoted by an asterisk. For the construction of items see the appendix. Results are based on the total weighted samples.

Education had quite different effects in the Russian setting. Access to higher education under the Soviet regime often involved an expressed commitment to the values of the Communist order, and many of the highly educated in our sample are located among the Chelyabinsk-70 residents. Nevertheless, Table 7.7 indicates that biocentric attitudes were distinctly more common among better educated Russians ($r = .25$). This suggests that the anthropocentric values of the Soviet regime were not supported by the very pool of better-educated individuals who provide the social and technological cadre of the nation. Neither the valuation of nature nor collectivist attitudes were strongly linked to educational level, although there was a weak relationship in the opposite direction for the latter dimension ($r = -.12$). The other social status measures available in the Russian survey generally replicated these patterns. For instance, biocentric attitudes were positively related to income ($r = .07$), and a measure of household conveniences (heating, plumbing, and cooking facilities) ($r = .14$); these measures were weakly and ambiguously correlated with the valuation of nature and collectivist attitudes.

The American findings are not quite so anomalous, however, if we focus our attention only on the city of Spokane, thus excluding the peculiarities of the Richland and Yakama Indian Reservation samples. Among Spokane residents there was a weak positive correlation ($r = .06$) between educational level and support for the NEP. In percentage terms, 32 percent of those with some college education scored very high (7 or more) on the NEP scale, compared to only 23 percent among those with a high school diploma or less.

At a single point in time it is difficult to disentangle the effects of historical change borne by generations and the modernization forces of social change represented by educational level because both factors overlap. The young are both young and better educated; the old are both old and less educated. In further analyses (not shown), we attempted to disentangle the separate effects of these variables and develop a more powerful definition of the social locus of environmental attitudes. When both factors are combined they define sharply polarized social groups: older and less-educated Americans and Russians were least likely to support environmental values linked to the New Environmental Paradigm, while younger and better-educated respondents were much more likely to support these values. In short, both factors seemed to be working independently to transform the values of both publics.

Political Location

A final way that we can locate respondents is in reference to their political positions, such as partisanship or ideology. Although partisanship in the American setting does not always reflect a liberal or conservative position, on many matters it is correct to say that Republican party identifiers are likely to be conservative in their views and Democratic party identifiers are inclined to be more liberal. Such is the case with the NEP—it has consistently been associated with the more liberal political forces and is more closely associated with the Democratic party. Table 7.8 presents findings for both surveys. With regard to Hanford respondents, there is virtually a monotonic increase in pro-NEP sentiment as one moves across the partisan spectrum from strong GOP through Independent to strong Democratic identifiers.

It is more difficult to talk about the same partisan contrasts in the context of Russian politics in the early 1990s. There was no clear equivalent to the left/right framework of Western politics, and thus no equivalent

Table 7.8
Partisan differences in environmental values

Hanford survey		Mayak survey			
Partisanship	NEP	Partisanship	Biocentric	Value nature	Collectivist
Strong Democrat	.93	Greens	.13	.10	.01
Democrat	.65	Democratic Russia	−.06	.00	−.04
Weak Democrat	.85	Communist	−.08	.16	−.02
Independent	.25	Patriots	−.21	.26	.05
Weak Republican	.08	No party	.06	.07	.02
Republican	−.94	Not sure	−.14	−.29	.00
Strong Republican	−1.59				
Eta	.28		.15	.17	.12

Note: Table entries are scores on the environmental values indexes expressed as deviations from the overall grand mean score on the indexes. Index construction is described in the appendix.

to the debate over the position of environmentalists in terms of this long-standing political cleavage. Yet Russia was racing headlong into electoral politics and at least quasi-democratic competition among political groups, even if the nature of these groupings remained fluid. Table 7.8 compares the environmental attitudes of Russians according to the partisan preferences they expressed at the time of our survey. Each of these groups was admittedly a loosely defined collective, short of the formalized parties and partisan attachments found in established Western party systems. We organized the party groups roughly along a reformist/conservative continuum, with nonpartisans at the bottom of the list.[15] Supporters of the Greens, a relatively large proportion of our sample (18 percent), were predictably the most likely to express biocentric attitudes and place a high valuation on nature. Adherents of the more centrist Democratic Russia placed themselves near the center of the three environmental dimensions. The two conservative groups, supporters of the Communist party and the Patriots, highlight the potential contrasts between different dimensions of environmentalism. These two political groups leaned toward an anthropocentric view of nature, scoring below the population average on this dimension, while simultaneously scoring highly on the traditional valuation of nature dimension.

Conclusion

Although Russia continues to experience massive social and economic dislocation, one of the most striking findings from our survey is the support for green attitudes found among Chelyabinsk residents at the start of this process in the early 1990s. We found broad support for many elements of what has been labeled the New Environmental Paradigm by social scientists studying environmental politics in the West. Chelyabinsk residents value nature very highly, express a certain skepticism toward technology, and are aware of the limits to humankind's use of natural resources. Residents of the Hanford region display even greater support for these same principles. Indeed, so common are these beliefs that one might question whether the pure industrial values of the Dominant Social Paradigm really existed as the defining characteristics of societal values even during the high point of industrialism (Kempton, Boster, and Hartley 1995).

At the same time, we would stop far short of claiming that support for a New Environmental Paradigm was widespread among Chelyabinsk residents. The breadth of support for green positions—for both Russians and Americans—partially came from a rejection of the DSP, but it more generally reflected an idyllic view of nature that coexisted with industrial values in the minds of many citizens, Russian and American alike. People in modern industrial societies tend to hold a positive view of rural life and entertain a romanticized image of humans living in harmony with nature; these sentiments likely represent a cultural legacy of our preindustrial past. Ironically, these idyllic views are often held by the most urban and educated of the public, and not by those who presently live in rural areas and are close to nature. These conservationist orientations imply an anthropocentric perspective on nature, where humans can protect the environment while also pursuing economic growth. This is the popular environmentalism of the World Wildlife Fund and the Nature Conservancy in the West, or the All-Russian Society for the Protection of Nature in the East.

Among Western environmentalists, however, a second form of environmental thinking has developed in conflict with the values of the DSP. A biocentric view of nature questions the principles of industrial societies' exploitation of nature and the pattern of social and political relations that grow out of these principles in these societies. These are the core values of the NEP. In the West these views are represented by political groups such as Friends of the Earth, Greenpeace, and Earth First! In Russia such elite-challenging views are apparently just developing. Chelyabinsk residents were clearly restrained in their support for biocentric environmental perspectives, and the cluster or reformist values that accompany these orientations. However, the social location of these NEP orientations was similar to the patterns evident in the West—that is, more common among the young, the better educated, and the adherents of postmaterial values.

Another notable feature of our findings was the lack of structure linking various environmental attitudes among the Russian public. Conservationist and biocentric attitudes tend to be interrelated among our American respondents, but these were distinct attitudinal dimensions for our Chelyabinsk respondents. There was virtually no relationship between both dimensions of environmental attitudes, and collectivist attitudes formed yet another dimension. Among Western publics these attitudes formed a tighter cluster of beliefs that comprise the popular base of environmentalism.

These results lead us to several observations about the nature of environmental thinking in Chelyabinsk, and likely in Russia as a whole. In many ways, our findings are similar to what might have been found in the West before the development of the environmental movement in the 1970s. Environmentalism in Russia, as was the early environmental movement in the West, was primarily a conservationist orientation coexisting with the DSP, rather than an alternative social paradigm. The apparent goal was to preserve nature or protect humankind from the worst effects of industrialization and urbanization. Advocates of more fundamental environmental reform are a distinct minority in Russia, and these views are not integrated into the public's general valuation of nature. This conclusion overlaps with Oleg Yanitsky's (1996) claim that the values of the NEP are underdeveloped in Russia, contributing to the erosion of the environmental movement throughout the 1990s. There has been a generation of debate in the West on what it means to be green, which gave rise to the political thinking embodied in the NEP. Such discourse is still beginning in Russia, and environmental thinking among the citizenry is therefore just developing.

The pattern of environmental attitudes also suggests a strategy of mobilization for Russian environmental groups. There is a large social base for environmental activism that taps the positive views of nature that are common among residents of the Chelyabinsk region. Thus campaigns that stress the destruction of nature and the need to restore the natural environment normally can count on greater support than campaigns that directly challenge the military-industrial complex at Chelyabinsk-65. As we noted in chapter 4, concerns about the environmental consequences of Mayak are unrelated to the biocentric attitudes of Russians—but the NEP values of Americans are strongly related to environmental concerns. Even if the goal of some Russian environmental activists might lean toward the broader social changes of the NEP, the popular base for such action is less extensive.[16] We explore these ideas more directly in chapter 10.

The modest support for the values of the NEP should not be taken as an unchanging element of Russian political thinking. Environmental attitudes are still relatively underdeveloped, and discourse on environmental issues and new ways of environmental thinking are just beginning. The core supporters of NEP orientations are, in fact, social groups who have the potential to influence the course of political change in Russia today—the better educated and the young—and who will guide Russia in the future. One of the major accomplishments of the environmental movement in the West has been to educate the public and influence the formation of public opinion on environmental issues. Russian environmentalists also need to foster new environmental thinking among the Russian public if environmentalism is to become a broad force of political change.

Western environmental groups face a similar dilemma; they can attract more popular support when they moderate their demands to tap the broad conservationist values of Western publics, while campaigns that directly challenge the values of the DSP attract less popular support (Dalton 1994). By combining both appeals, however, advocates of the NEP can press the process of reform forward in many particular circumstances. Furthermore, our results show that these beliefs are interlinked among the Americans we surveyed; thus the dichotomy is not being forced on them. More traditional, conservationist views of the environment are likely to be more common among the residents of eastern Washington, and among those who are concerned about Hanford's environmental impact because of its threat to the environment and the well-being of downwind populations. NEP orientations are more likely to be concentrated among environmental activists, and among those who are concerned about Hanford because of their general environmental consciousness. One implication of this finding is that a green identity can play a larger role in creating a basis for support of groups protesting the Hanford facility, and in guiding the behavior of these groups.

Notes

1. Scholars such as Anne Bramwell (1989), Andrew Dobson (1995) and Robyn Eckersley (1992) have described the various ideological currents and subcurrents that exist within the Western environmental movement, highlighting the diversity of green thought.

2. In a later publication, however, Yanitsky (1995, 5) openly speculates that Russians will not be receptive to the values of the New Environmental Paradigm.

3. Dunlap and Van Liere found a similar juxtaposition of the valuation of nature and human dominance of nature in their study of the NEP (1978, 13), though their American sample seemed to express more support for a biocentric view of nature than is true for the distinctly conservative Hanford area sample.

4. Only 39 percent saw these limits to population growth, but this percentage was affected by the large proportion who gave a "don't know" response to this question (40 percent, the highest of any question on this list). When calculated on the base of those who expressed an opinion, 63 percent agreed we are approaching the earth's population limits, a figure closer to the findings from the Hanford region (also see Dunlap and Van Liere 1978, 13; Milbrath 1984, 32–3).

5. The positive and negative loadings of items on this dimension reflect the wording of the separate questions. The pro-NEP worded responses load positively on this dimension, and questions worded in the pro-DSP direction load negatively.

6. The rotated dimensions act simply to clarify the loadings on these three dimensions rather than indicating an alternative structure.

7. Zimmerman's (1994) national study of Russian policy opinions similarly found that these political opinions are weakly structured. This appears a general pattern as citizens are learning to function in a new democratic environment.

8. This pattern may be a methodological artifact of response set for a public not familiar with opinion surveys; that is, some respondents fall into a pattern of agreeing or disagreeing with a general battery and do not notice that some items have been phrased in an opposite direction.

9. However, Oleg Yanitsky (1993, 49) and Boris Doktorov, Boris Firsov, and Viatcheslav Safronov (1993) have observed that Russia is still struggling to accomplish the economic affluence of materialism, and so the number of postmaterialists in Russia is quite limited.

10. See appendix B for the construction of these indices.

11. There is a modest negative relationship (−.17) between biocentric attitudes and collectivist attitudes, which contrasts with the pattern suggested by the NEP, but which is consistent with the support for a private economy expressed by those who hold biocentric attitudes.

12. In the case of "other minorities" (17 persons, two of whom are African Americans and the rest indicating "other" under ethnicity) there is a clear anti-

NEP inclination. Given the difficulty of knowing who these others are, we reserve judgment on the significance of this finding.

13. The generational grouping was defined as the period during which the respondent reached fifteen years of age. The definition of American generations was as follows: prewar (1945 and before), postwar (1946–1959), 1960s (1960–1970), Détente (1970–1979), and Reagan/Bush (1980 and later).

14. We created meaningful historical breaks for the Russian age groups, even though this produced slightly different time periods from the American study. The Russian historical periods were Stalinist period (1955 and before), Khruschev era (1956–1964), Brezhnev era (1965–1984), Gorbachev era (1985 and later).

15. We also compared environmental attitudes to ideological positions. Left/right attitudes were related to NEP support in the American survey ($r = -.06$), and reformist/conservative self-location was related to biocentric attitudes ($r = -.09$) in the Russian survey (see table 7.3).

16. One might find, however, that the values of the NEP are more important in motivating the environmental activists who comprise the movement, and thus are important in maintaining the commitment of these activists (see chapter 8).

8

Environmental Thinking among Environmental Leaders in Russia

Paula Garb

> The most active support for environmental issues comes from younger people and the better educated, those who have probably suffered the least from environmental degradation. Their motivation comes from a philosophical view that values nature.
>
> —Lydia Popova

This chapter examines environmental values, their roots, and the cultural and political context in which they have developed among leaders of Russia's post-Communist environmental movement. These orientations provide a framework for conceptualizing environmental issues that helps the leaders understand and interpret events, and guides their actions. The chapter determines whether leaders of the movement have developed a green philosophy and whether it resembles environmentalism found in Western advanced industrial nations or is dominated by its own distinct characteristics.

The results of this study, based on in-depth interviewing of environmental elites, show how these individuals articulate their beliefs and explain their development. These are people whose attitudes, in comparison with the public at large, are presumably better defined and articulated. Understanding their values and reasoning enables us to have a clearer picture of fluid trends in the Russian environmental movement, to comprehend the actions taken by the movement in its efforts to impact environmental policy, and to predict how future public opinion might be shaped by these influentials.

Theoretical Framework

We examined the environmental thinking of Russian environmental activists by looking at their attitudes toward nature and their approaches to the various economic policies that affect environmental conditions. For example, we determine whether they hold anthropocentric or biocentric views of nature. In addition, we focus on attitudes toward current economic reforms, particularly privatization, toward technological solutions to environmental problems, and toward the concept of limits to economic growth.

We chose these areas of analysis because they are indicators of beliefs that fall within the realm of the New Environmental Paradigm (NEP), the term used for environmentalism that challenges the prevailing sociopolitical norms of advanced industrial democracies (see chapter 7). For instance, according to research in the West, adherents to the NEP value nature for its own sake and have a biocentric view of nature that assigns equal value to humans, plants, and animals.

We should be cautious, however, of simply extrapolating Western belief systems to Russia. Oleg Yanitsky (1995) warns against mechanically applying Western paradigms to Russia, where "mentality was formed under conditions of forced and still unfinished industrialization under a totalitarian regime." This does not means, he stresses, that "the laws of nature operate differently in Russia. It only means that their perception and interpretation by Russian culture differs from those in the 'West' " (Yanitsky 1995, 50).

Douglas Weiner (1988) identified three trends in Russian environmental thinking of the 1920s that are helpful in analyzing the context for the evolution of contemporary Russian environmental beliefs. These trends were abruptly interrupted with the advent of Stalin and Lysenkoism in the early 1930s, but they were revived during the mid-1950s in the Khrushchev thaw and have influenced contemporary environmental thinking in Russia.

Weiner described a pastoralist orientation that was comparable to the biocentric thrust of the NEP. Pastoralists valued nature for aesthetic and moral reasons. "Repelled by modern industrial society—capitalist or socialist—its adherents sought to return to an idealized, organic, agrarian golden age when humanity had not yet despoiled the earth. . . . They emphasized that nature is valuable in itself, irrespective of its utility to humans, and that other living things have an equal right to existence" (Weiner 1988, 231).

Another trend in early Russian environmental thinking, according to Weiner, was the anthropocentric, ecological view, held almost exclusively by natural scientists. They believed that nature had a distinctive structure characterized by interdependence among its biotic components and by a state of relative equilibrium or at least proportionality. The ecologists emphasized nature's fragility. They were motivated less out of a concern for the survival of other life forms for their own sake and more worried about how a breakdown in natural ecosystems would impact civilization.

The third approach noted by Weiner was the utilitarian view, which was also anthropocentric. Utilitarians tended to define resources narrowly, based on their economic utility. They excluded recreational and aesthetic amenities from their concerns, as well as living or nonliving things whose economic value was dubious in their minds. In contrast, scholars find that Western environmentalists share a preference for qualitative values that are not easily accommodated by economic cost/benefit models; environmentalists also express skepticism about the ability of the market to regulate environmental issues. In the case of Russia, such postmaterialist values are cultivated by an extremely narrow circle of people, and some of them are eager to assimilate elements of Western civilization, such as the market and private property (Yanitsky 1993).

Another component of the NEP concerns distrust of technological solutions to environmental problems, and a belief in the need to limit economic growth. Among Western environmental leaders these views involve the conviction that people are in danger of losing control over the consequences of technological development, that the bad effects of technology outweigh its good effects, and that technological "fixes" will only perpetuate environmental damage.

Yanitsky (1993) also talks about the technocratic thinking among Russian environmentalists, and Weiner (1988) notes that this was a strong theme in early Soviet history. Technology, however, was often viewed in positive terms. The ecologists argued that their scientific expertise could ensure that growth would remain within the possibilities afforded by healthy nature, and proponents of the "utilitarian" view advocated the principle of sustained yield wherever applicable. They wanted to make resource use generally more efficient.

Weiner concluded that in prerevolutionary Russia the utilitarian and pastoralist tendencies flourished, whereas the Soviet period saw the gradual dominance of the ecological approach to conservation. According to Weiner, the Old Bolshevik intelligentsia supported the ecological approach because:

> they regarded socialism's double mission to be enlightenment and the rational organization of social and economic life on the basis of science. So they greeted ecological *zapovedniki* warmly. . . . By providing a materialist, scientific explanation for complex natural phenomena, ecological science would enlighten. And by establishing permissible and recommended parameters of economic activity for specific natural regions on the basis of etalon studies, ecologists would help promote a rational and self-sustaining economy. Above all others, the People's Commissariat of Education (*Narkompros*) became the institutional guardian of this sense of mission" (Weiner 1988, 231).

Beyond theoretical debates on the content of environmental attitudes, there are other difficulties in studying environmental attitudes in contemporary Russia. For instance, there has not been much continuity in a country with a history of deep crises, involving wars and revolutions. This cannot help but impact value formation. Yanitsky maintains that while the West is making a transition to a postindustrial society, in Russian society no such trend has emerged: "Indeed, how can such values be cultivated when most of the Russian population lives in poverty, and as in the past, acts as a bearer of the values of 'leveling' distribution, of executive bullying, and of provincialism—values that manifest themselves in archaic forms of social action" (1993, 49).

Second, the political situation is so fluid that opinions and values are shaped and reshaped much more rapidly than in more stable societies that are not undergoing social revolutions. It is inefficient to take one "snapshot" of these opinions and values and put the photo in a frame or a model developed on the basis of vast empirical information collected in other societies or in other times. Many more "snapshots," taken from different angles, will be necessry to draw more solid conclusions. Presented here is a "photo album" of the way people in Russia conceptualize environmentalism in the 1990s.

Methods

The primary source of information for this chapter came from working with, observing, and interviewing environmental activists over a total period of six months during eleven trips, between June 1991 and March 1998. In an effort to assess support for the New Environmental Paradigm among Russian environmentalists, we developed a questionnaire from surveys of environmental values in three Western democracies (Milbrath 1984). These questions were used in semistructured interviews and informal conversations with eight environmental activists in Moscow and twenty in the Chelyabinsk region, between June 1991 and March 1998.

Another source of qualitative data is from interview transcripts published by Oleg Yanitsky (1993). Most of the quotes taken from Yanitsky's book to illustrate relevant points are by environmental leaders in Moscow whom I have also interviewed.

The third source of data is our survey of residents in the Chelyabinsk region, conducted in 1992, and a survey of twenty environmental leaders in Moscow and Chelyabinsk. The survey questions were focused on determining whether there is support among environmental leaders and various other sections of the public for the New Environmental Paradigm.

Background of Informants

The vast majority of informants belonged to the largest and strongest of the Russian environmental organizations, the Socio-Ecological Union. They were either among the national leadership based in Moscow or in the leadership of affiliated organizations in the Chelyabinsk region. Other interviewees were leaders of the Moscow-based Russian Ecological Union, the Moscow Green Party, the Moscow Ecological Foundation, or the Chelyabinsk Association of Greens (the only organization contacted that was politically conservative and had links with Russian nationalists).

The environmental leaders in the sample were typical of environmental activists in terms of ethnic identity, socioeconomic status, gender, age, and other social characteristics. The ethnic backgrounds of environmentalists reflected the local demographics. In Moscow and Chelyabinsk they were ethnic Russians, as was the vast majority of the population. The exceptions were some villages of the Chelyabinsk Oblast, where Tatars and Bashkirs constitute a majority and were the leading environmental activists. Environmental leaders tended to be in their forties or older. They were fairly evenly divided along gender lines (although it appears that women were the most involved in the everyday activities). They generally had a higher education, usually in technical fields or the natural sciences.

Yanitsky (1993) refers to these professionals as the "marginal intelligentsia" who had not gained entry to the state, party, or other structures of power before perestroika. Usually they were capable of leadership positions but had avoided being part of the establishment or had been expelled for their criticism.

Environmental Attitudes

Our public opinion survey in Chelyabinsk indicated the absence of a cohesive environmental ideology, which may be similar to the environmental movement in the West at its inception. In other words, respondents may have supported one or more components of the NEP, but generally not the whole spectrum of beliefs. This is unlike the situation in the West where these beliefs fit into a single framework.

The Chelyabinsk data showed a high valuation of nature among all sectors of the public, but only a small minority held strong biocentric views. The survey also indicated (although the results were less robust than for the valuation of nature questions) some acceptance of the limits to economic growth and a hesitant distrust of technology.

The following presentation of statements taken from interviews with environmental leaders attempts to bring alive the survey data by providing informants with the opportunity to articulate in their own words the philosophy that motivates them in their activism.

Valuation of Nature

Environmentalists in Moscow commonly explained that their dedication to environmental issues was because "they always loved nature." Their love was not from books, but from everyday contact with nature at dachas, summer camps, and even urban houses with plots. Lyubov Rubinchik, in charge of the Socio-Ecological Union's information center, described her own early history this way:

> On the whole I grew up amid nature. Our little town was green then, there was a river nearby, and beyond it a forest. From the time I was ten years old my brother and I spent all our time on the river, in a boat (Yanitsky 1993, 199).

Other Russian environmentalists had a positive formative experience with nature later in life. Maria Cherkasova, another SEU leader, and a graduate of Moscow University's biology department, described how her environmental consciousness was raised by working at the Sayano-Shushenskoe reservoir, which was formed by building a giant dam on the Yenisei River:

> I saw how not only birds, but also tens of thousands of people were driven from that rich river valley that became the bottom of a monstrous artificial lake. None of them would ever be able to return. There I came to understand that both animals and people share a single fate. Neither animals nor people are being left with a place to live (Yanitsky 1993, 178).

Many environmental leaders grew up almost entirely in an urban environment from which they wanted to escape. Since the 1970s, Muscovite Yevgeny Schwartz has been a leader in the environmental *druzhiny* movement (student nature patrols), and came into the movement this way:

> As I remember I always felt oppressed in the urban environment, because it restricted my behaviour too much. . . . I turned to biology and environmentalism not so much from the urge to protect nature, as from the human need to find a refuge (Yanitsky 1993, 207).

Environmentalists who were educated in the student nature patrols of Moscow University's biology department were presented by their movement colleagues as prime examples of "environmentalists of the highest order." A common refrain I heard from veterans of the nature patrols was that these groups were established to protect nature, not other human beings.

The attitudes expressed by veterans of the student patrol movement that burgeoned in the 1960s and 1970s were shaped when the environmental orientations of the 1920s were being revived. The biology department of Moscow University (where many of today's leaders were educated) played a major role in this revival, which, I believe, was rooted in the "ecological" approach. However, there seems to be a stronger biocentric strain in that approach today.

Lydia Popova, a Muscovite leader in the SEU, thought that green philosophy in Russia, including her own, was derived from Russian cultural traditions:[1]

> People without an environmental philosophy leave the movement very quickly. This green philosophy in Russia comes from Russian traditions. It also comes from living in an urban society where the government did everything to destroy nature. The people are going back to a tradition they lost but that is still in them and is pushing them forward. The most active supporters are younger and more educated people, those who probably suffered less than others because they enjoyed the benefits of the system and could protect themselves. It is less an objective problem in their own lives, and more a philosophical motivation driving them.

Sviatoslav Zabelin, the leader of the SEU, did not idealize past traditions of an environmentally sound Russian rural cultural environment. He felt that the whole history of human beings has been marked by environmental destruction, but that, for the first time, a new consciousness was evolving worldwide: "The rise of this movement represents the onset of humanity's social maturity, as opposed to biological instincts" (Yanitsky 1993, 217).

In Chelyabinsk, unlike Moscow, there did not seem to have been an educational institution, department, or an environmental organization that fostered an environmental sensitivity like that of the Muscovite environmentalists that we interviewed. The focus of higher education was entirely technocratic. Only in 1993, had Chelyabinsk opened a university offering courses in the humanities and social sciences in addition to the hard sciences.

Natalia Mironova, chair of the Chelyabinsk Movement for Nuclear Safety and Cochair of the Democratic Green Party maintained that a green philosophy was not the primary motivation among the region's environmentalists:[2]

> Our movement was spurred by the health problems of our children. This is why our most active people are women. They are responding to a specific reality. They are mainly engineers, people who work at industries that are sources of the very pollution they fight. This is how their philosophy was developed.

Larissa Pilenen, one of the engineers Mironova was referring to, eloquently expressed both the mentality of the technocrat and the mentality of the New Environmental Paradigm in, perhaps, its Russian form:[3]

Our people have always been close to nature, to their source; they chose to spend time close to nature if they could. But since everything here was turned upside down by forced industrialization that affected not only nature but our consciousness, we changed psychologically. Despite my deep understanding of our industrial tragedy, having worked for 17 years at our metallurgical plant, I have to say that I love it, because I designed many of the units. My intellectual contribution is as dear to me as my children, so I love the place, even though it is horrible. Our people would have taken better care of nature if industrialization had not destroyed our consciousness. Today people want to go back to nature, but it is for the purpose of consumption. We want houses, dachas, because we can pickle cucumbers there. There is nothing terribly wrong with that, but this isn't the kind of motivation that enriches, enlightens people. What we don't have is the feeling that people are part of nature, *not* its master. Women understand this better because they bear children. Men physiologically can't understand this.

Pilenen said she began developing a green philosophy in 1984–85, when her daughter was stricken by debilitating allergies. That was when she remembered first hearing the word ecology:[4]

As an engineer at the metallurgy plant I was killing people with the same hands that were trying to give relief to my ailing daughter. All morning I looked after my daughter, but from 1:00 to 7:00 at night I designed a blast furnace. . . . People have to understand that they are a piece of nature, and nature has its laws, regardless of what we want. If my arm wants to live it has to follow my head. So people must learn from nature to see how to arrange our life so that it fits with other creatures. Other creatures don't burn wood to get warm. They find other ways.

Sergei Kalachev, Cochair of the Democratic Green Party of Chelyabinsk, was the only environmental leader there who so clearly articulated a biocentric view of nature, one that seemed to have evolved from early childhood. He told me:[5]

I have always loved nature, as far back as I can remember. Maybe God meant for me to be nature's defender. When I was younger I helped put out several forest fires. When the job was done the sight of the scorched trees was horrible. I have long realized that man is not king of nature, but part of nature. So it was only natural that I joined the environmental movement.

There are also movement leaders who were motivated primarily by their position as victims of the horrendous environmental damage caused by the weapons complex. Vladimir Skulatov was an active member of the Kyshtym-57 Foundation and of the Chelyabinsk Nuclear Safety Movement. He came from a village that was evacuated a year after the 1957 accident, and moved to another village that he believed was also exposed to radiation from Mayak. In 1990, when he came to work in Chelyabinsk as a driver, he started going to meetings of the local environmental movement:[6]

> No matter what, I would've eventually joined the environmental movement. I have always been close to nature. Since childhood I have hunted and fished. I have watched the environment die. There's such a big difference between the environment now and when I was a boy, not so long ago. I can't be indifferent to this.

Since 1989–90, when the public was finally informed about the nature of Mayak contamination and the related health risks, several Muslyumovo residents have maintained a loosely structured environmental group linked with the Movement for Nuclear Safety in Chelyabinsk. The Muslyumovo organization has been the only center of consistent environmental activism outside the oblast's administrative center.

The leadership of this group had no prior involvement in environmental activism or green philosophy. Our Chelyabinsk public survey showed that support for elements of the New Environmental Paradigm was weakest in Muslyumovo, where, for example, anthropocentric views of nature were clearly prevalent and the good effects of technology were seen to outweigh the bad effects (chapter 7). Semistructured interviews of the local population brought similar results, although this in-depth interviewing enabled informants to qualify their statements in a way that indicated greater acceptance of key principles of the NEP, especially the valuation of nature for nature's sake. Perhaps these rural inhabitants took too literally such questions about whether animals and plants exist to be exploited by humans.

Guzman Kabirov was one of three leaders of Muslyumovo's twenty or so environmental activists. His interview illustrated the typical response I heard from Muslyumovo environmentalists about green values. When I spoke to Kabirov about his attitudes toward nature in order to better understand whether his views were more biocentric or anthropocentric the conversation did not result in any "quotable quotes" until the next day. Apparently he went home thinking about my questions. The following afternoon he sought me out to tell me that he had never really thought about the issues I raised in such a focused way, and felt he had not adequately answered my questions. This is what I learned from him:[7]

> My grandmother used to tell me that Russians pray to wood (or icons), Mongolians to rocks, but that Tatars pray to God. My grandmother taught me that God rules everything—nature and man. When I thought about all that we talked about yesterday I came to the conclusion that if humans would suddenly die out, within five years nature would take over the earth again and no trace would remain of man-made things. I don't know if it really is possible for nature to flourish like that as long as man exists. I don't know if we can really live in harmony with nature.

Economic Dimension of Environmental Attitudes

Most informants found it difficult to articulate clear positions on some of the economic components of the New Environmental Paradigm. It seemed as though environmental elites in Russia avoided addressing economic and political issues that were related to the environment. Perhaps they felt that these were matters outside their field of expertise or even interests as greens. Ann Rubin, an American who worked in Moscow with the SEU from 1991–93, pointed out:[8]

> Environmentalists in Russia are clearly against the environmentally devastating aspects of the old command system of centralized planning. But they are also skeptical of what the market economy, as it is being developed in Russia, can do for the environment and for environmental restoration.

In the debate on political reforms in the early 1990s, privatization was a key issue. Environmental leaders in Moscow generally supported small- and medium-sized business ventures and distrusted "big industry and industrial groups."[9] In outlying regions, such as Chelyabinsk, privatization was endorsed wholeheartedly by local environmentalists as the primary solution to environmental degradation. Galina Nikitina of Chelyabinsk articulated her thoughts as follows:[10]

> Only privatization of all industry can solve our problems. A private owner will care about developing the business, ensuring profits, and seeing to it that production is environmentally safe.

This was how Vladimir Skulatov explained his convictions that privatization would improve the environment, and that the profit motive would not interfere with this goal:[11]

Anyone who owns a manufacturing business will be concerned about the enterprise and how it affects the environment of his children and grandchildren. When no one owned these businesses the management just didn't care what happened.

Sviatoslav Zabelin did not think that environmental improvements depended on whether industries were privately or state owned. As he explained, "technological modernization of industries is central to preventing pollution, and that can be accomplished under any form of ownership."[12]

A more controversial issue was government-owned versus privately owned nuclear waste management. Some believed that in the midst of the political and economic chaos of the early 1990s, the whole nuclear cycle needed to be controlled by the state to prevent nuclear proliferation. Others felt that as long as the state was overseeing its own activities in nuclear waste management, it could not be trusted to make the right decisions. Government supervision would be meaningful, they said, only if private companies were managing radioactive waste. Natalia Mironova explained how these matters are viewed in Chelyabinsk:[13]

The charter of the Democratic Green Party makes a general statement in support of a market economy, but does not elaborate. In Russia, as long as the military-industrial complex is owned by the state we cannot hope for normal environmental control. We discuss these issues only informally, among ourselves. But I think my opinion is prevalent in the group. We can't expect any ecological control unless the military-industrial complex is privatized. It's a problem throughout the region because 80 percent of all industry works for defense. Monitoring won't have much effect as long as the government is in charge of controlling its own industries' compliance with environmental regulations. This is because state-owned enterprises dictate to the government. It's especially evident in the nuclear complex. Whatever Victor Mikhailov [Minister of Atomic Power—] needs for the nuclear power industry, Yeltsin approves.

Lydia Popova in Moscow, in contrast, expressed a more skeptical view that was prevalent among the SEU national leadership:[14]

In this transition period in Russia, when the economy is very unstable and there is rampant corruption, the involvement of private capital will actually increase the corruption and will aggravate the environment. A company established by the Ministry of Atomic Power wants to destroy chemical weapons by underground nuclear explosions at Novaya Zemlya. The company wants to earn hard currency by providing services to other countries. This kind of privatization will only make things worse. We have to be cautious about privatizing the military complex. There should be two parallel courses—one of privatization and the other of elaborating strong legislation and regulation. But regulation should come first, then privatization. Otherwise the environment in Russia will be destroyed completely.

Although individual environmentalists took positions on various matters related to economic and political reform, their organizations had not elaborated platforms on these issues. When asked about their preferred economic system they seemed to be suspicious of both capitalism and communism. Their answers reflected a certain ambivalence, but then this may not be so different from the views of the general public who were still grappling with these complex issues.

The ambivalent attitudes toward economic reform were illustrated in an interview with Zabelin in September 1991. He said that some environmentalists had thought it was easier to be an environmentalist in the West, but he felt these impressions were shattered at an international environmental conference held in Moscow in March 1991. This was the first time that many key Russian environmentalists were confronted with specific firsthand information about environmental movements in Western countries. According to Zabelin, they learned that environmentalism has an uphill climb no matter what the political or economic system. They explained that this was because state structures, no matter what the system, oppose genuine environmentalism, and everywhere the power balance has to be maintained by environmentalists exerting constant pressure on the interest groups that show less concern for the environment. The one clear advantage Russian activists saw in the West was that legislation gives environmentalists more leverage over their governments.

Technology and the Environment

Another aspect of environmental values involves the approach toward technological solutions to environmental problems. It is clear that technocratic thinking has impacted Russian society. The whole thrust of Soviet efforts to "catch up with and surpass America," a slogan prevalent in the speeches of Soviet leaders and on huge building signs in every Soviet city, focused the educational system on training technocrats who could master and advance science and technology to help the country meet this all-important goal. In a society that was rebuilding its economy from the rubble of World War II, technology was seen as the only hope for recovery and progress. This was a common theme in the media throughout the 1950s and 1960s when careers in science and engineering were among the most prestigious, and when today's environmental leaders were choosing and pursuing their vocations.

Another reason for an orientation toward technological "fixes" was the belief that the severe economic crisis, with industries running on ancient and polluting technology, could be overcome only by modernizing technology. Many environmental leaders are scientists or engineers. These are professions with the strongest traditions of Soviet technocratism, so the views of our informants reflected this lifelong training that inculcated a continuing belief in the power of technological solutions.

Valery Sevastianov, coordinator of the Moscow Ecological Federation and deputy in the Moscow City Duma is an electrical power engineer. Chernobyl convinced him and other engineers that they should join the ecological movement, since they shared in the moral responsibility for Chernobyl. Their goal was to prevent another Chernobyl disaster by promoting safer, more modern technology. This was his view of technology's role:[15]

> Someday in the distant future there may be alternative energy sources. Right now we have to improve on what we have. In the long run, perhaps, in the next century, we will develop new technology, including biotechnology. Bionics will help us restructure industrial production and living conditions. . . . We shouldn't copy the United States, which has a worse record for ecological damage than other advanced countries . . . The model is in Western Europe, in some respects, Japan. We should copy the best advanced technologies of each of these countries. . . . We can have living standards like in the U.S. but by using different technology.

Oleg Stakhanov, who worked on publicity for the Socio-Ecological Union, articulated his view of technology this way:[16]

> Our technology is so old that any replacements will improve the environment. One of the fundamental reasons for our catastrophic situation today is that our petro dollars were not invested in technology. . . . Russians will have to advance technology somewhat to improve environmental conditions. African countries will have to advance even more. In the West people will have to regress, as it were, and use less energy per person. We have to neutralize the harmful effects of our backward technology by modernizing it. Meanwhile we have to promote ecological education to change the way people think about utilizing the environment.

When I asked Oleg what he meant by "utilizing the environment" he talked about habits of consumption that touched on another economic dimension of the NEP—limits to growth. He maintained that directing people to consume less for the sake of the environment was not a difficult task in Russia:[17]

Most people in Russia would rather live in a society that is just but less luxurious than in a more affluent one without justice. Justice and modesty are principles that are primary and acceptable to Russians. This is the way we've been raised. We're never going to catch up to America. People realize that now. Now Russians are oriented toward restoring justice and evening out living standards, and genuinely implementing the socialist slogan that distribution should be according to one's work.

Limits to Growth

Environmental leaders tended to support the notion that there are limits to economic growth, that the planet cannot support unlimited growth and resource use, and therefore people must restrain their consumption. They talked about refraining from pursuing American living standards on a worldwide basis, as a general philosophy rather than as a call to fellow Russians. For instance, while Zabelin believed that all previous history had been geared toward environmental destruction, he was encouraged by the green philosophy that he saw worldwide: "The main sign of this maturity is humanity's willingness to limit its needs" (Yanitsky 1993, 217).

Seviastianov had this view of how people will be persuaded to limit their consumption habits:[18]

If we look at the technological progress of such countries as the U.S. and Japan, the world will have limited resources, the planet will die. We must develop much less energy-intensive technology and exchange information more with each other. This will lead to a reasonable limit to consumption, not the return to nature that Jean Jacques Rousseau advocated.

Some leaders did not see how this approach was realistic for a Russia in deep economic crisis, with production levels plummeting. Lyubov Rubinchik was among them:[19]

If people don't change their attitude to consumption they won't change the way they utilize nature. . . . We can't get people to stop driving cars or flying airplanes in the foreseeable future, in our lifetime. You have to take the least of two evils.

It's easier to get people to understand that they don't need to drive so quickly and accept the slower speeds of electric cars rather than not drive cars at all. We have to be realistic about what people in this country are willing to do to protect the environment.

Rubinchik said the SEU advocated a plan for buying up inferior and therefore cheap land so that environmentalists would turn 30 percent of the world's land into models of environmentally sound land use. She maintained that the idea was possible to implement everywhere except Russia, where she believed much more work must be done to change the people's values about land use. Although she was not religious she looked to the Russian Orthodox Church as the institution most likely to instill in the people more environmentally sound, biocentric values:[20]

These values can be changed only by the Church. The Church, as an institution, developed anthropocentrism. Only the Church can turn it around. Environmentalists should open up dialogue with the clergy, who have understood many things too, are very strong, and are led by progressive people.

Conclusion

In the early 1990s, the leaders of Russia's environmental movement were in the process of articulating for themselves and the movement a "green" philosophy that bore an overall resemblance to the New Environmental Paradigm in the West. The in-depth interviews, excerpts of which are presented here, did not reflect a clearly defined set of NEP views. This may occur because the multiple choice questions in the survey helped informants see in concise form ideas that were still in their embryonic stage and that have not yet crystallized; these orientations were harder to articulate in the less-structured interviews. It was not easy for respondents to provide ready answers that fit neatly into the New Economic Paradigm, especially on questions related to the economic dimension of the NEP. It was as if they had either not thought about the issues very much, or had done so in slightly different terms that were shaped by the specific realities of Russia today. In addition, there was a perceptible difference between environmental leaders in Moscow and Chelyabinsk. Environmentalists in Moscow were more likely to express a more developed environmental ideology than their counterparts in Chelyabinsk.

The valuation of nature was most readily and clearly supported by informants in Moscow and Chelyabinsk. Their orientations tended to be in transition from anthropocentric to biocentric views. The roots of these views were found largely in the "ecological" trend of the 1920s, which later dominated the environmental scene, providing the main proponents of environmental protection and institutional support throughout the Soviet period (i.e., the Society for Nature Protection). The biocentric approach of the "pastoralists" of the 1920s appeared to become more dominant in the 1990s. Perhaps it could be traced directly to adherents of this trend preserved within the Soviet environmental movement, or perhaps it is related to the pre-Christian, pagan roots of these values in Russian culture. It is also possible that the growing biocentrism in the movement was influenced by contact with the green philosophy of counterparts in Western nations.

Attitudes to the economic dimensions of the NEP were much harder to define. Environmentalists were ambiguous about the economic and political particulars of reform, even in relation to environmental issues. Privatization was embraced more wholeheartedly in Chelyabinsk, whereas there was more skepticism of privatization in Moscow. In addition, there was a strong reliance on technological "fixes"—an outgrowth of the highly technocratic educational and value system inculcated throughout the Soviet period. On the limits to economic growth, environmentalists supported this idea in the abstract, but less when it came to Russia in the early 1990s, where slogans urging people to limit their consumption would sound absurd. If the question were related to a hypothetical postcrisis Russia, a common answer would be like that of Oleg Stakhanov, who maintained that Russian culture emphasizes modesty in consumption.

Perhaps, as the movement gains more experience and furthers its ideological discourse, environmental values in Russia will become structured more cohesively into a single framework of environmental thinking, and developed opinions will form. More widespread and meaningful contact with Western environmentalists will also be an important factor in shaping these ideas. However, it is possible that the cultural and historical differences, and differences in economic realities will prevent the formation in Russia of the Western NEP in its entirety. The achievement of this study is to set the groundwork for further observations of the process of value formation and the influencing factors domestically and internationally. Will what appears to be a burgeoning environmentalism resemble that of the West, or will it be molded quite differently in Russia? Three factors will dominate as the determinants—the length and depth of Russia's economic crisis, the role of Russian nationalism in culture and politics, and the quality of contact with Western environmentalists.

Notes

1. Lydia Popova, Coordinator of Nuclear Programs, Socio-Ecological Union. Conversation with Paula Garb, Russell Dalton, and John Whiteley, Irvine, California, June 1993.

2. Conversation with Paula Garb, Russell Dalton, and John Whiteley, Irvine, California, June 1993.

3. Conversation with Larissa Pilenen, member of the board of the Chelyabinsk Movement for Nuclear Safety, Chelyabinsk, September 1992.

4. Conversation with Pilenen, September 1992.

5. Conversation with Sergei Kalachev, Cochair of the Chelyabinsk Democratic Green Party, Chelyabinsk, May 1992.

6. Vladimir Skulatov, member of the board, Chelyabinsk Movement for Nuclear Safety. Conversation with Yevgeny Gontmacher, Chelyabinsk, December 1991.

7. Guzman Kabirov, leader of the Muslyumovo branch of the Chelyabinsk Movement for Nuclear Safety. Conversation with author, Chelyabinsk, September 23, 1993.

8. Ann Rubin, U.S. coordinator of the Moscow Environmental Clearinghouse. Conversation with author, Moscow, December 18, 1992.

9. Maria Ivanian, secretary of Green party. Conversation with author, Moscow, March 1992.

10. Galina Nikitina, member of the board, Chelyabinsk Movement for Nuclear Safety. Conversation with author, Moscow, September 1992.

11. Conversation with Vladimir Skulatov, December 1991.

12. Sviatoslav Zabelin, chair of Socio-Ecological Union. Conversation with author, December 26, 1992.

13. Conversation with Natalia Mironova, June 1993.

14. Conversation with Lydia Popova, June 1993.

15. Conversation with Valery Sevastianov, Moscow, December 20, 1993.

16. Conversation with Oleg Stakhanov, Moscow, December 20, 1993.

17. Conversation with Oleg Stakhanov, December 1993.

18. Conversation with Valery Sevastianov, December 1993.

19. Conversation with Lyubov Rubinchik, coordinator of Information, Socio-Ecological Union, Moscow, December 20, 1993.

20. Conversation with Lyubov Rudinchik, December 1993.

9

The Mobilization of Public Support for Environmental Action

John C. Pierce, Russell J. Dalton, and Ira Gluck

> The main difference between Eastern and Western environmentalists: in the West they are fighting mainly to protect nature, in the East they are fighting to protect themselves.
>
> —Igor Altshuler

Citizen complaints about even the most extreme conditions of environmental degradation and public expressions of the most fright-filled feelings of personal risk remain an unanswered claim if discontent is not mobilized into political action. Why do individuals band together to express their feelings and to engage in action designed to address the conditions producing those feelings? Are the reasons and objectives of involvement similar or different between residents near the two nuclear weapons facilities?

This chapter provides some insight into these questions about the sources of political mobilization against the facilities at Hanford and Mayak. To what extent do the citizens of the various American and Russian cities near the nuclear weapons complexes express their discontent in action? In particular, are there any differences in the political mobilization of Americans and Russians in this regard? Within each country, do the variously situated communities differ in the character of their political activity?

This chapter does more than merely describe the relative levels of political mobilization of citizens in the vicinity of Hanford and Mayak. We also probe the underlying reasons for environmental action across nations, communities, and types of individuals. To what extent are Americans or Russians mobilized by a perceived threat to themselves as victims in which they act on the basis of their shared grievances? Some non-Western (and especially Japanese) environmental movements are mobilized by perceptions of a specific threat to human health that makes victims of a particular population group (Lewis 1980; Pierce, et al. 1989, 72–94). The Hanford and Chelyabinsk settings are theoretically interesting because of the exceptional magnitude of the environmental problems in both areas, which might reasonably stimulate political action. An alternative explanation is that people are mobilized in major part by their political values, such as a general concern for the environment. Indeed, cross-national comparisons of environmental mobilization in the United States and other Western democracies often finds that environmental action grows out of a particular ideological paradigm, the New Environmental Paradigm (NEP) (see chapter 7; Dunlap and Catton 1979; Dunlap and Van Liere 1978).

This chapter compares a victimization or grievance model, a value-based model, and a social location model as sources of environmental mobilization against the Hanford and Mayak nuclear weapons plants, along with other possible explanations for that political activity. We begin with an exploration of citizen feelings about environmental groups as channels for political mobilization in both nations. We examine the American and Russian publics' approval of environmental interest groups, trust in the information the groups provide, actual membership in these groups, and perceptions of the groups' political effectiveness. We then examine the correlates of these orientations, searching for evidence about alternative explanations of mobilization. Finally, we explore some contextual effects, looking at the degree to which the correlates of mobilization differ across communities, assessing the extent to which these contexts themselves had an effect beyond the preceding independent variables.

Support for Environmental Groups

Earlier chapters described the emergence of public protest against the environmental consequences of nuclear weapons production at Hanford and Mayak (chapters 5 and 6). Subsequent to the formation of protest groups in both regions during the mid-1980s, there have been ebbs and flows in their level of activity and public visibility. Moreover, it was especially difficult to gauge popular support for environmental groups in Chelyabinsk because of the instability in the larger political climate. The vast majority of the public supported the environmentalists in a March 1991 regional referendum opposing the construction of a new nuclear power plant at Mayak; but ongoing involvement in group activities has been limited (and decreasing). Public support for environmental groups represents an important measure of the general potential for the transformation of citizen disaffection into political change. Our parallel surveys of public opinion in both regions thus provide a valuable method to gauge the degree of popular support for environmental groups, then to examine the bases of that support.

Our study employed four separate indicators of the public's support for environmental groups. The first item measured approval of groups that are active on environmental matters. The second item tapped citizen trust in environmental groups as a source of information about the facilities. Although approval and trust both reflect general support for environmental groups, they are conceptually distinct. Approval measures the respondents' affective feelings about groups; as such, many analysts consider approval items to be a "soft" measure of support. Trust represents opinions that are at least partially based on evaluations of group performance and is consequently a somewhat "firmer" measure. A third indicator of support taps the respondents' membership in (and attitude toward joining) an environmental group. A behavioral measure, such as membership, typically provides the strongest test for environmental group support because activism requires a real commitment of time and energy. Finally, we asked about the perceived efficacy of environmental groups; that is, respondents indicated whether environmental groups were seen as actually influencing government policy. Groups that are viewed as more effective should be better able to maintain membership and morale, as well as expand in membership and scope of action.

Our analysis begins with our measures of the public's general approval of environmental groups for each of the communities in the study.[1] As in other chapters, we have paired sites across the two countries according to their relationship to the facilities: Richland and Chelyabinsk-70 are "host" communities where people who work at the facilities live;

Spokane and Chelyabinsk City are the major "downwind" urban centers; the Yakama Indian Reservation (White Swan, Toppenish, and Wapato) and Kyshtym/Muslyumovo are small, largely indigenous minority communities adjacent to the facilities; Wenatchee and Chebarkul are the control sites located just beyond the immediate region of the facilities.[2]

The findings in figure 9.1 reflect several noteworthy differences, both across countries and among the American and Russian communities. In each cross-national pair of cities, the Russian sample was much more likely to express complete approval of environmental groups than was the American counterpart sample. The largest difference in approval was between the two control cities of Chebarkul and Wenatchee, with 72 percent and 9 percent respectively giving their complete approval. It

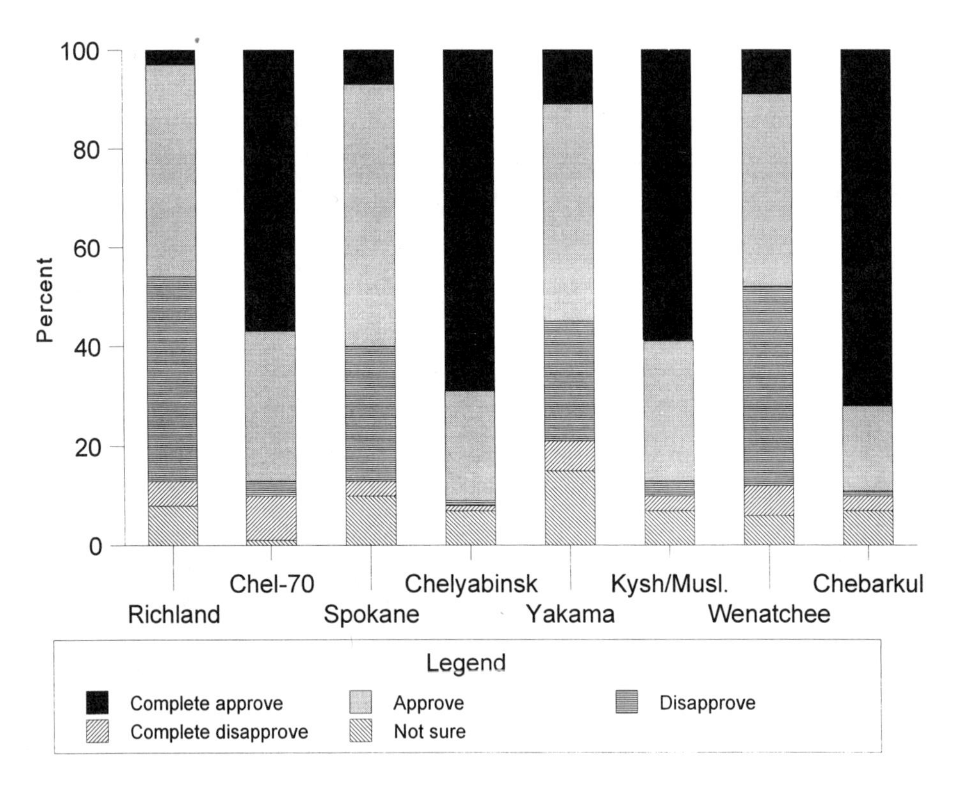

Figure 9.1
Approval of environmental groups

is not that Americans did not support environmental groups—they did—but they expressed their support with the more modest category of "mostly approve."

Significant differences also existed among the cities within each country. For example, disapproval of environmental groups was higher in Hanford's hometown of Richland and in the control site of Wenatchee than in any of the downwinder locations. Richland residents were evenly divided between those who approved of environmental groups (46 percent) and those who disapproved (46 percent). Most residents of Spokane (60 percent) and the towns on the Yakama Reservation (55 percent) expressed approval of these groups.

The geographic distribution of sentiments toward environmental groups is roughly similar in Russia. Environmental approval was lowest in the closed host city of Chelyabinsk-70, the functional equivalent to Richland, where most residents had occupational links to the Mayak facility. The ambivalent feelings among employees of Mayak are illustrated by the comments of one official from the facility:

> I am absolutely certain that environmental movements are necessary. Only a strong and active environmental movement can force the technocrats to think about the environment. This happened in the West a lot earlier. In the United States the turning point was in the 1950s. In our country it happened during the period of glasnost. . . . Unfortunately, our environmental movement is destructive. It can only think in terms of banning everything.

The greatest support for environmental groups came from the city of Chelyabinsk, where the local green movement had its political base. When combined, the approval levels in the downwind communities of Kyshtym and Muslyumovo were close to those of Chelyabinsk-70, but this reflects a mix of two different reactions. Kyshtym residents were indirectly linked to the facilities and only 49 percent expressed complete approval of environmental groups. In contrast, Muslyumovo residents had felt a heavy negative impact from the facilities (see chapter 4) and were much more sympathetic toward environmental groups (72 percent expressed complete approval).

The Russian survey contained an additional question that asked about approval of groups specifically active in protesting the environmental consequences of Mayak.[3] Approval of Mayak protest groups roughly paralleled approval of environmental groups in general. For instance, 64 percent of the total Russian sample completely approved of environmental groups in general; 63 percent of those aware of the Mayak protest groups completely approved of them. The two questions also tracked each other across the various Russian sites. These similarities are important because it is possible that the contentious nature of protests against the nuclear weapons facilities evoked less support than general opinions toward environmentalists. At least in the Russian case, approval of environmental groups spanned these two different aspects of the movement.

Trust in environmental groups was another important measure of support because it taps public willingness to rely on these groups as a source of information or as an influential actor in the political process. Table 9.1 presents the publics' trust in environmental groups as a source of information. The figure displays the percentage "mostly" or "completely" trusting environmental groups and other political actors as sources of information about the facilities.[4]

Again, several potentially revealing cross-national and intra-national patterns appear in the figure. In the two major downwind cities, Spokane and Chelyabinsk, there was a striking similarity in trust ascribed to environmental groups and most other political actors. Residents of these two cities considered environmental groups and the media (TV/radio and newspapers) as the most trustworthy sources of information about the facilities. For instance, 54 percent of Spokane residents trusted environmental groups, as did 57 percent of Chelyabinsk residents. At the other end of the continuum, officials from the facilities and the government agencies responsible for both facilities (U.S. Department of Energy and the Russian Ministry of Atomic Power) were given little credibility by the residents of both cities; both downwinder publics were also generally skeptical of local government. The greatest discrepancy between these downwind cities was in their evaluation of certain national and international agencies. Spokane residents were relatively trustful of the EPA, while Chelyabinsk residents were distinctly less trustful of the Russian Ministry of the Environment. Conversely, people in the city of Chelyabinsk had more confidence in the International Atomic Energy Agency (IAEA) than did Spokane residents.

Table 9.1
Trust in various actors as a source of information about the facilities

Status	Richland	Chelyabinsk-70	Spokane	Chelyabinsk	Yakama Reservation	Kyshtym/ Muslyumovo	Wenatchee	Chebarkul
TV/radio	38	43	56	66	56	60	43	57
Environmental group	32	44	54	57	55	37	41	59
Newspapers	36	25	52	52	52	59	38	44
International Atomic Energy Agency	46	42	27	44	33	30	24	30
Facility scientists	87	45	38	33	48	29	39	43
EPA/Ministry of the Environment	62	31	47	35	53	33	41	41
Local government	50	24	30	20	31	29	28	30
State legislature/ government deputies	31	35	26	30	45	17	34	22
DOE/Minatom	52	25	26	20	42	20	34	23
Facility officials	66	30	21	19	34	15	22	22

Note: Table entries are the percentage that "completely" or "mostly" trust each actor as a source of information about the respective facility. Missing data were included in the calculation of percentages.

A markedly different pattern occurs when we compare the results from the two areas where plant employees are concentrated: the host communities of Richland and Chelyabinsk-70. Richland residents were highly trustful of the scientists from Hanford, officials at Hanford, and the Environmental Protection Agency; they were openly skeptical about environmentalists and the media.

These statistics are evidence of what is widely described as the "Richland family" phenomenon—a pervasive sense of confidence in the facilities and in the government's nuclear weapons program. Such sentiments were typified by the following comments registered by Richland residents on their returned surveys. A fifty-two-year-old Hanford employee wrote:

> I think we have too many "bleeding hearts." If the protestors would get off their backsides and "get a life," then our whole country would be better off.

A forty-five-year-old Hanford technician echoed these same feelings:

> I work at Hanford in the area of radioactive waste remediation [and I] do know that most of the information about Hanford that comes from environmental pressure groups, politicians, and the media is fertilizer. The problems at Hanford are mostly caused by people who will not let us do our job, because if we did there would be no problem and their source of income would dry up. These are mostly government consultants, DOE (Washington, D.C.), and environmental pressure groups (HEAL, Heart of America, Sierra Club, etc.). We can do our job safely and cost-effectively if we are not micromanaged. The storage vaults at Hanford, with very minor exceptions, pose no problem to the public or the environment.

Many residents of Richland viewed themselves as the true environmentalists, trying to clean up the accumulated by-products of Hanford's years of plutonium production—environmental interest groups were seen as grandstanding and trying to politicize the cleanup process.

Many residents of Chelyabinsk-70 also were scientists and technicians whose careers were closely interwoven with the nuclear operations at Chelyabinsk-65. Even so, residents of Chelyabinsk-70 displayed considerable ambivalence about the representatives of those facilities. They trusted the scientists at Chelyabinsk-65, but they were quite skeptical of plant officials and representatives of the Ministry of Ecology and the Ministry of Atomic Power (Minatom). Perhaps the most striking evidence of these doubts was the equal levels of trust they placed in environmentalists and Chelyabinsk-65 scientists!

The comparisons between the downwind communities on the Yakama Reservation and the communities of Kyshtym and Muslyumovo were even more complex. Trust in environmental groups was relatively high in the American downwinder and indigenous peoples communities—possibly reflecting a view that environmental groups were their advocates, representing them as victims of environmental degradation. Residents from the three towns on the reservation were also relatively trustful of the media, but more skeptical of local government and the Department of Energy.

In the Chelyabinsk region, trust in environmental groups as an information source appeared low in the downwind indigenous communities. Again, however, the combined Kyshtym/Muslyumovo statistic blends two divergent patterns; only 22 percent of Kyshtym residents said they trust environmental groups. Kyshtym's links to Mayak prompted some feelings similar to the "Richland family" and Hanford. For instance, a deputy of the Kyshtym local assembly was quite critical of environmental groups: "At the beginning, the parents of sick children and those who suffered themselves participated in the ecological movement. As for now, only the obsessed participate, those with the reputation for schizophrenia." In Muslyumovo, however, trust in environmental groups is much higher (55 percent).[5] The media were a relatively trusted information source by residents of both communities, just as they were skeptical about Chelyabinsk-65 officials and the Ministry of Atomic Power.

A third question in the survey tapped perceptions of political effectiveness of environmental groups. Ned Muller and Karl-Dieter Opp (1986) have shown that this is an important dimension of social movement support, with potentially strong links to the individual's willingness to engage in political action. The distribution of those perceptions is shown in figure 9.2, and there are striking cross-national differences on this question.[6] More American respondents than Russians saw environmental groups as being politically effective. Even in Richland, for instance, 66 percent of the public believed that environmental groups were at least fairly effective. The Richland statistic was higher than for the American public as a whole; a contemporaneous national survey found that 38 percent of Americans said environmental groups are politically effective

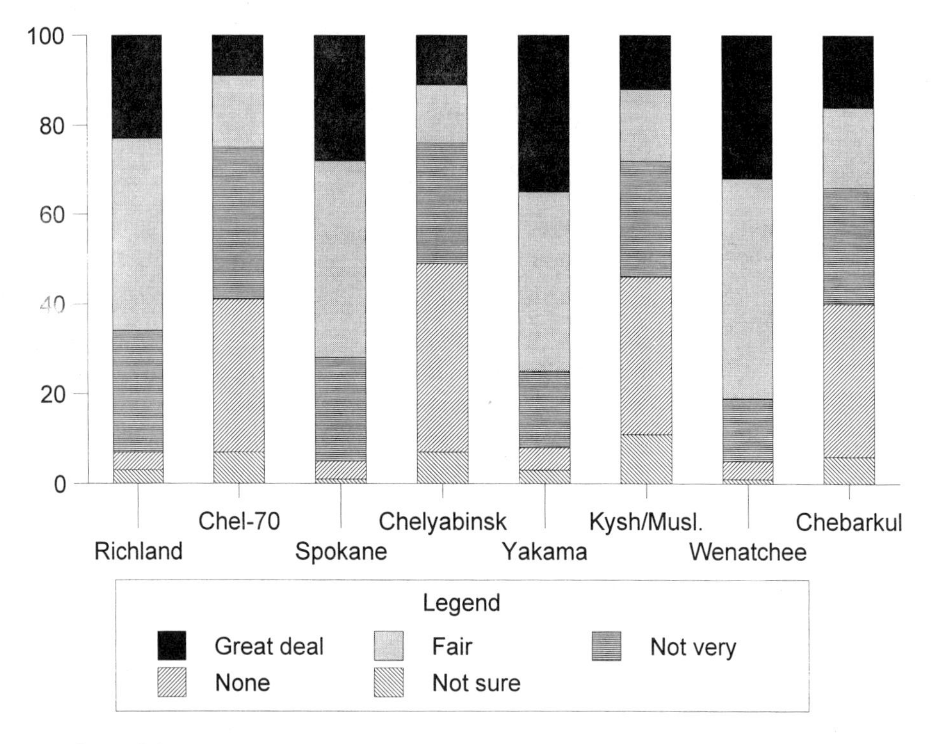

Figure 9.2
Perceived effectiveness of environmental groups

(Dunlap, Gallup, and Gallup 1993). This provides potent evidence that the effect of these groups has been felt within the facilities.

In each pair of communities, Russian residents were much less optimistic than Americans about the ability of environmental groups to influence policy outcomes. This holds even though the Nuclear Safety Movement had been relatively effective over the previous several years in stimulating debate about Mayak and blocking the construction of the Southern Urals Power Station at Chelyabinsk-65. In the city of Chelyabinsk, the political base of the environmental movement, only 24 percent of those surveyed thought environmental groups were fairly effective. Furthermore, when compared with the Russian national population, residents of the Chelyabinsk region were even more pessimistic than the Russian national population about the effectiveness of environmental groups. Across the Chelyabinsk region, roughly one-quarter of our respondents said environmental groups were at least fairly effective, compared to almost half in a recent Russian national survey (Dunlap, Gallup, and Gallup 1993).

Two factors might explain the difference between our Chelyabinsk sample and the Russian national findings. As mentioned earlier, environmental groups were fairly weak in the Chelyabinsk region at the time of our survey, having declined from an earlier peak in the late 1980s. Without an effective organization and strong membership base, residents of the region were unlikely to see these groups as being able to act effectively against the government. The second factor concerns the regional government itself. Because of the long-standing security concerns and influence of the military industrial complex in Chelyabinsk-65, the regional government was more conservative than the Russian national government (chapter 11). Weak environmental groups, combined with a strong governmental body, can explain much of the pessimism found in Chelyabinsk.

The final measure of support for environmental groups is behavioral. Abstract approval or trust of environmental groups was relatively easy to express in a public opinion survey. A more demanding, and possibly more accurate, measure of support entails a willingness to translate positive feelings into real behavior. The survey asked whether respondents were members of a group protesting the activities at Hanford or Mayak, or were willing to join such a group.[7] As we expected from William Schreckhise's (chapter 5) and Paula Garb's (chapter 6) descriptions of the membership base of environmental groups in both regions, only a minority of those surveyed reported that they belong to a group—roughly 2 percent of the weighted Hanford and Mayak samples (table 9.2). Membership rates in the Hanford area were lower than national membership statistics in the United States, although national statistics also include membership in a wide variety of green groups.[8] Russian membership rates were consistent with the results of national surveys, which implies that the type of focused local activism occurring in Chelyabinsk paralleled the broad boundaries of the current Russian environmental movement. This is consistent with our expectation described in chapter 1 that Russian mobilization would be more grievance based, while the American movement would include ideologically mobilized participants.

Table 9.2
Membership in environmental groups protesting the facilities by city

Membership status	Richland	Chelyabinsk-70	Spokane	Chelyabinsk	Yakama reservation	Kyshtym/ Muslyumovo	Wenatchee	Chebarkul
Respondent member	2.1	2.2	2.0	1.4	4.1	3.4	1.2	1.0
Member in family	1.7	.7	1.0	1.4	4.1	2.0	.6	1.0
Willing to be member	7.0	30.2	11.7	42.8	15.2	47.3	11.0	44.7
Not willing	66.8	40.3	50.5	32.2	38.6	18.5	49.4	17.5
Not sure	22.4	26.6	34.9	22.2	37.9	28.8	37.8	35.9
Total	100%	100%	100%	100%	100%	100%	100%	100%
(N)	(527)	(139)	(505)	(369)	(145)	(205)	(172)	(103)

Note: Column totals may not equal 100% because of rounding. When used in subsequent analyses as a dependent variable, membership is recoded into a three-category index; respondent membership, family membership, and willingness to be a member were combined into a single category. In the Russian survey this question was only asked of respondents who had heard of the groups protesting Mayak.

Variations across the pairings of cities were relatively similar. In both countries, the lowest reported group membership was in the "control" cities of Chebarkul and Wenatchee because, among other reasons, no active local environmental movement existed there in which they could participate. Likewise, in both countries, the highest reported membership was in the several indigenous/minority locales. In the downwind village of Muslyumovo, 6 percent of those who responded to this question said they or a member of their household belong to an environmental group. The facility host cities (Richland and Chelyabinsk-70) and the downwinder cities (Spokane and Chelyabinsk) fell between the other two types.

The major divergence between the two countries surfaces in the next two categories—those willing to be members and those unwilling to become members. In all pairings of Russian and American communities, the Russian respondents were more likely to express a willingness to join an environmental group than were the Americans. In this instance, then, one would conclude that if the costs of action were otherwise relatively equal, the Russian groups would have a greater potential for acting as agents of political mobilization than would their American counterparts.

Obviously, the mobilization process is more complex than simply assessing the expressed willingness of citizens to join a protest group (see chapter 1). Mobilization for public interest objectives most likely depends on several factors: the capacity and freedom of an individual to expend the resources to become active, the availability of appropriate groups to join, the distribution of group incentives that cement members' allegiances and provide the foundation for making claims upon their loyalty, the institutional context for political action, and the capacity of the group itself to make a difference in the political outcomes (McFarland 1984, 195–205). The latter effectiveness depends both on the character of the group and its leaders, and on the degree to which the system itself admits to influence via the interest group channel of interest articulation (Kitschelt 1986; Knoke 1990).

Recall from Figure 9.2 that Americans were much more likely than Russians to see such groups as being effective. Thus, despite the willingness to join and participate, there obviously existed a powerful disincentive in Russia relating to the perceived ineffectiveness of these groups. Again, although this perception of political effectiveness was only one of several potentially salient incentives to potential members of these groups, it may be worthwhile for Russian groups to focus more intently on selective material incentives than on the proposition that the group itself can have a significant effect on current policy outcomes.

Predicting Support for Environmental Groups

The literature on the mobilization of citizen participation in protest movements offers several explanations of why citizens might support environmental action (see chapter 1; Rohrschneider 1990). Three of these models are most relevant for our project. Most basic is the social location model, which argues that support for protest movements is centered among certain demographic groups, such as the young or certain class groups. The victimization (or grievance) model, in contrast, explains public support for environmentalism by individuals' direct and personal exposure to environmental problems. In other words, those who have direct experience with local pollution are more likely than citizens lacking this experience to support environmental groups. The values model reflects the assumption that citizens support environmental groups largely because of their personal values and beliefs, which may exist independent of the specific wrongs linked to the nuclear facilities. We will examine each of these models separately in the sections that follow, and then combine them into a multivariate analysis.

The Social Location Model

Research on Western publics has shown that support for environmental groups is substantially greater among certain demographic groups than others. For example, education and income are potential predictors of environmental support in two ways. First, prior research has shown that individuals in a shared position in the social structure also may share a political interest or perceived stake in environmental policy (Olson and Landsberg 1973). That shared perspective may provide the foundation for the political mobilization required to influence policy regarding that interest. In fact, the better-educated and more affluent are more concerned with noneconomic quality-of-life issues, such as the state of the environment (Cotgrove and Duff 1980; Milbrath 1984; Dalton 1994).

Second, social structural variables may reflect differing capacities and potential for mobilization. Political mobilization requires individual and collective resources in order to organize and express demands on the political system. The better-educated and more affluent are more likely to possess the personal resources and political skills that enable them to become involved in political groups, such as those associated with the environmental movement. Another potential correlate of environmental support is age. Younger-age groups are more interested in environmental issues and are more active in the movement. These results are common in research in the West, and analysts often describe similar patterns for Russian environmentalism (DeBardeleben 1992).

We correlated these three social characteristics with the two most discriminating measures of support for green groups: approval of environmental groups in general (from figure 9.1) and membership in groups protesting the facilities (table 9.2). These analyses are based on the weighted American and Russian samples that combine respondents from the various sampling sites; this allows us to compare the overall differences across nations.[9]

Table 9.3 displays the correlations between the social structural variables and the measures of environmental group support. In the Hanford study, social status—measured by education and income—had only a modest influence on either approval of environmental groups or membership in an environmental group opposing Hanford. This attenuated relationship might partially arise from the peculiar demographics of the

Table 9.3
Social location model of environmental group support

	Hanford		Chelyabinsk	
	Approval	Membership	Approval	Membership
Education	−.01	.10*	.02	−.03
Income	.12*	.04	.04	−.01
Age	.23*	.15*	−.02	.01

Note: The measures of group support are approval of environmental groups in general (figure 9.1) and membership in environmental groups (table 9.1). Table entries are Pearson r correlations; coefficients significant at .05 level are denoted by an asterisk. Results are based on the weighted samples in both nations.

communities we are studying. Richland, where support for environmental groups was lowest, was an area with an unusually high proportion of better-educated and affluent respondents—many of whom have links to the Hanford facilities. Conversely, there was considerable support for environmental groups in the Yakama Indian Reservation communities, whose residents tended to occupy lower social status levels.

Social status also appeared to have little impact on Russian attitudes toward environmental groups. Again, however, this may reflect the unusual nature of the communities we are studying. Chelyabinsk-70 had a high proportion of upper social status respondents because it was a scientific research center; these residents were generally more skeptical about environmentalists and less likely to join an environmental group than other citizens. Conversely, the least-affluent and least-educated sample was found in Muslyumovo, where support for the movement is the highest.

Age was significantly correlated with approval of and membership in environmental groups among the Hanford area residents. Furthermore, the same basic age gradient occurred in each of the American communities we surveyed. Surprisingly, significant age differences did not appear in support for environmental groups among the Chelyabinsk area residents. As other observers have noticed, the environmental protest movements in Russia were often led and staffed by the middle-aged, which tends to flatten the age recruitment pattern of environmental groups.

In summary, the environmental movement protesting Mayak lacked the type of youthful, middle class constituency that modestly defined the core of the environmental movement pushing for change at Hanford (and the type of constituency generally found for Western environmental groups). This evidence provides a first sign that the mobilization appeal of the anti-Mayak groups followed a different pattern than generally observed in the West. In social terms, at least, the Russian movement had a more diffuse constituency. The consensus about the poor state of the environment in the Chelyabinsk region is so pervasive (chapter 4), that environmental concerns may cross all social divides. Victimization-based grievance may disguise, and perhaps make irrelevant, potential social/structural sources of environmental mobilization. This problem is compounded by the rather peculiar demographic makeup of these regions in relationship to environmental issues.

The Victimization Model

Research on Western environmental groups has featured a debate regarding the putative rational basis of environmental action. In his *Logic of Collective Action,* Mancur Olson (1965) provided a framework for judging "rational" political action. Olson maintained that in large group settings, rational individuals only undertake activities from which they derive selective benefits (available only to participants) that exceed the costs of participating. If the benefits resulting from action are insufficient to counterbalance the cost, then it is rational for individuals to refrain from becoming active. In many cases, if not most, Olson would maintain that the costs of action exceed the benefits, hence most people will remain inactive. Furthermore, the rationale for inaction is exceptionally strong when the benefits resulting from action are shared goods (both participants and nonparticipants benefit), as is the case for most environmental policies. In these instances, most people should be "free-riders"; that is, people can be expected to leave political action to others to avoid the costs of involvement, while reaping the fruits of the others' actions when those collective benefits become available.

Environmental action often seems to violate the tenets of Olson's rational-actor model. Environmental groups generally provide collective benefits (such as clean air or less pollution) rather than the selective benefits of other interest groups (such as the employee benefits obtained by unions or the agricultural subsidies obtained by farmer associations). The rational individual, then, might presumably act as a free-rider, avoiding the costs of action, while enjoying the environmental benefits reaped by those who become active. Because environmental action often yields only incremental benefits that are widely shared, the personal cost of action relative to the direct benefits received might easily dissuade a rational actor from participating. As a group, scholars studying environmental politics have been rather critical of Olson's logic of collective action because there often has been politically significant environmental action in circumstances where Olson's logic would predict inaction (Mitchell 1979; Walsh 1988; Melucci 1980).[10]

However, the scale of environmental problems at Hanford and Mayak are so great that a rational self-interest approach may be appropriate. The mobilization of activists is normally greater in communities where ecological damage has occurred or is threatened. This can be as dramatic as major nuclear accidents (Walsh 1988; Walsh and Warland 1983), protests against pollution in Japan (Lewis 1980; McKean 1981), or protests against specific projects that threaten local residents or their quality of life (the NIMBY syndrome). In Chelyabinsk especially, the scope of environmental damage is extreme, dwarfing the scale of typical environmental problems. Because of the extensive radiological damage and the long-term failure of the Soviet government to inform and treat those exposed to this hazard, environmental activism in the Chelyabinsk area may be explained at least in part by the "victimization" model. Chelyabinsk is, after all, widely known for being the most polluted place on earth—the benefits of environmental action thus can be equally great. Although the environmental damage at Hanford is relatively less severe than at Mayak, the radiation releases and environmental contamination are still extreme by American standards. By all accounts, of all the nuclear weapons product sites in the nation, Hanford is the most polluted and largest-scale cleanup problem for the U.S. Department of Energy. Thus self-interest born of victimization may be an important explanation for environmental action against both facilities.

Our survey used four distinct indexes to tap respondents' feelings of victimization. The first was a measure of perceptions of national or global environmental conditions. This measure reflects the extent to which respondents viewed national and global environmental problems as very important. We asked each respondent to rate the quality of the environment in the nation and in the world as a whole.[11] We combined negative ratings ("fairly bad" or "very bad") into a three-category index, such that a higher score means greater concern about the environment (see chapter 4 for an analysis of these items). Consequently, a positive correlation indicates that a concern about the environment is associated with greater support for environmental interest groups.

The second measure tapped perceptions of local environmental conditions. We asked respondents to rate the seriousness of ten specific local environmental problems (see appendix B for index construction). Again, a positive correlation means that the greater the perceived problems, the greater the support for environmental interest groups.

The third index focused explicitly on the facilities and their perceived environmental impact. We asked about respondents' perceptions of past environmental problems connected with the facilities. The Hanford and Mayak questionnaires inquired about four separate problems: off-site leakage of radioactive gases; the leakage of liquid wastes; problems with the long-term storage of nuclear wastes, and past health or safety problems for the respondent's family. To create an index we counted the number of items the respondent felt constituted "major problems."

Not all environmental concerns are based on actual problems; often the concern is with the potential for environmental or health damage. Thus the Russian questionnaire contained an additional battery that asked whether residents of the region were worried about potential environmental problems at Mayak in the future. We asked respondents about five separate possibilities: their personal safety or that of their family, leakage of radioactive gases, leakage of radioactive liquids, the storage of nuclear wastes, and the likelihood that future generations will be harmed by radioactivity from Chelyabinsk-65.[12]

The correlations of these indexes with support for environmental groups are shown in Table 9.4. Support for environmental interest groups

Table 9.4
Victimization model of environmental group support

	Hanford		Chelyabinsk	
	Approval	Membership	Approval	Membership
National/global concern	.28*	.14*	.03	−.02
Local concern	.24*	.16*	.08*	.04
Problems of nuclear facility	.32*	.28*	.13*	.16*
Risk of future problems	—	—	.15*	.13*

Note: The measures of group support are approval of environmental groups in general (figure 9.1) and membership in environmental groups (table 9.2). Table entries are Pearson r correlations; coefficients significant at .05 level are denoted by an asterisk. Results are based on the weighted samples in both nations. The risk variable was not included in the U.S. survey.

is more strongly based on grievance-related environmental concern among Americans than among Russians. Broad environmental concerns (i.e., the nation and the world) and local environmental concerns are more linked to support for environmental groups and potential membership among the Hanford survey respondents. Surprisingly, neither national nor local environmental concerns had much influence on support for the environmental movement in Russia. Environmental concerns were more narrowly focused in the Russian survey. Only the perceived problems of Mayak and the risk of future problems at the facility were significantly related to support for environmental groups. For instance, perceptions of past problems at Chelyabinsk-65 showed a moderate relationship with both approval (r = .13) and membership (r = .16). Expressed in percentage terms, among those who mention three or more problems at Chelyabinsk-65, 70 percent completely approved of environmental groups; among those who report no problems at the complex, only 39 percent completely approved. Even here, however, the strength of these relationships is much greater among the American respondents.

It is still too early to conclude that these findings reflect a victim orientation among Americans. The weak correlations in the Russian sample may occur simply because the local conditions were so bad that there was not enough variation to affect support for green groups. Alternatively, since environmental interest groups were seen as less effective in the Russian setting, this belief may lead residents of the Chelyabinsk region to consider these groups as being largely irrelevant to environmental remediation.[13] In addition, chapter 7 showed that Russian attitudes were less developed on environmental questions, which may attenuate all correlations in the Russian survey. Finally, the measures of perceived problems at each facility were the overall strongest predictor in both nations. This suggests that in both countries a sense of being a potential and/or past victim may be an important motivator to mobilization.

The Values Model

A theoretical alternative to the victimization model argues that support for environmental action is based in major part on the values and ideology of the public. This approach presumes that certain value systems are associated with a high level of support for environmental groups based on "principles" rather than on narrow self-interest or perceptions of immediate environmental conditions. Some students of environmental politics in the West, for example, have linked environmentalism to the rise of "new" political values and a new type of political thinking among mass publics (Inglehart 1990; Milbrath 1984). Similar arguments are found in writings about Russian environmentalism (Yanitsky 1993; DeBardeleben 1992).

One of the core concepts in the environmental literature is the idea of a New Environmental Paradigm (NEP). Developed and applied in diverse settings by Dunlap and his colleagues (Dunlap and Van Liere 1978; Catton and Dunlap 1980; Milbrath 1984), the NEP presumes that public values toward society and the environment are changing in advanced industrial societies. Adherents of the NEP see humans and nature as participants in an integrated biosystem, view humans and nature as having coequal rights, and believe that the earth exists as a spaceship with a finite supply of natural resources.

Chapter 7 compared respondents in our two surveys according to their adherence to the elements of the New Environmental Paradigm. We identified three separate dimensions of environmental attitudes among the Russian respondents. The first, "biocentric" dimension reflected a high intrinsic value being placed on nature in relationship to the needs of humans (e.g., "plants and animals exist primarily for human use"). The second, "valuation of nature" dimension represented a belief that humankind and nature should live in harmony. The third "collective values" dimension reflected a concern with collective rather than individual rights in the relationship of humans and nature. In the American sample, all three dimensions combined into a single measure of adherence to the New Environmental Paradigm. In the Russian sample, in contrast, each dimension represented a separate and distinct element of environmental thinking.

Another measure of political values is Ronald Inglehart's (1990) index of postmaterialism. Postmaterial values reflect the individual's freedom to focus on higher-level needs, such as an aesthetic appreciation of the environment, compared to materialist values reflecting more basic needs such as security and safety. Research on Western environmental groups routinely finds that support for the environment generally is higher among postmaterialists than among materialists (Inglehart 1990, ch. 8; Dalton 1994, ch. 3). Similarly, several Russian specialists have discussed postmaterial orientations in the context of Russian environmentalism (Yanitksy 1993; DeBardeleben 1992; Doktorov, Firsov, and Safronov 1993).

Another relevant political value involves a broader sense of ideological orientation. A considerable amount of literature argues that political location in left/right terms is a good predictor of environmental group support (Inglehart 1990; Dunlap and Mertig 1992). Western analysts often view environmentalism as a liberal position. While this is by no means an absolute, there is a clear pattern of leftist support for environmentalism. Inglehart, for example, showed that those who identified themselves as leftists were about twice as likely as rightists to voice approval of environmental groups or become a member of a group (1990, 54). In the Hanford survey, we included a question that asked respondents to position themselves along a liberal to conservative continuum.[14]

Similarly, Russian scholars routinely place environmental groups at the forefront of the political reform movement (DeBardeleben 1992; Yanitsky 1993). However, the value and meaning of concepts such as "left" and "right" were more tenuous in the context of Russian politics, especially in the early 1990s when our surveys were conducted. These ideological terms were confounded by their association with conflicting symbols from the Soviet and post-Soviet eras, and thus did not share a common use or meaning among the Russian populace. To tap comparable attitudes to the liberal/conservative orientations of Americans, we asked our Russian respondents to position themselves on an eleven-point scale, with "reformers" at one end and "conservatives" at the other. Because the reformers/conservatives terms were still somewhat ambiguous, we also assessed opinions on a variety of items that tap elements of liberal, reformist thought in more precise terms—in particular, attitudes toward a private economy and antimilitarism.[15] In addition, because environmental groups were among the vanguard of forces pushing for democratization in the former Soviet Union, we expected that supporters of environmental groups would share democratic reform values.[16]

Table 9.5 presents the correlations between these various measures of political values and support for environmental groups. Several clear

Table 9.5
Ideological model of environmental group support

	Hanford		Chelyabinsk	
	Approval	Membership	Approval	Membership
New environmental paradigm				
Biocentric attitudes	.41*	.27*	.02	−.04
Value nature	.49*	.25*	.12*	.02
Collectivist values	.29*	.13*	.06*	.06
Summary NEP	.53*	.32*	—	—
Postmaterial values	.24*	.22*	.02	.06
Reformist attitudes				
Left/right	.34*	.24	—	—
Reformer/conservative	—	—	.07	.10*
Private economy	—	—	.00	.01
Antimilitarism	.27*	.18*	.05	.08*
Democratic attitudes	—	—	−.11*	−.14*

Note: The measures of group support are approval of environmental groups in general (figure 9.1) and membership in environmental groups (table 9.2). Table entries are Pearson r correlations; coefficients significant at .05 level are denoted by an asterisk. Results are based on the weighted samples in both nations. The variables marked with dashes were not included in the survey.

patterns emerge from the table. Overall, the values and ideology measures were much more highly correlated with environmental group support in the Hanford area than in the Chelyabinsk region. Each of the subdimensions of the New Environmental Paradigm was strongly related to both approval of and membership in environmental groups among Americans, and the summary NEP measure had a dramatically high correlation ($r = .53$) with approval of environmental groups. Postmaterialism, too, was strongly related to support for environmental groups in the American sample. One might claim that these relationships approach a tautology: support for NEP environmental values correlates with support for environmental groups. Yet this "obvious" relationship did not appear within our Russian sample. The correlations between the NEP indexes and group support ranged from no relationship to very weak ones. Even biocentric attitudes, the dimension most closely linked to the NEP in the Russia sample (chapter 7), displayed negligible correlations with environmental group support.

Similar cross-national differences appeared for the relationship between reformist attitudes and group support. Reformist or liberal political positions were strongly linked to environmental group support in the American setting. The correlation between liberal/conservative position and approval of environmental groups is .24. In contrast, the correlation between reformer/conservative orientations and group support was considerably weaker among Russians (r = .10). The other reformist measures—privatization of the economy, antimilitarism, and democratic orientations—displayed equally weak correlations. Indeed, the strongest and most consistent relationship between reformist attitudes and group support in Chelyabinsk is between democratic attitudes and group support. However, this relationship runs in the wrong direction; those who were least democratic in orientation were more supportive of the environmental movement.[17]

In summary, there are several notable patterns in these data. We found a clear contrast between the citizens near Hanford and the citizens near Chelyabinsk. American orientations toward environmental groups were strongly buttressed by the citizens' adherence to the New Environmental Paradigm, their postmaterial values, and their liberal political identification. In other words, the critics of Hanford and the supporters of the environmental movement were at least partially mobilized by a general set of political values that transcend the specific environmental problems and controversies of Hanford. More than a movement of self-interest, environmentalism at Hanford in good measure also reflected the changing values and culture of Americans.

These same values played a surprisingly weak role in predicting Russian support for environmental groups. This may be symptomatic of the general ambiguity of the political and ideological underpinnings of Russian environmentalism at the time of our survey. Russian nationalism has historically combined environmentalism with notions of "Greater Russia." There is a deep connection with the land in many Slavic writings, most notably those of Dostoyevsky, such as is expressed in *The Brothers Karamozov*. In the Chelyabinsk region, the Association of Greens represented this right-wing environmentalism. At the same time, groups like the Nuclear Safety Movement might have attracted support from reformist elements of the population. As both these messages existed in the same political arena, it means that the character and ideology of Russian environmentalism has not been shaped yet. In a sense, environmentalism resonates differently with contemporary political cultures in the two countries (see chapter 1). Endorsing this interpretation, chapter 7 showed that while there was considerable support for elements of the NEP among Russians, these opinions were not unified into a cohesive belief structure to the extent that they are among Americans. Thus it is premature to suggest that there exists an integrated value or ideological structure underlying Russian environmental mobilization in the Chelyabinsk region.

A Multivariate Analysis

Until this point, we have looked at separate theories to explain support for environmental groups, and potential membership in such groups. In actuality, of course, the decision to support or become active in the environmental movement is a multistaged process in which individuals consider numerous factors. Some people are motivated by a feeling of victimization, others are motivated by adherence to alternative political values, and still others are empowered by their social location. In most instances, however, a mix of influences likely comes into play. Moreover, it is probable that there are other factors entering into this equation. The crucial task is to determine the overall configuration of this decision-making process, and to estimate the relative weight of the elements of the process.

We see this as a multistep process. The decision to become active in a group protesting Hanford or Mayak is most immediately dependent on the citizen's image of the environmental movement. Approval of the movement or specific groups is obviously an important aspect of this image; people are unlikely to become active in a movement that they do not support. In addition, we have stressed that considerations of effectiveness can play an equally important part. Especially in the Russian context, citizens commonly approve of environmental groups but they tend to question whether these groups could be effective in the quasi-democratic environment of contemporary Russian politics.

Further back in the causal chain, we have argued that victimization and value orientation both can come into play in determining the attitudinal basis of environmental action. We have found that both factors seemed important to residents around Hanford, while neither carried great weight in the Chelyabinsk region. We need to determine the relative weight of each explanation when considered jointly, and in comparison to other potential influences.

Finally, social location provides an empirical starting point for determining attitudes toward the facilities and determining the likelihood of political action. We would include social status and age as social-location variables, as well as the contextual influence of residing in one of the communities in our study. Although one's residence in Muslyumovo or Toppenish, for example, might provide the basis for an individual's relationship to the facilities and define the context in which opinions about the facilities are formed, our causal framework suggests that the effects of social location will largely be mediated by intervening attitudes and opinions.

The final stage of our analysis thus assesses the relative effects of the various causal sources of mobilization that we discussed earlier in this chapter. The dependent variable is membership or potential membership in a group protesting the facilities. The potential predictors were grouped into four categories: image of environmental groups (approval and effectiveness of environmental groups); self-interest or grievance (the number of perceived local environmental problems and the number of perceived problems associated with the facilities); values (the New Environmental paradigm, left/right attitudes, and postmaterial values); and social location (educational level, age, and community of residence). We entered these predictors into separate analyses for the Hanford survey and the Mayak survey.

Table 9.6 displays the results of these two regression analyses. In the American survey, all four sets of variables exerted some direct influence in mobilizing participation in groups opposing the Hanford facilities. As expected, general approval of environmental groups had a strong direct effect (ß = .14). In addition, concerns about the past environmental problems of the facilities are directly related to group support (ß = .10). All three value-based measures also had significant effects. Finally, social conditions—education, age, and residence on the Yakama Indian Reservation—exerted a direct influence on group support, even when controlling for the attitudinal measures in the analysis. The analyses also show

Table 9.6
Multivariate model of membership in environmental groups

	Hanford	Chelyabinsk
Environmental group image		
Approve of environmental groups	.14*	.15*
Effectiveness of environmental groups	.02	.15*
Self-interest		
Local environmental problems	.04	−.08
Problems at facilities	.10*	.11*
Values/attitudes		
New environmental paradigm	.14*	.02
Left/right (Reformer/conservative)	.06*	.06
Postmaterial values	.08*	.04
Social conditions		
Educational level	.11*	.00
Age	−.07*	−.05
Richland/Chelyabinsk-70	.00	−.09
Spokane/Chelyabinsk	−.03	−.03
Yakama/Muslyumovo	.13*	.09
Kyshtym	—	.08
Multiple R	.42	.32

Note: The dependent variable is membership in environmental groups (table 9.1). Table entries are standardized regression coefficients; coefficients significant at .05 level are denoted by an asterisk. Results are based on the unweighted samples in both nations, using pairwise deletion of missing data.

that the sharp differences we previously observed between Richland and Spokane residents can be explained by their contrasting attitudinal profiles. Holding attitudes constant, residence in these communities per se did not significantly affect group membership.

The bases of environmental mobilization were more narrowly defined among the Russian sample. The image of environmental groups emerged as a strong predictor of (potential) membership in the groups protesting Mayak. Moreover, as we have noted in several points of our analysis, the perceived effectiveness of environmental groups was an important influence in the mobilization process in Russia (ß = .15). In simple percentage terms, 79 percent of those who believed that environmental groups could greatly affect policy were members or potential members of these groups, compared to only 53 percent among those who felt these groups had almost no effect. In comparison, perceptions of effectiveness had virtually no impact on American mobilization patterns, either in the multivariate analysis or in a simple bivariate analysis.[18] The importance of Russian environmentalists' image is perhaps understandable as Russia and the Chelyabinsk region were undergoing a process of democratization where autonomous citizen groups and direct democracy were unproven (and somewhat suspect) methods of action.

If we take a step back in the causal chain, Russians also placed greater weight on grievance and self-interest over values in deciding to join a group protesting Mayak's activities. Concerns about the environmental problems of the facilities exerted a direct impact (ß = .11) on environmental activity that went beyond the effects also represented by the image of environmental groups. In contrast, none of the values measured—the NEP, left/right attitudes, or postmaterial values—had a significant direct impact on environmental mobilization among residents in the Chelyabinsk area. Even the connections between values and both group approval and group effectiveness were relatively weak. Finally, the impact of social location on environmental action was initially quite weak (Table 9.3), and the multivariate analyses show that these effects are mediated by attitudinal measures. Although there remains a slight tendency for Kyshtym (.08) and Muslyumovo (.09) residents to be more active and Chelyabinsk-70 residents to be less active (−.09), these were not statistically significant differences.

Hanford, Mayak, and Environmental Action

This chapter has examined public perceptions of environmental groups in the Hanford and Mayak regions, and assessed the potential for these groups to mobilize citizen protest against both facilities. The analysis began by gauging the extent of public support for environmental interest groups, considering overall approval, assessment of group effectiveness, level of trust in environmental groups as a source of information, and willingness to join a group as indicators of public support.

There were several significant patterns in our data on the relative levels of support for environmental interest groups in the two nations. First, approval of these groups was greater in the Chelyabinsk region than in the region around Hanford. Within nations, approval was highest in the downwind cities of Spokane and Chelyabinsk, as well as in the towns of the Yakama reservation and the Russian downwind village of Muslyumovo. At the same time, Americans tended to see environmental groups as being more effective advocates than did Russians. Finally, Americans in the Hanford area were less willing than Russians in the Chelyabinsk area to join environmental interest groups. In summary, there was greater affective support for these groups among the Russians we surveyed than among their American opposites, but there was a more frequent perception that environmental groups were more effective in the United States than in Russia.

There are several factors that might account for the broad differences we uncovered between the Americans and Russians we surveyed. One answer involves the relative maturity of the democratic processes in the two countries. On the one hand, Americans' more extensive experience with interest group involvement in the policy process may have influenced their survey responses regardless of what individuals felt about environmental groups per se. On the other hand, the public in newly democratic Russia may have seen interest groups as an opportunity to mobilize and influence policy in the face of severe degradation, but in the early 1990s they lacked the experience to know whether such groups could be effective in the Russian political forum.

This chapter also attempted to model the bases of environmental support in both settings. We found that support for environmental groups among residents of the Chelyabinsk region represented relatively diffuse orientations; most relationships were weak relative to the American results. As we have noted repeatedly in these empirical analyses, Russian attitudes toward the environment were still in a process of formation in the early 1990s, so public opinions were still relatively unstructured and thus difficult to predict.

We examined three broad mobilization theories—social location, victimization, and values—to explain support for environmental groups. For residents of the region around Hanford, we found that all three of those theories had some support in the results. Social position was important in defining support for the movement, as well as the values of the citizenry and their perceptions of problems related to the Hanford facilities. Environmental mobilization against Hanford was thus motivated both by grievances (real or perceived) linked to the facilities and by a broader belief structure that stressed postmaterial values and a concern for nature.

The distinctive aspect of the Russian findings was the breadth of support for environmental groups during this early phase of environmental mobilization, and the limited applicability of our theoretical models in explaining support for the movement. The groups protesting against Mayak drew from a diffuse social base, so that movement support was not particularly concentrated among specific demographic or social sectors. Similarly, general political values were not significantly related to group support, even when the logical connection seemed obvious (as with the New Environmental Paradigm). Two factors seemed to condition group support among residents of the Chelyabinsk region. The first was a concern about the specific environmental problems of the facilities. Because the environmental damages from Mayak had been so great, a grievance or self-interest model of political action seemed justified there. Our data support this conclusion.

In addition, perceptions that environmental groups could influence policy were significantly related to the mobilization potential of the environmental movement in Chelyabinsk. This, too, is understandable. In a context of great political uncertainty during the transition to democracy, many Russians questioned the ability of environmental groups to bring about real policy change. Chapter 6's account of the movement's travails during this period justify these concerns. When Chelyabinsk residents believed that the groups could be effective, they were more likely to participate. Without policy potential, people were inactive even in the face of the horrendous environmental damage wrought by Mayak. Moreover, the political retrenchment that has occurred since the time of our fieldwork (see chapter 6) has likely further eroded the perceptions that environmental action can influence policy making.

What do these overall patterns say about the relative capacity of environmental interest groups to act as vehicles for the mobilization of citizen protest in the two countries? The answer to this question is rather complex. If one looked only at the levels of support for environmental interest groups, the potential for mobilization seemed greater in the Russian setting than in the American. The more favorable views of environmental groups and the greater willingness to join them were important forces, although a low assessment of group effectiveness acted to blunt the effects of such positive affect. It may be that what environmental interest groups needed in Chelyabinsk (and Russia generally) were more opportunities to demonstrate their effectiveness and to capitalize on their other affective attributes. Those officials in government who favored environmental causes could have helped greatly in this respect by providing opportunities for substantive participation by environmental group representatives in planning and program development.

The significantly greater ability to use statistical models to explain citizen attitudes toward environmental interest groups in the Hanford area suggests the potential for greater mobilization in this case. That is, the fact that citizen views of environmental groups were linked to attitudes, values, and social/structural characteristics provided a powerful tool to organize resource solicitation and mobilize individual involvement for environmental interest groups. Shared values and perceptions broaden the range of incentives that the groups can offer to potential supporters and active members. Moreover, many of those values constitute stable social characteristics, and thus provide a foundation for sustaining mobilization beyond an immediate crisis of victimization. The lack of such an ideological base among Russians is likely one factor that has contributed to the fluid and uncertain nature of the environmental movement in Chelyabinsk, and in Russia more generally.

At the same time, in both the Chelyabinsk and Hanford settings, perceptions of the degree of degradation and the potential for further damage to the environment, especially the potential risk to public health, were central influences on support for environmental interest groups. That potential presumably can be activated in times of crisis if citizen mobilization is desired.

In sum, there seem to be at least three primary elements to the potential for environmental mobilization: positive views of environmental groups; a shared set of resources, conditions, or values that can motivate individuals to affiliate with environmental groups; and the political context that legitimizes that mobilization. The two countries shared in the first two elements, although these commonalities are not necessarily manifest in the same form. If anything, the Russian environmentalists in the Chelyabinsk area had the advantage on the first, while the environmental groups at Hanford had the advantage on the second. The third element, though, clearly favored the American environmental movement, with a higher probability of green, group-based advocacy activity being politically tolerated and legally protected.

Even though Hanford was shrouded in secrecy for nearly fifty years, it remains an acceptable part of the American political culture for protest to occur and to be articulated via citizen groups. Moreover, even though environmental groups were not liked as much or trusted as fully in the American setting as in the Russian, they were viewed as being much more effective in their advocacy. The demonstrated record of interest group success—whether the result of their actions is preferred or not—makes clear to many citizens that if you want to get something done in politics, citizen action is a viable means of influence. The fact that environmental groups were seen as less effective in Chelyabinsk surely placed a damper on the political fires of citizen protest. And events since our survey have probably further eroded the public image of the environmental movement in Chelyabinsk.

As Russian citizens tried to find ways to make their lives better in the aftermath of the fall of the Soviet regime, the new and relatively untested democratic system for deciding the course of public policy will face many occasions where the short-term desires and actions of politicians may retard the development of citizen-based environmental groups. Yet the long-term health of the democratic process requires that such groups form and become effective in their advocacy. The constructive channeling of citizen concerns in circumstances of many tough choices is a major challenge facing the young Russian democracy in the coming years. The way in which environmental groups act, and are reacted to, could play a significant part in how the Russian democratic state develops.

Notes

1. The U.S. questionnaire asked the following question: "There are a number of groups that are active on environmental and nature protection issues in the United States today. Generally speaking, do you completely approve, mostly approve, mostly disapprove, or completely disapprove of these groups?"

2. See appendix A for additional information on each locale.

3. If the respondents had heard about the groups protesting against the facility, they were asked: "In general, can you tell me whether you approve or disapprove

of these groups overall?" Then, if approve/disapprove: "Is that completely or mostly?"

4. The U.S. question reads: "Please indicate how trustworthy you believe each of the following sources are in terms of providing honest and accurate information about Hanford." In addition to environmental interest groups, the citizens in both countries were asked about a set of other information sources.

5. A consistent finding from the Russia survey was the lack of support for environmental groups among Kyshtym residents. This town's name is synonymous with the infamous 1957 explosion at Mayak and stands downwind of the facilities and its discharges. Yet Kyshtym's residents expressed the lowest level of approval or trust of environmental groups of the five Russian sites studied. The most likely explanation for Kyshtym's deviance from the general pattern is the relationship between Mayak and the town. Mayak has enjoyed many benefits from the national government. Although relatively few of these benefits have trickled down to Kyshtym, the town's residents harbored the hope that the community would be incorporated into Chelyabinsk-65 in the future. Victor Fetisov, former director of the Mayak plant, was elected as a deputy of the Russian Federation in 1994 partially on the basis of that promise. In addition, while many in Kyshtym were concerned about the environmental consequences of Mayak, many former residents had accepted lucrative employment within one of the closed cities. Consequently, there were many family and personal ties linking Kyshtym residents and employees at Chelyabinsk-65 or Chelyabinsk-70. While residents of Kyshtym understandably resented the environmental damage that the plant had inflicted upon the town (see chapter 4), many also saw employment at the facilities as a source of upward mobility and financial security.

6. The U.S. question reads: "In your opinion, how much of an effect can individual citizens and citizen groups have on solving our environmental problems? A great deal, a fair amount, not very much, almost none at all, not sure."

7. Respondents were asked the following question: "Can you tell me whether you or someone else in your household is a member of one of these groups? (If not a member) Would you be willing to join one of these groups, or would you certainly not join?" In the Russian survey this item was only asked of those who said in a previous question that they had heard of these groups.

8. Approximately 2 percent of the weighted Hanford sample said they were members of an environmental group, compared with 11 percent in a comparable American national survey. By comparison, 2 percent in the weighted Chelyabinsk sample said they belonged to a group, compared with 3 percent nationally. See Dunlap, Gallup, and Gallup (1993) for the national statistics.

9. Another reason for combining samples was to maximize the variation on our measures. Our preliminary analyses show that perceptions of the environment and perceptions of environmental groups varied sharply across communities. We will, however, examine community differences in a multivariate model that follows these analyses.

10. Olson and other researchers (Muller and Opp 1986) have noted that there can be a variety of "rational" explanations for activism that can raise the expected

utility of participants above the threshold of inaction. Some of these explanations include entertainment value, social affiliation rewards, and the costs of nonaction. Thus many environmental groups, such as Greenpeace and the Sierra Club, provide direct selective incentives, such as magazines, discounts, "ecotourist" travel and merchandise, and bumper stickers to encourage membership and maintain affiliation.

11. See figure 4.1 in chapter 4 for more details on this measure.

12. The U.S. questionnaire did not include this battery. We counted the number of items that respondents rated a "very great risk." This index ranges from zero (no risks perceived) to five (each area poses a risk).

13. Separate analyses showed that of the four support measures, the perceived effectiveness of environmental groups seemed the least affected by concerns about the quality of the environment among Americans. Among Russians, however, the number of serious local environmental problems was linked to perceived effectiveness in three of the five communities (Kyshtym and Muslyumovo are the exception).

14. The question is: "In political matters, people talk of "liberal" and "conservative." How would you place yourself in terms of these political categories?" Respondents were given a string of numbers running from 1 to 10, with the following labels: 1) "liberal," 5) "middle of the road," and 10) "conservative."

15. In Russia, reformists favored privatization of the economy, highlighting the problems of making simple and direct translations of Western political terms to Russian politics.

16. The democracy measure includes six statements: 1) "there is too much democracy in Russia today"; 2) "a group that tolerates too many differences of opinion among its own members cannot exist for long"; 3) "it is better to live in an orderly society than to allow people so much freedom that they can become disruptive"; 4) "free speech is just not worth it if it means we have to put up with the danger of extremist political views"; 5) "society shouldn't have to put up with political views that are fundamentally different from the views of the majority"; and 6) "because demonstrations frequently become disorderly and disruptive, radical and extremist political groups shouldn't be allowed to demonstrate." The index is a count of the number of prodemocracy responses given on the six items.

17. In a more detailed analysis of the Russian survey, Gluck (1993) suggested that the findings for the democracy index may be a methodological artifact of index construction. An alternative measure of democratic attitudes showed a relationship in the expected direction, although the magnitude of the correlations remains very weak (.06 and .02).

18. Among American respondents who think environmental groups have a great deal of influence, 29 percent were members or potential members of a group; among those who say environmentalists have little influence, 14 percent were members or potential members.

V

The Government Response

10

The Governmental Response at Hanford

Nicholas P. Lovrich and John M. Whiteley

We must realize that when we broke the atom apart and released its energy we changed the history of the world. It is essential that the United States in this area of national strength and national vigor should be second to none—and on this river, in these reactors, by your effort, that great objective has been maintained.

—John F. Kennedy
Hanford, Washington
September 1963

American Exceptionalism: The Strong Democracy, Weak State Heritage

The story of the governmental response at Hanford is one of a major agency of U.S. government, initially empowered by broad-based social trust in its highly significant national security role, losing the public's trust and endeavoring to reestablish itself as a credible and efficacious agent of the people's business. This effort is set in the broader context of a society wherein the official agents of government are viewed with considerable suspicion, and where the overall level of public suspicion and mistrust of government has been rising since the late 1960s (Nye, Zelikow, and King 1997). In America, quite by design, public officials operate within the strict bounds of an adversarial legal system that offers aggrieved citizens, interest groups, and even other public agencies a broad array of political and legal avenues to challenge public authorities (Sandel 1996). A general inclination toward suspicion of government, the jealous protection of civil and property rights, and a strong devotion to federalism all inevitably came into play when the national government—a weak government confronting a strong democracy context—took on a task at Hanford that touched the lives of American citizens in a profound way.

To add further to the political drama at the Hanford nuclear complex, the presence of separate, aggrieved "sovereign" (treaty-based) tribal governments in the vicinity of Hanford contributes to the story of Hanford's effort to reestablish trust in the public. The capacity of parties who oppose Hanford's policies to question the activities of the national government is greatly enhanced by the presence of tribal governments, whose "trust relationship" with the federal government provides special grounds for requesting access to information that can be highly valuable to Hanford's critics (Pevar 1992).

A fiercely independent judiciary protective of private rights, a strong state (subnational) government authority, and significant national and state-level legislation directed toward the protection of the environment—administered by an activist Department of Ecology of the state of Washington and the U.S. Environmental Protection Agency—similarly are all noteworthy factors in comprehending the governmental response to post–Cold War developments at Hanford. This political environment facilitates the formation and growth of nongovernmental organizations (NGOs) and interest groups that use the relatively porous seams of American government to question putatively dubious governmental policies and activities that merit a broader public review (Rimmerman 1997; DeLeon 1997). The fiercely protected and longstanding tradition of a free and independent—even at times predatory—press provides further fuel for the fires of elite-challenging political action witnessed in the Hanford setting (Fallows 1996). Weak national political parties (vis-à-vis state, local, and candidate-based organizations) and strong local and substate political/governmental powers insure that geographically concentrated problems occasioned by national policy, such as was the case at Hanford, will beget significant attention in the national legislative process.

The governmental setting described here, featuring a hyperpluralistic collection of actors enjoying relatively easy access to established societal, governmental, jurisprudential, and mass-communication channels, reflects a uniquely American understanding of democracy (Lipset 1996;

Riggs 1998). The American exceptionalism which Seymour Martin Lipset discusses features a setting in which democratic values are strongly held within the public and are localized (e.g., elected school boards, many local elective offices, etc.). In contrast, national public institutions tend to be weak and subject to frequent challenge, and require extraordinary feats of leadership to accomplish major policy initiatives. Lipset's characterization of American political life provides a fitting backdrop for our assessment of the efforts of Department of Energy (DOE) officials at Hanford to reestablish the public trust they once enjoyed with the citizens they served.

In addition to these broader considerations of politics in America, the American West—long known for its fierce sense of independence and alienation from federal government authority (most recently evident in the Sagebrush Rebellion and County Supremacy movement)—provides even further inspiration for local actors in the Hanford area to take political and legal actions to challenge the motives and actions of officials in far-off Washington, D.C. (Alm 1993; Cawley 1993). The American West is a region of sparse population and vast tracts of federal lands, hence federal agency decisions and activities—particularly with respect to natural resource and environmental policies—are frequent targets of vocal and organized regional opposition. Such was the case in the prestatehood days of the Oregon Territory, and such is the case now in the contemporary Washington State setting for Hanford activities (Dietrich 1995). The U.S. Department of Energy "bureaucrats" and "beltway bandits" can be painted with the same satirical, condemning brush as were the "carpetbaggers" of yesteryear (Sheldon 1988). While not all social scientists would agree with the characterization of the political culture of the Pacific Northwest constructed from Lipset, Alm, Cawley, Dietrich, and Sheldon (see Smith 1993 for an alternative view), most probably would find this description of the political backdrop to events at Hanford of heuristic value in this analysis.

John M. Whiteley's account of the history of Hanford in chapter 2 documents accurately the "fall from grace" experienced by Hanford authorities as repeated revelations of hidden accidents, public deceit, and purposeful deception came to light during the past decade. Federal officials at Hanford initially were seen as preserving the global peace and assuring the nation's safety, and as such were accorded unquestioned loyalty. Furthermore, they made vast public investments in the regional economy, and the Tri-Cities area (Richland, Pasco, and Kennewick) became a prosperous community because of the Hanford operation. In time, however, a significant proportion of the region's people became painfully aware of having been the unsuspecting victims of both accidental and intentional exposures to some of the most harmful forms of environmental contamination known to mankind. Some area citizens disclaim many of the criticisms voiced against Hanford's DOE managers, and have remained loyal to the governmental authorities notwithstanding evidence of failures and official deception. This is particularly the case in the immediate vicinity of the plant populated by the families of the highly paid and well-educated scientists, engineers, and technicians for whom Hanford has represented a special type of "family." We saw compelling evidence of this strong sense of loyalty to Hanford in the surveys conducted in Richland.

Many other citizens of the region, however, now mistrust official versions of past and present activities at Hanford. Some of these aggrieved citizens have sought compensation via individual lawsuits or a class action downwinders' lawsuit for injury to health and/or property they believe resulted from the presence of the nuclear weapons complex. Others have placed the events in the broader context of American environmental politics, and have mobilized the considerable network of national and regional environmental groups into political and financial support for area-based critics of the facility. Bill Schreckhise's research (chapter 5) describes a number of the more prominent groups, and sets forth the character of their principal activities. Finally, a significant number of disaffected "insiders" have exited the Hanford "family" of faithful supporters to become quite troublesome, legally protected "whistleblowers." Hanford whistleblowers periodically pass on to critics and inquisitive state, tribal, and local governmental authorities knowledge of how their interests and the well-being of their constituents were placed in jeopardy by Hanford authorities.

Against this broad backdrop, what follows is an account of how Hanford authorities have sought to reestablish a measure of public trust in their stewardship over Hanford. Among the several elements of American exceptionalism of particular importance in the Hanford drama were the following: the vitality of American federalism; the legal framework for questioning federal administrative actions provided for in the Administrative Procedure Act of 1946; strong tribal government powers provided for in Indian law; protections for whistleblowers; increasingly exacting standards of public access to governmental records in the Freedom of Information Act; and the framework for public participation in environmental policy decisions established in the National Environmental Policy Act (all developments reflective of America's populist, individualist, and egalitarian values). Each of these "strong democracy/weak state" features of American political institutions, and several of the more modern-day progressive/populist governmental practices requiring public access to information and a concrete role in policy deliberations by stakeholder groups, are given special weight by an ever-present independent judiciary (Rohr 1986). In combination, these elements of the political context posed significant challenges to the federal authorities at Hanford whose "cover of secrecy" was ripped away from them at the close of the Cold War.

Having set this institutional and legal stage, we move on to ask: Who were the most significant actors pressuring Hanford to "come clean" on its history and change the way it does its business? What were the ultimate results of these efforts? Has the DOE really changed how it does its business at Hanford? How are the region's citizens reacting to the Hanford authorities' newly announced mission of site cleanup and environmental restoration? After reviewing the record of events, we should be able to answer this key question: Should the DOE be viewed as a truly reformed and newly "reinvented" agency that deserves broad public trust, or is it most properly viewed as having accomplished a master stroke of artful deceit and Machiavellian self-preservation through the crafty co-optation of its critics and the initiation of efforts at environmental stabilization and restoration? In coming to this judgment we will offer some lessons that might be drawn from this assessment of governmental responsiveness at Hanford that might be instructive to environmentalists and democratic reformers in Russia.

Legal and Institutional Environment for Hanford's DOE Managers

After the U.S. Department of Energy had taken over responsibility for the management of the nation's wartime-era nuclear weapons development program from the military and the Atomic Energy Commission, the DOE generally followed a pattern of operation that had arisen around the "top secret" wartime atomic bomb development effort. The agency created a highly centralized, intensely security-conscious organizational structure to manage the many research, production, and testing facilities that maintained the American postwar nuclear deterrent. DOE had a virtual exemption from the normal rules of public agency accountability (e.g., public budgets and published plans of work, provisions for intergovernmental coordination, publication of records available for public inspection, etc.) for its nuclear weapons development and production responsibilities.

DOE's ample budgets were accompanied by very strong pressures from the president and Congress to keep ahead of their nuclear rival, the USSR—both in terms of our basic nuclear science knowledge and in the steady enhancement of our stockpile of nuclear warheads. As these pressures born of the intense competition inherent in the nuclear arms race ebbed and flowed, the historical record clearly shows that Hanford officials on numerous occasions succumbed to the pressures of the moment and engaged in activities which, had they been subject to public scrutiny, these officials might never have undertaken.

The historical record also clearly shows that the Nuclear Regulatory Commission, in its careful and thorough effort to establish a record of safe operation to convince a cautious public of the value of nuclear-generated power, had developed a much more demanding and efficacious system of risk assessment, risk management, and safety auditing than that in place in the DOE (D'Antonio 1993). In due time these differences between the two agencies would become chief points of critical comparison for the opponents of Hanford's DOE authorities. The mantle of secrecy enjoyed by the DOE (and its predecessors) absolved the agency's nuclear weapons site managers from keeping up with the changing nature of public sector management practices clearly in place in the NRC (Rees 1994).

When time for dramatic change came for the DOE's nuclear weapons complex managers in Washington, D.C., and at Hanford (and similar facilities), the agency's powerful "family" culture and established standard operating procedures proved "highly resistant" to change (Reed, Lemak, and Hesser 1997). While the DOE's nuclear weapons management bureaucracy kept itself exempt from the significant changes sweeping over the rest of the federal government's natural resource and environmental agencies, those agencies were all trying to adapt to change. For example, American state governments since the 1970s have fought quite effectively to gain a key role in the administration of many federal programs as national legislation on environmental standards and natural resource management developed.

The relative newcomer, the Environmental Protection Agency, early on accepted provisions for delegating its oversight authority to states capable of and willing to commit to the enforcement of national statutory provisions. The long-established agencies, such as the Forest Service, Bureau of Land Management, Fish and Wildlife Service, Department of Transportation, and Department of Agriculture, found new ways to work with related state agencies on a broad range of issues (Clarke and McCool 1996). The Intergovernmental Personnel Act provided for the interchange of thousands of federal and state agency employees over the years, and the networks of public service professionals, academic researchers, and environmental activists across the country created many informal linkages between and among federal and state environmental and natural resource agencies, and between those officials and university-based researchers. In contrast, very little such networking and legislative activity came about in the administration of the nation's nuclear weapons plants. When pressures for change inevitably came to the DOE managers responsible for nuclear weapons facilities, change came only after the most stubborn resistance of these managers was overcome by the persistent efforts of a wide array of interests, as we discuss later in this chapter.

It should be recalled that the American Constitution scarcely makes reference to the federal bureaucracy and how it should be managed. Part of America's weak-state heritage was the presumption that the bulk of public life would be managed by state and local governments. Federal responsibilities would certainly be important, but relatively few in number and hence would require relatively few employees for Congress and the president to manage (Barry and Whitcomb 1987). With the growth of the nation into a continental country, the assumption of world power status, and ultimately the creation of a welfare state infrastructure in the New Deal years, the harsh reality finally hit home—namely, we have a substantial federal apparatus without having a legal/political rationale by which it should be managed (Spicer 1995). In a legal setting where a long tradition of strong property rights comes into conflict with a large, unplanned state empowered with sovereign immunity, what are the rightful limits of federal action? What claims made by citizens against the federal government should law provide for in its balancing of collective and individual values? When most decisions affecting the liberties and property of individuals are not made in the halls of Congress but rather in administrative agencies, what provisions for public input and public scrutiny of agency plans, regulations, and practices should be in place?

These are indeed knotty questions, and they are the province of administrative law in the United States. Key to this area of public life is the enactment of the Administrative Procedure Act (APA) of 1946 (Davis 1969); most states have comparable statutes setting out the proper powers of public agencies, and identifying just grounds for civil action against agents of the institutions of the state. Importantly, these state administrative procedure acts typically prescribe the public processes to be followed by state agencies in carrying out the duties delegated to them by elected legislative and executive officials. The cardinal principles of the APA include the concepts of open public scrutiny and accountability for agency actions, prior publication and timely notice of intended actions, official public hearings on the record, provision of neutral hearing processes for complaints and appeals, and ultimately the right of appeal to civil courts for the adjudication of cases under the APA (and comparable state statutes). The idea underlying the act was that the Congress (and state legislatures) would be permitted to delegate substantial authority to executive agencies to make regulations "binding as law." This lawful delegation would be permissible notwithstanding the constitutional prohibition of delegation of powers. Importantly, the courts would be given the ultimate authority in monitoring (via adjudication of telling cases) the behavior of public agencies to check any excessive intrusion into constitutionally protected civil and/or property rights (Stillman 1987).

Over the years relatively clear standards for public notification and hearing participation have been established, the conditions required for individual tort claims and class action suits have been elaborated, and the standards for the hearing of administrative appeals and complaints have been put into practice (Caldwell 1976). All major public agencies operating in the strong democracy, weak state setting are cognizant of these provisions, and all major interest groups are similarly aware of the rules of the game by which public agencies must play. Again, by virtue of the national security cover protecting the managers of American nuclear weapons complexes, the part of the DOE that dealt with nuclear weapons production was not required to play by these same rules until the end of the Cold War.

What happened when the DOE was forced to accommodate these features of democratic administrative processes in the operation of its nuclear weapons development and production facilities? How was the drama of the fall from grace (i.e., exemption on the grounds of national security) played out at the Hanford facility? In what follows it will be clear how the strong democracy, weak government setting of American politics came to structure events at Hanford as the agency sought to convince the public of its veracity and efficacy through a broad range of policy initiatives. The challenge facing the DOE entailed confessing to deception and deceit in the past, making appropriate public amends, assuring a more open and trustworthy mode of operation in the future, and assisting the victims of Hanford's past mistakes and misdeeds. Finally, the agency needs to convince a skeptical public that it has shifted its core mission from nuclear weapons production to environmental cleanup, a task that it can perform with alacrity, appropriate expertise, and economic efficiency.

The following sections set forth the track record of DOE managers at Hanford with respect to these goals and objectives, reflecting the "new realities" faced by DOE nuclear weapons site managers. The Tri-Party Agreement joining Hanford DOE officials, the EPA, and the state of Washington serves as the focal point of the assessment of the governmental response at Hanford. As noted in chapter 2, this legal agreement establishes a structure within which DOE managers have to play by the same rules as other natural resource and environmental agencies—namely, maintaining open records, facilitating public participation in decision making, providing for stakeholder involvement in problem assessment and policy option development, and active coordination with state and tribal governments. In this case, the state of Washington is granted the status of a regulator, sharing responsibilities with the EPA. With this focus for our assessment, it will be clear how the elements of American exceptionalism discussed earlier come to the fore. After a careful consideration of the record of events occurring after the fall from grace, we will proffer an overall assessment on the governmental response at Hanford.

Emergence of Differing Realities at Hanford

In tracing key developments in the old reality at Hanford since 1943, and in noting the emergence of new realities over the past decade, it is apparent that various observers have quite different perspectives. Michele Gerber has observed that "more historical data are publicly available about the Hanford site than are available about any other nuclear defense site in the world" (Gerber 1992c, 19). Writing in the same issue of *Perspective,* however, the founder of the Hanford Education Action League, Reverend William H. Houff, drew attention to a lasting legacy of operation by secrecy and deception, "a paired dynamic that taints and corrupts every aspect of America's nuclear weapons production establishment . . . nuclear technocrats became so accustomed to misrepresenting the reality and risks of their activities that they inevitably ended up deceiving themselves as well" (Houff 1992, 3). For Houff, nuclear experts were working behind a curtain of secrecy and deception and focused "mainly on the urgency of making nuclear weapons, they largely ignored other factors—moral and practical" (1992, 24). In his view, the historical record provided by the DOE under the compulsion of Freedom of Information Act provisions remains hopelessly incomplete.

A more tangible issue than the relative completeness of the "official" historical record is the DOE compliance record with the remediation milestones of the Tri-Party Agreement. John D. Wagoner, manager in 1992 of the Richland office of the Department of Energy, drew attention to the fact that "it takes three to seven years of investigation and study before a final cleanup decision can be made on a (nuclear waste) unit. Multiply that by 78 units and you get a sense of the magnitude of the job" (Wagoner 1992, 4).

In characterizing its compliance with the Tri-Party Agreement, *A Look at Hanford: 1992 Progress Report* (Richland Office, DOE, 1992) struck an optimistic tone and indicated that DOE and its contractors have: 1) met 239 of 243 Tri-Party Agreement milestones on time; 2) 4 milestones are incomplete while DOE change requests are under review by the regulators; 3) 35 of the original 160 milestones have been renegotiated; and 4) An additional 203 new milestones were added through the end of fiscal year 1992 (Richland Office, DOE, 4). The Tri-Party Agreement was characterized as the guiding force of the Hanford cleanup—setting the pace and assigning priorities for the future. Tri-Party Agreement milestones are what environmental remediation work is centered around, and they provide a principal means by which to bring Hanford into compliance with federal and state environmental laws monitored by the U.S. EPA and the Department of Ecology of the state of Washington.

Both the DOE nuclear experts who wrote and approved the sections on compliance with Tri-Party Agreement milestones, and the critics of the DOE record from the environmental community, agree on the importance of compliance with TPA timelines and milestones. It is the degree of compliance performance being achieved by the DOE that is the subject of some sharply differing interpretations of current realities at Hanford.

A critical view of the DOE compliance record is provided by Todd Martin of the Hanford Education Action League. Writing about the same general time frame, Martin reviewed the record on twenty major milestones. His conclusion was that the "TPA becomes one more agreement that DOE manipulates to fit its own needs, rather than the mechanism for accountability to provide for Hanford cleanup" (Martin 1993, 16). Martin's critical review noted numerous major problem areas, such as the grout program for mixing tank waste with cement-like material and the pretreating of tank wastes in order to separate the "low-level" and "high-level" fractions (the low-level would go for disposal to grout, the high-level to vitrification). Martin criticized provisions for the interim stabilization of single-shelled tanks and the pumping of free liquid in them into double-shelled tanks and the retrieval of tank wastes in a circumstance where the commercial technology for getting the waste out of the tanks does not yet exist. He was also skeptical about whether future deadlines could be met, such as the initiation of a tank farm closure demonstration project by the year 2004 (an integral part of the final disposition of the tanks themselves, tank piping, contaminated soil, and tank waste), the closure of all 149 single-shell tanks by the year 2018, and the cessation by June of 1995 of further discharge of liquid waste streams directly into the soil. Finally, Martin doubted that by May of 1996 Hanford would have submitted all the plans and permit applications designed to bring all facilities at Hanford into compliance with the Resource Conservation and Recovery Act of 1976 (RCRA).

The formal framework for dialogue between these differing environmental realities is the Tri-Party Agreement. This dialogue involves issues that are new to the twentieth century and that are largely unsolved from a scientific and technological point of view. The various realities existing at Hanford can be seen in the different perceptions of the problems associated with one of the single-shelled tanks: Tank 101-SY, which was filled with over 1 million gallons of high-level waste. Up to 10,000 cubic feet of hydrogen gas builds up roughly every three months, which raises the level of the crust. The crust then returns to its static level when there is a great "burp" of the explosive gas mixture. The hole in the top of the tank is 42 inches across, making it very difficult to service the highly radioactive tank by remote control. The short-term waste management plan was to install a pump to stir the mixture continually so the hydrogen bleeds off routinely.

After the signing of the Tri-Party Agreement, state of Washington inspectors had access to the Hanford facility. In 1992, a state inspector found that three separate leak detection systems for Tank 101-SY were inoperable.[1] Clearly there was a long way to go to solve these problems from the past. In *A Look at Hanford: 1992 Progress Report,* the Richland office of the Department of Energy identified Tank 101-SY as the DOE's "top safety issue at Hanford" (DOE Office of Communication 1992, 4). The reason for concern was that potentially flammable gases accumulate within the tank and vent, or "burp," about every 100 days.

The DOE took specific steps in 1992 to improve the situation. They included the installation of new monitoring equipment to provide more information on what happens inside Tank 101-SY between ventings, the installation of a frame to support a new pump that will circulate the waste to keep gas from accumulating, and modification of the tank access openings to accommodate the pump and improve monitoring. The design of a new pump for the tank was listed by DOE as part of "numerous" 1992 achievements. It was finally installed on July 4, 1993.[2] To environmentalists this is a record of delayed reaction; to DOE it is a record of successful technical innovation.

The DOE commentary on the safety problems faced with high-level waste in twenty-four older single-shelled tanks is also worth noting: "The number-two tank safety issue (after Tank 101-SY) is waste in 24 older tanks which contain varying amounts of ferrocyanide compounds, nitrites and nitrates. Ferrocyanide was added to these tanks between 1954 and 1959 in an effort to cause radioactive cesium to settle to the bottom. However, if this mixture is allowed to become extremely hot, a runaway reaction or explosion could occur" (Richland Office, DOE 1992, p. 6).

As an historical note, it was high-level radioactive waste that became extremely hot and caused an explosion near Kyshtym in Russia in 1957. The explosion sent 20 million curies over the landscape of the Southern Urals. Substantial portions of that vast landscape remain unusable, and will be unusable for many years to come.

Long-Term Future of Environmental Remediation at Hanford: A Shift in Federal Priorities Forcing Renegotiation of TPA Benchmarks

At the time of the dramatic expansion of Hanford in the early 1950s, the major environmental protection legislation still lay in the distant future. Furthermore, the Atomic Energy Commission and its successor agencies, including the Department of Energy, steadfastly maintained that they were exempt from compliance with U.S. environmental legislation as it evolved over time. One consequence of this exempt status is that neither the AEC nor the DOE (nor the other organizations with interim stewardship over Hanford) established disposal practices in the early years that would have minimized radioactive and chemical contamination of the environment. It is clearly much easier to contain hazardous waste as it is generated than to clean it up once it has been released into rivers, settling ponds, wells, or other outlets.

The nearly ten years that have elapsed since the Tri-Party Agreement was signed have led to much greater awareness of the extent of environmental problems remaining at Hanford, and awareness of the lack of existing technology to solve some of these problems. A more focused debate has emerged over what goals are reasonable to seek (containment and stabilization versus creating green fields and kindergarten sandboxes). Similarly, the debate has narrowed over who should be responsible for obtaining them (scrapping the TPA versus some level of reformulation of it), and to what extent the environmental restoration effort by the DOE should be affected by the effort to reduce the federal deficit. The later point, which was such a priority for the 104th and 105th Congresses (elected in 1994 and 1996), became a new, stark reality in part because of the electoral loss of thirty-year House veteran Thomas S. Foley (current Ambassador to Japan). As Speaker of the House, Congressman Foley assured the continued flow of funds to Hanford. With his loss, the assumption of power by the GOP majority, and an increased emphasis on balancing the federal budget, the politically privileged status of Hanford was no more.

In the wake of the 1994 election, a number of core assumptions regarding the amount of federal commitment required to restore Hanford to an adequate level of remediation were challenged by a series of analytic (some would say ideologically inspired) reports.[3] Each of these documents represented a challenge to "business as usual" at Hanford as it had evolved since the signing of the Tri-Party Agreement. Taken in combination, the documents questioned the status, accomplishments, and the adequacy of the Tri-Party Agreement itself. They each reassessed what was possible as opposed to desirable under the rubric of environmental "cleanup." Each report also refined the characterizations of environmental contamination present at the site, especially in relationship to considerations of the adequacy of current technology required and the minimization of risks to worker health and safety.

Regarding the level of continued funding required, there have been two elements to the challenge. The first suggests that instead of the thirty-year cleanup time frame envisioned by the Department of Energy in 1989, expenditures should extend well into the twenty-second century at a lesser standard of cleanup. Second, instead of using the Tri-Party Agreement milestones as a determinant of funding levels for the cleanup efforts (with civil and criminal penalties for not doing so), the fiscal policy debate in Washington, D.C., has shifted to concerns about national priorities and balancing the federal budget. To the detriment of Hanford managers and the victims of Hanford's past activities, the urgency of environmental restoration in eastern Washington State appears to have waned seriously in the corridors of decision making in Washington, D.C. Following is a brief characterization of each of the four key studies that have precipitated the fundamental rethinking of cardinal assumptions about the future of Hanford.

Train Wreck along the River of Money

Train Wreck along the River of Money: An Evaluation of the Hanford Cleanup was prepared by two independent consultants to the U.S. Senate Committee on Energy and Natural Resources; it was commissioned in September 1994 and delivered in March 1995 (Blush and Heitman 1995). The report characterized the cleanup at Hanford as the largest civil works project in world history. The evidence reviewed in the report produced a number of conclusions, some of which have precipitated a lively debate. One conclusion was that:

> . . . the current process for cleaning up the Hanford site is not working effectively, the site is spending more money on the cleanup than it can justify, and no reliable data exists [sic] for completing the cleanup. We believe Congress must act decisively to salvage the program and prevent further taxpayer dollars from being squandered. (Blush and Heitman 1995, 1).

The report also concluded that very little cleanup had occurred despite the expenditure of $7.5 billion from 1989 through 1995, with a projected expenditure of $1.5 billion per year for the foreseeable future.

The report defined cleanup as not only environmental remediation, but all activities necessary to contain, treat, store, and ultimately dispose of the hazardous wastes present at Hanford. By this definition, the Tri-Party Agreement is fundable by Congress "only if it is willing to forego [sic] appropriating money for other needs that almost certainly have a higher national priority" (Blush and Heitman 1995, 3).

The report also found what it deemed to be critical flaws in both the schedules based on the Tri-Party Agreement and the substantive provisions of the Tri-Party Agreement. The report contended that the time schedules of the TPA are "unworkable, disjunctive, lack scientific and technical merit, undermine any sense of accountability for taxpayer dollars, and most importantly, are having an overall negative effect on worker and public health and safety" (Blush and Heitman 1995, 3). This harsh indictment was based on the contention that Hanford was floundering in a legal and regulatory morass.

According to the authors of the "Blush Report," as it is affectionately known in Hanford circles, the foundation of the legal and regulatory morass occurring at Hanford was to be found in the unworkable nature of the Tri-Party Agreement's approach to bringing the site into compliance with state and federal laws governing the cleanup and management of hazardous waste sites. The selection of a thirty-year time frame for cleanup, and the requirements of milestones associated with that time frame were not realistic in the view of the Blush Report authors.

Another controversial conclusion of the report was that none of the parties to the Tri-Party Agreement appreciated the "difficulty of combining EPA and state regulations with those of DOE." They claimed that the signatories failed to appreciate the level of scientific and technical uncertainty associated with the cleanup, as well as the implications of combining DOE, federal, and state environmental regulations to tackle Hanford's enormously complicated technical, safety, health, and environmental risks, and the effect these unknown factors would have on the scope, schedule, and cost of the cleanup (Blush and Heitman 1995, 3).

In subsequent testimony before the U.S. Senate Committee on Energy and Natural Resources, Blush and Heitman elaborated upon their basic conclusions as they related to public policy choices.[4] The authors believed that the federal government had a clear duty to stabilize the hazardous materials stored at Hanford and to provide an appropriate level of spending to clean up the site. However, choices must be made to strike a proper balance among competing social and public health needs, and a new framework was necessary. They cited a recent EPA and Washington State strategy paper that indicated that the 100 area (along the Columbia River) would not be cleaned up until the twenty-second century.

The central conclusion of the Blush and Heitman report was that the Tri-Party Agreement constituted a failure that Congress needed to replace with a more workable and affordable approach to cleanup. They called particular attention to the important relationship between assumptions about future land use at Hanford, reasonable levels of cleanup, and the scope and cost of the work. At present "the cleanup is not proceeding rationally toward any coherent set of cleanup goals or in accordance with any reasonable schedule or budget."[5]

The authors noted the paradoxes inherent in choice of a time frame and standards for the Hanford cleanup. If the standards were set too high, then it would take DOE and its contractors virtually an indefinite length of time to clean up the site. If the time frame selected was too short, then the costs and risks to the workers, to the public, and to the environment would be unacceptable to the American people. According to Blush and Heitman, the Tri-Party Agreement did not address satisfactorily the trade-offs between costs, public safety, and environmental risk.

The hearings proposed a set of changes to address the problems identified with the Tri-Party Agreement: 1) reform the legal and regulatory framework for cleanup; 2) resolve the question of the intended level of environmental cleanup; 3) require DOE to undertake and maintain an integrated risk assessment of the Hanford site as the primary basis for budgeting; 4) require DOE to defer any actions that further near-surface disposal of large quantities of long-lived radionuclides at Hanford until the National Academy of Sciences completed a comprehensive study of the scientific basis for such a policy; and 5) require DOE to produce and periodically update a long-range plan for site cleanup that integrates all aspects of waste management, facility cleanup, and environmental restoration.

The authors of the Blush and Heitman Report concluded that after five years of operation under the TPA, a subsequent five-year agreement should be negotiated. They further suggested that DOE should produce and update periodically a long-range plan that integrated environmental restoration, waste management, and facility cleanup into a coherent system of activity. In what was obviously a controversial and confrontational recommendation, the authors stated that "to the extent that the state of Washington aspires to have Hanford cleaned up to standards that exceed those promulgated in accordance with the recommendations above, Congress should formulate a new arrangement that results in the sharing of financial responsibility for cleanup between the Federal Government and Washington State."[6] Obviously, while the members of Congress receiving the report were heartened by the observations and recommendations of Blush and Heitman, the Washington State authorities, tribal governments, downwinders, environmental interest groups, and Hanford DOE officials were shocked and greatly angered by the report.

Closing the Circle on the Splitting of the Atom

In 1995, a second report prepared by the Office of Environmental Management within the DOE addressed Hanford's status: "Closing the Circle on the Splitting of the Atom: The Environmental Legacy of Nuclear Weapons Production in the United States and What the Department of Energy is Doing About It." The report was intended for citizen education, and it focused on a range of environmental, safety, and health problems at DOE facilities. It described nuclear weapons production from uranium production to final assembly, and characterized the sources and types of wastes and contamination generated in the process. One intent of the DOE publication, according to former Secretary of Energy Hazel O'Leary, was to advance an effort "to earn public trust and foster informed public participation."[7]

In the introduction, Thomas P. Grumbly, DOE assistant secretary for environmental management, made several observations worthy of close attention. First, he observed that dealing with the waste and contamination resulting from nuclear weapons production was a problem that, for the most part, had been postponed at the time the waste was produced. This was, in retrospect, a very costly decision. Many records were either not kept or remain inadequate for the purposes of guiding cleanup activities. Second, the report acknowledged that the challenges facing the Department of Energy "are political and social as well as technical. . . ." In his opinion, however, DOE was meeting those challenges as well as could be expected. He acknowledged that there were numerous critics who would disagree with his characterization of DOE's record of performance. In support of his contention, Grumbly argued that his agency was "reducing risks, treating wastes, developing new technologies, and building democratic institutions for a constructive debate on our future course."[8]

Critics of the DOE, primarily congressional fiscal conservatives as well as the usual supporters of environmental protection, noted that the conclusions of the Blush Report were made six years after the negotiation of the Tri-Party Agreement. To them, the notion that key decisions still remained to be made in the future was quite astounding. They argued that several key questions remained unanswered despite the enormous expenditure of funds and the passage of substantial time. Most essential among those questions were: Where and how will we treat and dispose of the backload of wastes from nuclear weapons production? How clean is clean? And, should we exhume large volumes of contaminated soil in order to allow for unlimited use of the land in the future?

Depending on how these questions were answered, and in what time frame, estimates of the cost of cleanup have ranged upward to $1 trillion, and in a time frame ranging from thirty years to well into the twenty-second century. To pose such fundamental questions in 1995 indicated how unresolved some of the most basic considerations remained at that time.

Another striking feature of *Closing the Circle on the Splitting of the Atom* was the discussion of the interrelated tasks falling under the rubric of environmental cleanup. Since "cleanup" had become a generic term with little precise meaning, it is instructive to review this discussion. The environmental management mission was presented as being considerably broader than cleanup. Stabilizing and maintaining the large number of nuclear materials and facilities were of the utmost urgency, and these represented the most high-risk activities to be undertaken. This statement of cleanup priority applied to facilities, many of which are now over forty years old, which had exceeded the useful time for which they were designed. Stabilization was essential to protect the safety of workers. The sensitive materials involved were vulnerable to a list of potential catastrophes such as leakage, explosion, theft, terrorist attack, or avoidable exposure of radiation.

Managing and containing waste was a second task. The report stated that providing safe storage for the enormous quantities of waste was itself a monumental challenge. Hanford was specifically cited as requiring vigilance over underground tanks in order to minimize the risk of explosion. From 1989 through 1995, there had been a substantial increase in annual expenditures on this problem; they had grown from $50 million to over $500 million dollars annually.

The third task was environmental restoration, characterized by the DOE report as what is usually thought of as cleanup. It included both institutional (organizational) challenges and technological challenges. The root of the technical challenges was that no safe or effective technology was yet available to address—or even fully understand—the contamination problem. Included under the restoration task were stabilizing contaminated soil; pumping, treating, and containing groundwater; decontaminating, decommissioning, and demolishing process buildings, nuclear reactors, and chemical separation plants; and exhuming sludge and buried drums of waste. The cleanup objectives featured the goals of avoiding additional problems, minimizing hazards to the public or to workers, and minimizing cost and risk for the future.

The fourth task set out in the DOE report involved the development of new technology. The challenge in this area was to develop either lower-cost or substantially more effective approaches to the various cleanup, waste management, or stabilization problems identified in the previous three task areas.

The links among these four areas of the environmental management mission come together in the sheer scope of the time frame involved (thousands of years) and vast amount of radioactive materials present in a context where "we lack effective technologies and solutions for resolving many of the environmental and safety problems; we do not fully understand the potential health effects of prolonged exposure to materials that are both radioactive and chemically toxic; and, we must clear major institutional hurdles in the transition from nuclear weapons production to environmental cleanup."[9]

Estimating the Cold War Mortgage

The third 1995 report, "Estimating the Cold War Mortgage: The 1995 Baseline Environmental Management Report" was issued by the Office of Environmental Management of the Department of Energy.[10] Mandated by the U.S. Congress, it was the first of what are now annual reports that provide tentative schedules, life-cycle cost estimates, and the activities that are necessary to complete the DOE Environmental Management program.

Estimating the Cold War Mortgage contains a number of assumptions, estimates, and projections that were intended to promote public understanding of Hanford and its future. As a site, Hanford was projected to require the expenditure of 21 percent of all future funds allocated to DOE's environmental management effort. The mid-range estimate for departmental expenditures over a seventy-five-year period was $230 billion. Twenty-one percent of that overall figure would amount to over $48 billion for Hanford. In comparison to expenditures made since 1989, Hanford received an average annual allocation of approximately $1.5 billion, or $7.5 billion from 1989 through 1995. If that $1.5 billion level of funding were continued for the originally projected thirty years, the total expenditure would have been $45 billion (relatively close to the $48 billion projected over seventy-five years, assuming constant 1995 dollars for the purposes of comparison). Unfortunately, there were at least two problems with these DOE figures. First, the time frame projected in estimating costs in the report was seventy-five years, not the thirty years planning horizon employed in the Tri-Party Agreement. Such a time frame would reduce the per-year expenditure by well over 50 percent. The figure of $48 billion spread over seventy-five years would amount to slightly over $600 million a year. Such a major decrease in funds available would result in major layoffs in the near term, and substantial delays in some major categories of expenditure.

Second, severe curtailment in the short-term availability of funding would negatively affect progress in the four tasks of the DOE Environmental Management program. The report projects percentage expenditures by task as follows: 49 percent for waste management, 28 percent for environmental restoration, 10 percent for nuclear material and facility stabilization, 5 percent for technology development, and 8 percent for all other activities.[11]

Technology development was being counted on to help solve problems that were not currently solvable, and to reduce long-term costs by increasing efficiency and productivity. The problem of facility stabilization also was especially acute at Hanford. It does not appear to lend itself to a short-term containment in available funds without raising substantial risks to health and the environment. Given the vast volume of waste to manage at the Hanford site, it is not clear that the substantive discussions had occurred that would lead to a workable consensus that some expenditures could be deferred without risk of further environmental contamination. Finally, the topic of environmental restoration was what most observers equate with the term "cleanup." There was absolutely no interest in a massive deferral of environmental restoration costs among the key stakeholders affected by the Hanford site.

The estimates in *Estimating the Cold War Mortgage* explicitly excluded the costs of cleanup where no feasible cleanup technology existed. Most contaminated groundwater is included in that category of exclusion. Since contaminated groundwater is one of three top priorities at Hanford because of the risk of migration to the Columbia River, this is a major concern to the relevant stakeholders.

In a disclaimer, the report stated that "Projected costs significantly exceed current budget targets. Bridging this gap will require renegotiating compliance agreements and some statutory changes in addition to productivity improvements."[12] Given both the consultative obligations and political complexity of reaching decisions about the Hanford site, this statement certainly proved to be highly controversial for TPA stakeholder groups.

One of the government's purposes in issuing *Estimating the Cold War Mortgage* was to establish a more disciplined inventory of the problems and the potential liabilities at Hanford so the report could be used as a management tool. At the same time, the report identified the federal government as the authority which, in operating Hanford during the Cold War, assigned an overriding priority to weapons production. The report noted that: "Because of the priority on weapons production, the treatment and storage of radioactive and chemical waste was handled in a way that led to contamination of soil surface water, and groundwater, and an enormous backlog of waste and dangerous materials."[13]

Later in the report the central question is posed, "What Does the Nation Want to Buy?" After taking the position that fundamental choices had not yet been made, the report indicated that three influences on making those choices will be congressional mandates, adequate stakeholder input, and regulatory direction. Alternative choices were presented as having substantial cost and environmental implications:

> The cost and environmental implications of alternative choices can be profound. For example, many contaminated sites and facilities could be restored to a pristine condition, suitable for any desired use; they also could be restored to a point where they pose no near-term health risks to surrounding communities but are essentially surrounded by fences and left in place. Achieving pristine conditions would have a higher cost, but may or may not warrant the economic costs and potential ecosystem disruption or be legally required. Resolving such issues will depend on what the Nation wants to buy.[14]

In addition to the cost and environmental choices noted above, the cost of the environmental management program depended on how other key questions were answered in the future: What level of residual contamination should be allowed after cleanup? Should projects to reduce maintenance costs (i.e., high storage costs pending ultimate disposition of materials) be given priority over certain low-risk cleanup activities? In other words, how should cost affect priorities? Should cleanup and waste management proceed with existing technologies, or is it prudent in some cases to wait for the development of improved technologies? What criteria should guide decisions on this issue? Should waste treatment, storage, and disposal activities be carried out in decentralized, regional, or centralized facilities? And finally, how are issues of equity among states factored into configuration decisions?

All such cost estimates assume an acceptable level of accuracy of the data on which projections are based. It was not at all clear that the cost estimates set forth in *Estimating the Cold War Mortgage* were rooted in accurate data. Further, given the lack of availability of proven technology to solve some of the problems that were currently unsolvable, the projection of seventy-five years was not particularly realistic.

One especially valuable section of the report pertains to a critique of previous cost estimates of environmental liabilities and the problem of estimating cost in the face of large uncertainties. Some previous estimates before the Cold War ended assumed indefinite operation and had provisions for the renovation of some facilities. This new report noted that little emphasis was placed on more expensive activities, such as remediation at inactive sites.

Estimating the cost of future activities was therefore presented as a highly uncertain process. The source of the uncertainty included the lack of information for most of the 10,500 hazardous substance release sites of the DOE weapons complexes, the lack of knowledge about what remedies will be effective, the lack of knowledge about what remedies will be considered acceptable to regulators and the public, uncertainty about what level of protection for human health and the environment is possible through the various remedies, and uncertainty about the future land use of the various facilities.

The report also excluded a number of factors from the cost estimates that hinder comparisons with some previous estimates. For example, for the Hanford site it excluded large-surface water bodies, such as the Columbia River and most groundwater. Perhaps most important, the report excluded projects with no current feasible remediation approach. The reason for this exclusion was rather straightforward; because no effective remedial technology could be identified, no basis for estimating cost was available. Also not included were quantitative evaluations of risks to public health, workers, and the environment associated with environmental remediation activities that were included. Environmental management costs associated with the disposition of surplus weapons-grade plutonium were also excluded.

Cost estimates presented in the report also assume that the Yucca Mountain repository for high-level waste and spent nuclear fuel in Nevada will open in the second decade of the twenty-first century, and that the Waste Isolation Pilot Plant in New Mexico will open soon. There are strong reasons to believe that the Yucca Mountain facility will open on schedule.

One final matter concerning the DOE report merits comment. *Estimating the Cold War Mortgage* analyzed the Clinton administration's budget for fiscal year 1996 (which was supposed to take effect in October of 1995) through fiscal year 2000, using a base of constant 1995 dollars.

This budget reduced the target for fiscal year 2000 to $5.5 billion from the projected $6.6 billion for fiscal year 1996. The commentary on this is extremely noteworthy:

> A shortfall remains between the Base Case cost estimate (the estimated cost of meeting the Department's compliance agreements) and the Fiscal Year 1996 funding request and outyear targets. For the high Base Case estimate of $350 billion, this shortfall would be about $100 billion over the next forty years.[15]

Since Hanford receives about 21 percent of departmental allocations, it is reasonable to consider that it would receive a proportional share of the shortfall, or $21 billion short of "*meeting the Department's compliance agreements* [italics added]."[16]

Estimating the Cold War Mortgage argued that the cost effects of an assumed 20 percent productivity improvement would mitigate the effect of the reduced allocation. The report appeared, however, before the full budgetary consequences of the new Republican-controlled 104th Congress was apparent. The report also appeared before the political debates surrounding the 1996 presidential election changed the argument on a balanced federal budget from whether there should be a balanced budget to when and how it would occur.

It is not likely that the partisan turnover in the congressional delegation of the state of Washington will lead to disproportionately larger allocations to Hanford. The former Speaker of the House, Tom Foley, was a Democrat from the Fifth District in eastern Washington State (with Spokane as the principal city). The subsequent Speaker, Newt Gingrich, staked great personal prestige on accomplishing a balanced federal budget in seven years. None of his constituents in Georgia are known to be personally affected by environmental pollution from the Hanford Nuclear Reservation, nor are those of his successor from the state of Illinois, J. Dennis Hastert.

A second volume of *Estimating the Cold War Mortgage* reported specific data about the Hanford site. The report noted that early planners at Hanford expected that waste cleanup would start in the peaceful years following World War II:

> What followed, however, was the Cold War and an arms race continuing for decades. Plutonium production was always the top priority, and not until the late 1980s did environmental management finally receive serious consideration. As

a result, radioactive and chemically hazardous waste has been accumulating at Hanford since 1943. It accounts for approximately two-thirds of all the nuclear waste, by volume, in the DOE complex.[17]

The sheer magnitude of the contaminated soil and groundwater present at Hanford was staggering: 64 million cubic meters of soil, and 2.7 billion cubic meters of groundwater. In yet another index of the sheer magnitude of the contamination problem, the report stated that the full extent of soil contamination was not known yet, but about 200 square miles of groundwater are known to be contaminated. After a detailed location-by-location description of the current program, the section dealing with Hanford closed with cost estimates and a presentation of completion dates for major activity milestones. These milestones were broken down by categories of environmental restoration (through 2057), waste management (through 2035), storage and handling (through 2049), and nuclear material and facility stabilization (through 2047).

The report featured three categories of costs: a nondefense funding estimate of $502 million from 1995 through 2050; a program management cost estimate of $14.8 billion from 1995 through 2060; and a defense funding estimate of $72.6 billion from 1995 through 2060. The amounts were estimated in constant 1995 dollars.[18] Another question involves the source of defense funding estimates of $72 billion from 1995 through 2060. Assuming that Hanford consistently receives 21 percent of DOE Environmental Management funds, this would appear to project a total expenditure across all of the DOE complexes well in excess of $350 billion, excluding one category of program management costs. Including other listed management expenses, these calculations lead to a total budget for Hanford of $86 billion over sixty-five years, and thus a total for all DOE facilities well in excess of $400,000 billion. It should be remembered that the first volume also noted many exclusions in estimating costs.

The Galvin Report

One of the recurring themes of *Estimating the Cold War Mortgage* was the importance of developing new technology for use in the DOE's Environmental Management program. One key source of technological development is the complex of DOE national laboratories.

With the end of the Cold War, an intensive debate has arisen over the future role of the nuclear weapons laboratories. Former Chairman of the U.S. House of Representatives Science, Space, and Technology Committee, George E. Brown, Jr., introduced a bill in 1992 that would have required the phased consolidation of the various national weapons laboratories. Action on this legislation was deferred while a task force investigated alternative futures for the DOE's national weapons laboratories.

Known informally as the Galvin Report, after the chair of the task force, Robert W. Galvin, the report made a number of specific recommendations on the development of new technology for the DOE Environmental Management mission (GAO 1995). The Galvin report outlined in considerable detail the constructive role that the national laboratories could and should play in environmental remediation technology development. In the process of rendering their recommendation, the authors of the Galvin Report offered a devastating critique of how the environmental cleanup has progressed to date. In a bluntly worded assessment of the history of the DOE weapons complex cleanup effort, the report stated that the DOE was slow to adapt to the over three dozen pieces of environmental legislation enacted by the early 1990s, putting itself ten years behind in achieving compliance with mandates from the Environmental Protection Agency. Furthermore, DOE announced in 1989 that it would have the cleanup completed by 2019, but gave inadequate attention to the feasibility of the commitments it made. The report also observed that DOE was poorly equipped to deal with federal and state legislation and regulation. Finally, the study concluded that little actual cleanup had resulted from the $23 billion invested in environmental management since 1989 (GAO 1995, 27–29).

The "Hanford Syndrome" concept was introduced in the report. This characterization of Hanford's management history was identified as a risk aversion approach common to large bureaucracies whose symptoms are "an unwillingness to alter familiar behavior patterns, to stick with unproductive or failing procedures, to enhance tendencies for excessive resource allocation and regulation, and to oppose innovations. It is an important element in sustaining unproductive patterns of work" (GAO 1995, 29).

The Galvin Report also singled out the Tri-Party Agreement for special criticism as a constraint on the environmental remediation effort at Hanford (GAO 1995, 30). The report stated that the TPA was unrealistic and lacked proper input from those experienced in cleanup. The report claimed that the milestones force some make-work activities, and argued that some milestones should be abandoned altogether, including their penalties for noncompliance. Finally, the Galvin Report advised that some activities should be delayed until less costly and more effective technologies are developed.

The study held that an important reason for the slow pace of cleanup and assessment was the low quality of science and technology that was being applied in the field. The "pump and treat" approach to remediation of contaminated groundwater was offered as an example. This example is particularly relevant given the vast amount of contaminated groundwater requiring treatment at Hanford.

The Galvin Report also criticized DOE management for devoting an increasing proportion of resources to management oversight as compared to fieldwork. A Congressional Budget Office study concluded that 40 percent of cleanup program funds were devoted to administrative and support activities.[19] Furthermore, in the opinion of the Galvin Report, DOE management had made the absolutely wrong response to the problem. According to the report, when DOE learned that its costs were 40 percent above the private sector for environmental services, their response was to add between 1,200 and 1,600 full-time-equivalent people to management and oversight staff for the remediation effort.

Yet another level of criticism related to the projected costs of the DOE environmental management effort. The Galvin Report cited the estimates from DOE officials offered at different times as ranging from $300 billion to $1 trillion. These figures are, of course, difficult to compare since there are major cost exclusions in such previous estimates as those in *Estimating the Cold War Mortgage*. The assessment of the *Galvin Report* was that the "vast flow of funds into the program acts as an anesthetic, numbing the Department, State regulatory agencies, and affected stakeholders, hindering and delaying beneficial change" (GAO 1995, 37).

In an effort to bring about beneficial change, the Galvin Report offered a series of analyses of what they called "disconnects" that hindered progress, potentially discordant sets of activities that the Department has been incapable of harmonizing. Three disconnects were singled out for special consideration: in the areas of science/engineering and practical applications; in regulatory oversight and compliance; and in goals, objectives, and means.

The phrase "Valley of Death" was used to refer to the distance between those actively involved in the remediation effort and the insights and findings of the research community. One negative consequence of the "Valley of Death" disconnect was that waste remediation challenges across DOE sites had not been fully characterized, nor had applicable and available technologies been identified to deal with them. What was commonly called the "new-technology chain" was described as being seriously broken within DOE. This observation conveyed the view that there was little basic research being conducted, and little systematic analysis of field experience being done. The root deficiency, which made the various pertinent science/engineering applications disconnect a persistent problem, was the absence of a sustained, superior-quality scientific/technical review capacity at a high level within DOE. The lack of such leadership, and the poor management of the science/engineering-operational interface, exacerbated this problem at Hanford and the other nuclear weapons production sites.

A second disconnect was called the "Regulatory-Oversight-Compliance Management Disconnects." This type of problem concerned the self-inflicted, complex and frequently contradictory or redundant regulations and requirements that had become enormous obstacles to the remediation effort. In many circumstances there were harsh noncompliance provisions, and personal and civil legal penalties for failure. DOE officials and contractors were often intimidated, and this occasioned an excess conservatism in problem solving, sometimes bordering on abject inaction. There was little dispute that these circumstances aggravated inherent tendencies toward risk aversion. The process of compliance itself was presented as capable of being "quite burdensome, expensive and frequently fails to improve the affected activities" (GAO 1995, 32).

The third disconnect was the "Goals-Objectives-Means to Stakeholders' Interests Disconnect." The basic critique was that DOE had not clearly set the goals it should pursue in concert with affected stakeholders.

Examples included such basic questions as whether waste-contaminated soils were to be removed, or what was to be done with the large quantity of tritium-contaminated groundwater. A consequence of this disconnect was that longer-range difficulties were neglected due to concern for immediate acute problems such as tank leaks or potential explosions. While acknowledging that immediate acute problems must be dealt with, the critique pointed to a poorly conceptualized priority list that led to expensive, bad choices.

In contrast to the Manhattan Project, the report observed that the scientific challenge today is less profound, but the managerial challenges are more so. On this point, the Galvin Report indicated:

> We now have a poignant situation, for technology known to senior scientists and engineers both in the national laboratories and in the country's universities is in the wings that, appropriately applied, could dramatically alter the current prospects. (GAO 1995, 33)

The national laboratories were seen as having a critical role to play in this regard. Industrial site contractors are not inclined to undertake basic research because that effort might put at risk the meeting of compliance deadlines.

In summary, what is significant about all four of these studies is the questioning in 1995 of the basic assumptions of the 1989 Tri-Party Agreement and the actions that had been taken under this agreement up to this time. Most of the critical studies claimed that a consensus on the environmental goals for cleanup and the technical means needed to reach these goals had not been reached in sufficient detail to lead to effective remediation. It will become apparent in a later section that certain key stakeholders believed that such fundamental choices had already been made, and that these decisions were embodied in the legally obligating (if not binding) Tri-Party Agreement.

In addition, each report signaled a retrenchment in spending on environmental cleanup at Hanford and the other nuclear weapons facilities. Since these reports were issued, the annual federal appropriations to DOE Environmental Management have been curtailed substantially; the shortfall between estimated costs and available expenditures is going to amount to tens of billions of dollars at Hanford alone, irrespective of the cost calculation framework chosen. The basic point being made here is that 1995 marked the first appearance of official documents, which, taken together, foretold a massive shortfall between anticipated (and in many cases legally mandated) expenditures at Hanford and the funds that would actually be made available by the federal government. This inescapable conclusion applies irrespective of the time frame that is used for making funding estimates. Within six years of the new era of environmental cleanup at Hanford under the Tri-Party Agreement, and before many of the most pressing environmental problems had been addressed, the agreement was itself under attack.

Hanford Nuclear Reservation: A Public Hearing

The public debate over Hanford's future and the TPA continued at a public hearing before the Committee on Energy and Natural Resources of the United States Senate.[20] Held on March 22, 1995, the hearing provided a forum for some of the legislative stakeholders as well as the top DOE official in charge of the Environmental Management program and the two authors of *Train Wreck along the River of Money.*

The new Chairman of the Committee on Energy and Natural Resources, Senator Frank H. Murkowski, Republican of Alaska, began by posing three questions, and answering "No" to each of them himself: 1) Are existing environmental laws and regulations designed to protect the environment as well as the human health and welfare and safety, and are these being applied in the best interest of the cleanup, the workers at the site, and the people of the states of Washington, Oregon, and other states that might be affected? 2) Further, is the cleanup plan guided by a strategic vision and one that is likely to lead to success? 3) Do we have a clear idea of what will constitute success and achievement?[21]

In commenting on *Train Wreck along the River of Money,* Senator Murkowski concluded that in charting a future for Hanford, "we simply cannot get there from here under the current regulatory framework. . . ."[22] He called for creating exemptions to existing regulations, and establishing new priorities. The former chairman of the Committee on Energy and Natural Resources, J. Bennett Johnston, Democrat of Louisiana, echoed the theme, noting that Hanford had absorbed a disproportionate share of the federal budget: "Perhaps we would not shrink from spending such sums if they were necessary to protect the public health and safety or the environment, or if the money was being well spent."[23] Senator Johnston called for legislation that redefines the regulatory framework and establishes fiscal responsibility, a more realistic time frame, better standards, and a more clearly defined mission for the cleanup. Senator Johnston also noted the impossibility of the task facing the DOE. Acknowledging the presence in the hearing room of Thomas Grumbly, a lead administrator from DOE, Senator Johnston remarked:

> We have given him an impossible job. We have ordered him to meet standards he cannot attain, to use technologies that do not exist, to meet deadlines he cannot achieve, to employ workers he does not need, and to do it all with less money than that for which he has asked. If he fails, we have threatened to put him in jail.[24]

The principal alternative to existing policies appeared to be to vitiate the Tri-Party Agreement. A diametrically opposed point of view was expressed by Senator Patty Murray, Democrat from the state of Washington. Senator Murray observed that the federal government made the mess at Hanford and owed it to the citizens to clean it up under the framework of the Tri-Party Agreement. She argued that people residing in the region around Hanford deserved a voice in what happens, and that "The TPA is their voice . . . the Tri-Party Agreement must not be scrapped."[25] Senator Slade Gorton, a Republican from the state of Washington, similarly maintained that people living near the Hanford site should have the most influential role in that determination, and this meant the preservation of the Tri-Party Agreement.

The governor and attorney general of the state of Washington submitted statements for the official record. Governor Mike Lowry disputed the contention in *Train Wreck along the River of Money* that little progress had been made, citing the interim actions on Hanford's unstable and potentially explosive tank wastes, and the containment of mobile contaminants in groundwater. The defense Governor Lowry made of the Tri-Party Agreement foretold core arguments that will be offered in its behalf in the future when the state of Washington takes action to force the federal government to honor its contractual commitments made in the TPA.

. . . the Agreement has proven itself a viable, and when necessary, flexible framework for Hanford cleanup efforts. It has been substantially amended in recent years to reflect our growing technical knowledge, public concerns, and fiscal considerations. To boost efficiency, we have also negotiated a billion dollar cost reduction program and included new provisions to help integrate and streamline work under several regulatory regimes.[26]

Governor Lowry pledged to work to eliminate state regulatory processes that are of no value or are redundant.

The attorney general of the state of Washington, Christine O. Gregoire, made two essential points in opposition to those who wished to jettison the TPA:

Nothing could be more harmful to Hanford cleanup and to the confidence of the public. Washington State willingly entered a partnership with the federal government to ensure our nation's security. Why would our federal partner turn away from that partnership now? The message to our citizens will be a simple one—the federal government believes that the laws which apply to business, local and state government and private citizens alike should not apply to our federal government. And perhaps more distressing the message will be that our federal partner can't be trusted to fulfill its commitments.[27]

The twin themes of rule of law and trust are powerful ones for the stakeholders in the Pacific Northwest in the context of the forty years during which trust was breached in secrecy and the rule of law did not apply.

Reflecting DOE's position, Thomas P. Grumbly, assistant secretary for Environmental Management of the Department of Energy, called for substantial structural change in the approach to cleanup, advocating a fundamental revisiting of the relationships and regulatory structure between the federal government and the states. He argued that the Tri-Party Agreement did not lend itself well enough to balancing future land use plans, the realities of available funds, and the problem of ranking and assessing risks. Grumbly stated: "The fact of the matter is that DOE will have less money in the next five years than it has had in the last five years. There will not be enough funding beginning in fiscal year 1997 to fully comply with all of our agreements."[28]

Thus these hearings signaled growing public doubts in Washington and within the DOE about the Tri-Party Agreement as a framework for resolving Hanford's environmental problems.

Legislative Initiatives in Search of a Constituency

The 104th Congress elected in November of 1994 brought new Republican majorities in both the U.S. House of Representatives and the U.S. Senate. The new Congress took up the challenge to address the environmental legacy of the Cold War. Senate bill S.871 was cosponsored by the former chairman of the Committee on Energy and Natural Resources, J. Bennett Johnston, Democrat of Louisiana, and the new chairman, Frank H. Murkowski, Republican of Alaska.

Senate bill S.871, known as the Hanford Land Management Act, was intended to provide for the management and disposition of the Hanford Reservation. The bill revealed that the vision of Hanford's future reflected in the Tri-Party Agreement and the optimism reflected in DOE's initial thirty-year restoration initiative did not have fundamental support in the United States Senate. Even if the ideas reflected in Senate bill S.871 never became law, there does not appear to be interest in allocating the resources necessary for meaningful environmental restoration at Hanford.

Key ideas in this legislation involved the areas of alternate land use, risk assessment standards, criteria for materials and waste management, site restoration remedy selection, and work force restructuring. With respect to land use, the draft legislation allowed Congress to dispose of particular parcels or designate them for other uses. The bill also stated that proposed risk assessment procedures shall not exaggerate risk by inappropriately compounding multiple, hypothetical, conservative policy judgments. Cost-effectiveness was explicitly factored into environmental management activities. On waste management, the secretary of energy was directed to consider the risk to workers from exposure as part of environmental management, as well as the costs involved. Further, there was to be an explicit consideration of reducing risk to workers by interim storage of waste until it decays or decomposes, or waiting until more cost-effective treatment or disposal technologies are developed and available. On remedy selection, the reasonableness of cost was given emphasis. Finally, the secretary of energy was directed to reduce substantially the number of employees of the Department of Energy and its contractors.

In the U.S. House of Representatives, H.R. 2110 was introduced by Congressman Hastings from the state of Washington, with the cosponsorship of representatives from the states of Washington and Oregon. House of Representatives H.R. 2110 is known as the "Enhanced Environmental Cleanup and Management Demonstration Act of 1995." This draft law stated that the Department of Energy had achieved too little progress for the significant amount of money spent; contracting procedures were outmoded and lack competition; the sources of confusion and inefficiency included the lack of a clear chain of command and cumbersome bureaucratic process; internal orders issued by DOE interfered with compliance with environmental laws and add unnecessary costs to environmental restoration; DOE was accountable to several regulatory agencies, and regulatory requirements can be complex and redundant; and future land use issues are not necessarily related to cleanup decisions.

To improve the efficiency of management practices, the bill would have authorized a "site manager" with extensive authority. Remedies were to be selected after analyses of incremental costs and incremental risk reductions. The fostering of economic development was authorized as a key criterion in remedy selection. The site manager may disregard, in many instances, internal orders from DOE. In the case of Hanford, the state of Washington may exercise the authority vested in the administrator of the Environmental Protection Agency under the Comprehensive Environmental Response, Compensation, and Liability Act of 1980 (42 U.S.C. 9601–11050, West 1983 and Supp. 1990). Dispute resolution would be the means of resolving differences between the site manager and the state of Washington under CERCLA. The governor of the host state may overrule the site manager on remedy selection after a process of mediation. The bill also proposed a Land Use Council to develop a future land use plan. In developing the plan, the Land Use Council could consider land uses such as industrial, commercial, residential, agricultural, recreational, or open space. Technological considerations and costs of remediation were to be factored into decisions, along with risks to human health and environment. Technology development was explicitly authorized by the establishment of a program to foster remediation, waste characterization, and environmental technology testing using the actual Hanford site. In response to the various criticisms about the high costs associated with DOE's current approach to contracting and site supervision, H.R. 2110 encouraged market-based management and practices. The draft law authorized such usual private-sector business practices as implementing performance-oriented, incentive-based contracting and procurement practices.

Neither S.871 nor H.R. 2110 became law; nevertheless, it is evident from the development of this legislation that some influential members of the U.S. Congress believed that it was time to rethink the structure of the current Hanford approach to environmental management. The two legislative initiatives also reflect the differences between the interests of legislators who have a defense nuclear facility in their district or immediate geographical area, and those who do not.

In the 105th Congress, Representative Hastings, as well as other members of Congress, have continued to sponsor new legislation to restructure the government's relationship with Hanford. None of this legislation has yet been enacted. The Tri-Party Agreement continues to be the legal framework for the Hanford cleanup—albeit in a continually altered form. Reflecting these political pressures, in early 1995 the signatories to the TPA executed an amendment called the "Cost and Management Efficiency Initiative" (CEI) that committed Hanford to savings of $1 billion over a five-year period (1994–98). In May 1995, DOE representatives participated in a workshop with Hanford stakeholders and agreed to an additional $1 billion in savings (the current goal is a $2.8 billion savings over the 1994–98 fiscal years). These commitments were documented in the "St. Louis Blueprint for Action and Cost Control at Hanford" (Richland Office 1996). Hanford officials then briefed other groups concerned with the Hanford cleanup through a presentation to the Hanford Advisory Board.

The changing orientations of Congress have been clearly indicated in the budget allocations made for Hanford (figure 10.1). Budget authorizations (in current dollars) peaked in 1995. As the budget priorities and deficit sensitivities of the new Congress changed, the total funding for Hanford decreased in each subsequent year. Similarly, employment data show a decrease in the Hanford work force from 15,200 in 1995 to 10,250 in 1998. The DOE office at Richland acknowledged in their annual reports that since 1995 the projected cost of annual planned work has exceeded the budget authorized by Congress. These shortfalls posed

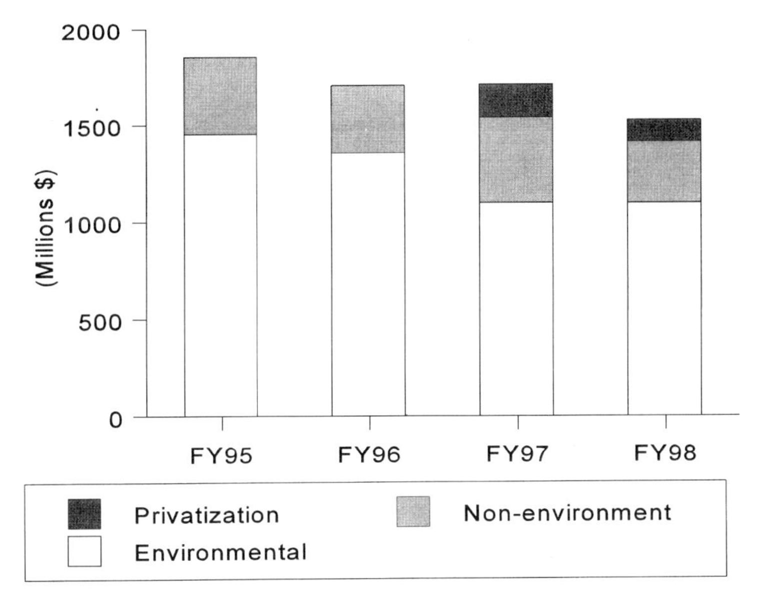

Figure 10.1
Hanford total funding summary, 1995–1998
Source: Department of Energy, Richland Operations Office, *Fiscal Year 1997 Annual Report,* p. 6.

difficulties in meeting the TPA targets for each year. One response was to identify tasks that would not be done, and another was to seek increases in the efficiency and effectiveness of ongoing cleanup efforts. In addition, DOE has sought to privatize portions of the cleanup effort, believing that commercial contractors would be more efficient than government-managed agencies in some cleanup activities. For fiscal years 1997 and 1998, these privatization activities amounted to roughly one-tenth of the environmental management budget.

These stopgap methods have taken an increasing toll on Hanford's cleanup efforts. In the 1997 Fiscal Year Report, the Richland Office of the Department of Energy admitted "the projected compliance gaps for FY 1998 and 1999 are significant and will be very difficult to overcome without additional funding (Richland Office 1997, 31). In practice, if not formally, the federal government and Congress have decreased their commitments to the timetables and cleanup plans embodied in the Tri-Party Agreement.

The Lessons of Hanford

This chapter focused on a central question: Was the DOE able to reestablish the sense of trust in its motives and in its capacity for effective operation that it enjoyed at the beginning of Hanford's history? (Leslie 1993; C. Thomas 1998). Relatedly, were DOE officials able to overcome the ubiquitous suspicion of government present in the U.S. political environment, particularly in the American West? Were the tribal and local and state governments brought into an effective partnership to accomplish a reasonable cleanup of the mess made of this land in the service of national security?

A good answer to these questions comes from a message written to accompany the most recent draft budget request for fiscal year 2000 operations. The annual process of formulating a budget request for Hanford involves a meeting of regulators, tribal governments, and stakeholder groups. Among the participants in the February 26, 1998, final session were the Washington Department of Ecology, the EPA Region 10, the Yakama Indian Nation, the Confederated Tribes of the Umatilla Indian Reservation, the Oregon Office of Energy, and the Hanford Advisory Board (HAB). The HAB was created in 1994 to provide stakeholder input to the Tri-Party Agreement implementation process. It includes representatives of local and regional governments, Native American tribes, regional business interests, Hanford workers' unions, the state of Oregon, environmental organizations, public health agencies, public interest groups, and the public at large. That meeting produced agreement on a budget request for $961 million. In commenting on the process of board deliberations and interaction with DOE officials, Merilyn Reeves, chair of the board (representing the Oregon League of Women Voters on the board) penned remarks for *Hanford Update: A Bulletin on Hanford Cleanup and Compliance* produced by the Tri-Party Agreement for dissemination to a mailing list of over 10,000 groups and citizens in the region. Ms. Reeves observed the following: "At the February meeting of the Hanford Advisory Board, members and alternates of the Board expressed frustration that there continues to be no firm national commitment for construction of facilities to treat, vitrify, and immobilize Hanford's high-level radioactive and hazardous tank wastes" (Reeves 1998,

3). Four pages of signatures of 42 members and alternates were attached to a statement emphasizing that facilities must be constructed to treat the tank wastes. The message was sent to DOE Secretary Federico Peña, members of the Northwest congressional delegation, and the governors of Washington and Oregon (table 10.1).

Other events reinforce the image drawn from this HAB report. Judith Jurji, president of the Hanford Downwinders' Coalition and a member of the Hanford Health Information Network (HHIN) Resource Center Advisory Board, has recently written a pessimistic assessment of the lessons learned from ten years of Hanford activism (Jurji 1998). On the plus side, she was positive about the public's ability to discuss and examine the complex issues involved in the Hanford cleanup. On the negative side, she felt that the issue lacked urgency for most Americans and politicians, and that insufficient progress was being made in addressing Hanford's environmental problems. Furthermore, a group of downwinders had sued the federal government to force action, and the lawyer representing this effort stressed the continued deception by government agencies in his public statements concerning downwinder litigation (J. Thomas 1998).[29]

A strong echo of American exceptionalism is heard in these observations—the claiming of individual rights from government, the suspicion of government and other elite elements of society, and the respect for populist understandings of democracy. Perhaps only in the United States would one expect to find such a strong belief in the capacity of the individual citizen to take active part in the highest-stakes issues of science, technology, and national politics. Sad to say, present here, as well, is the hard judgment against a government that is seen as still not fully worthy of trust.

The voices of the local people who have been intimately involved in Hanford affairs make it clear that the governmental response at Hanford has been far less than successful. While billions of dollars have been spent on the "characterization" of problems at Hanford, and millions have been spent on determining the levels of exposure citizens have suffered (and will need to document for independent tort claims against the government), the on-site problems at Hanford remain largely unresolved. While the federal government has spent a great deal of money to determine how much more money it will need to clean up the mess at Hanford,

Table 10.1
Statement to the Department of Energy by Members of the Hanford Advisory Board

HANFORD TANK WASTE TREATMENT CAPABILITY:

HANFORD ADVISORY BOARD MEMBERS DEMAND ACTION

Hanford has the most high-level radioactive and hazardous waste in the United States. However, no treatment facilities exist to retrieve and immobilize these wastes. Other sites have treatment facilities.

Postponing Tank Waste Treatment Is Costly. Hanford tanks are the greatest mortgage in the DOE complex, costing approximately $200 million per year to maintain a semblance of safe storage. Nearly seventy tanks have already leaked. It is easier to remove plutonium and uranium processing wastes from the tanks than from soil and groundwater. Adequate funds must be provided now to construct treatment plants for these wastes.

Unfulfilled Commitments. The federal government's commitments to treating Hanford's waste have consistently been unfulfilled—treatment has always been delayed. Risk assessments have shown that both a catastrophic tank failure and continued leaking pose unacceptably grave risks to the health of the Northwest's citizens, the environment, and agricultural economy. Delays only increase these risks. The nation has a legal and moral obligation to begin funding for retrieval and treatment of these extremely radioactive and hazardous Cold War wastes.

Get On With Cleanup and Build Treatment Plants Now. Members of the Hanford Advisory Board understand the problems and needs of this site. Further delay is not acceptable. The signers of this statement urge and demand assurance from the DOE, the administration, and Congress that there will be progress now and until the job is done:

• Provide safe, reliable, and fully operational treatment capability for Hanford's tank wastes now.

• Comply with the Tri-Party Agreement. The Hanford site needs to obtain treatment capability without further delay.

• Provide adequate life-cycle funding to build and operate treatment facilities at Hanford now.

• Establish sound management and technical expertise to successfully produce vitrified glass logs that permanently immobilize Hanford's tank wastes. The DOE, the administration and Congress must carry the program to success.

Source: Reeves (1998).

the federal government is showing definite signs of backing away from the solemn promises made in the Tri-Party Agreement to remediate Hanford to the fullest extent possible. Citizens knowledgeable about the situation at Hanford do not feel assured that corrective action is forthcoming. State, local, and tribal government leaders do not believe that the DOE and the federal government will make good on their promises of adequate funding. Downwinders and other stakeholder groups do not believe that the DOE has been fully forthcoming with evidence of human experimentation and accidental releases (see Steele 1995).

Our overall assessment of the governmental response at Hanford must be one of limited accomplishment in relation to the most daunting of environmental challenges. In the view of DOE critics, the primary beneficiaries of the $7.5 billion spent on cleanup since signature of the Tri-Party Agreement have been the contractors and the scientific and technical experts whose efforts to characterize the problems existing at the site are needed for appropriate remediation work. Unfortunately, in the view of Hanford's neighbors, relatively little reduction of threat to the people and the environment has been accomplished, despite the vast amount of money spent. Despite progress in a number of areas, many of the TPA targets have not been reached and others appear to be unreachable. Martin's earlier (1993) skepticism of the TPA timetable now seems well founded.

The activities of Hanford officials continue to confirm the frustrations and suspicions of neighbors and critics. In May 1997, DOE once again found itself subject to scathing criticism for the way it handled an accident when a containment lid was blown off a small storage facility at the Plutonium Finishing plant. The accident exposed workers to a brown plume of noxious gas, later found to contain trace amounts of plutonium (Geranois 1997a). Hanford officials responsible for the area in question ignored their own safety protocols and failed to notify emergency workers for hours. The workers exposed to the plume had to drive themselves to the hospital, and many developed skin rashes later (Geranois 1997b). The DOE fined the site operator, Fluor Daniel Hanford Company, $140,625 for the "level two" safety violation (potential for criticality), an action that Gerald Pollet of Heart of America Northwest characterized as "a light slap on the wrists" (Paulson 1998b).

The record of actual accomplishment at Hanford is a matter on which there are deeply held differences of opinion. For example, the DOE (1998c) has identified an initiative known as Enhanced Work Planning (EWP) as an important contributor to five successful demonstration projects. At the PUREX plant, deactivation was finished four months ahead of schedule and under budget by $10 million. Overall, a rough estimate of dollars saved or redirected by the EWP management initiative is $20 million.

On another endeavor not part of EWP, the B plant processing facility achieved a final "lights-out" deactiviation four years ahead of schedule in September 1998, at a savings to taxpayers of $100 million (DOE 1998d). The manager of the Richland office of the Department of Energy, John Wagoner, stated that "the successful deactivation of B plant represents our continuing commitment to cleaning up the Hanford site efficiently and cost-effectively" (DOE 1998d, 1).

On the critically important problem of protecting the Columbia River, the Department of Energy (1998e) announced a new, integrated plan with full public participation under the direction of undersecretary of the DOE Ernest Moniz. An Environmental and Molecular Sciences Laboratory has been commissioned to address in-ground treatment of contamination and the improvement of groundwater barrier technologies. Moniz also plans to enlist the expertise of the National Academy of Sciences to understand and address groundwater contamination better. The commitment from Undersecretary Moniz is "we are going to bring to bear on this environmental legacy of fifty years of weapons production our nation's finest scientific talent. We will have a much stronger role for the Pacific Northwest National Laboratory and other DOE national laboratories" (DOE 1998e, 2).

The examples that DOE offers to substantiate progress include a number of important themes. First, by turning to the National Academy of Sciences and increasing the role of the national laboratories, there is an enhanced effort to raise the level of scientific collaboration on solving Hanford's environmental problems. Second, there is a focus of these collaborative initiatives on selected key problems such as Columbia River contamination. In a fiscal climate in which there are not enough funds to address all problems, this focus on the most significant challenges and on broadening consultation is offered up as representing a constructive approach to Hanford cleanup.

At the same time, in 1996 the Department of Energy announced that one Hanford facility, the Fast Flux Test Facility, might be restarted to produce nuclear weapons material requested by the Department of Defense (Foster 1997; 1998a). This policy decision upset many of the groups' members, especially because the money to pay for the standby status was coming from Hanford's already shrinking cleanup budget (Foster 1998b). This announcement came on the heels of an announcement by Hanford officials that they were forced to cancel a registry and screening program for area residents during the early years of Hanford when significant amounts of radiation were being released. The program was being canceled because of federal budget cuts (Simon 1997).

Perhaps the saddest aspect to the drama of Hanford lies in the fact that the same federal government authorities who came out to the Pacific Northwest to use its water and land to produce nuclear weapons materials at a time when the nation was at serious risk are prepared to turn their backs on the citizens of the region now that the true extent of the damage done to it is more fully known. There is particular irony in the fact that progress on the Hanford Environmental Dose Reconstruction Project study and the Hanford Thyroid Disease Study coincide with the first clear evidence of the federal government's laying out a strategy for vitiating the Tri-Party Agreement.

In June 1998, Governor Gary Locke of Washington announced that the state was preparing to sue the U.S. Department of Energy for the slow pace of cleanup occurring at Hanford; the DOE has missed many milestones agreed to in the Tri-Party Agreement. To make matters worse, the DOE was forced to announce in early 1998 that waste from the high-level nuclear wastes storage tanks had reached the water table and was beginning what is projected to be a twenty-year migration to the nearby Columbia River (Paulson 1998a; Jamieson 1997). These developments could very well strain the relationship between DOE and Hanford area environmental groups that developed in the TPA "era of good feelings," and could signal the coming of a new period of intense and conflicted interaction. At the end of 1998, however, Washington State and DOE settled this lawsuit via the "Hanford Site High-Level Radioactive Waste

Tank Interim Stabilization Program and Interim Stabilization Consent Decree." This agreement is enforceable through the U.S. District Court for Eastern Washington.

Setting aside matters of constitutional jurisprudence for the moment, it is clear that the governmental response at Hanford has tended to heighten rather than ameliorate citizen suspicions of government (Tolchin 1996). The strong democracy expectation of citizens and interest group representatives and local, state, and tribal governments in the Hanford area have met with a weak and embattled federal agency that does not seem able to map out and maintain a reasonable course to environmental remediation and public health safety, despite the availability of enormous financial resources.

Perhaps the most ironic element of the drama of Hanford is that it is not Hanford DOE officials, but rather the stakeholders brought together on the Hanford Advisory Board who are the most effective voices advocating for DOE's Hanford budget! Whether this situation reflects an agency that is by and large down and out, or reflects one wherein the "domestication of dissent" has been fully accomplished remains to be seen. The ultimate judgment on the governmental response at Hanford will be possible if (and when) the mountain of nuclear waste sitting in leaking tanks at Hanford is processed into a safe form and securely stored, when the storage basins for spent fuel along the Columbia have been stabilized, and when the migration of contaminated groundwater toward the Columbia has been halted.

Careful observers of the Hanford case in Chelyabinsk might learn an important lesson from the events that transpired in this tragic set of circumstances all too similar to their own. Government promises of long-term commitments to address the real costs of nuclear arms production are easily made, but they cannot really be counted on by the people most directly affected. Even in the case of a wealthy country allowing the open advocacy of regional interests, these promises must be monitored continuously—and even then a change in national priorities can place the victims of government deception on the extreme back burner of official public concern. Perhaps keeping all stakeholders informed of events in both Hanford and Chelyabinsk can shorten somewhat the long odds of eventual restitution. Mutual concern has its own rewards, if only for one's moral and psychological health.

Notes

1. *Orange County Register,* A dangerous case of indigestion, July 5, 1993, p. 8.

2. Testimony of Thomas Grumbly before the U.S. Senate Committee on Energy and Natural Resources, March 22, 1995, p. 34.

3. The key studies are discussed below. They are: Steven Blush and Thomas Heitman, *Train Wreck along the River of Money: An Evaluation of the Hanford Cleanup.* A Report for the U.S. Senate Committee on Energy and Natural Resources, March 1995; U.S. Department of Energy Office of Environmental Management, *Closing the Circle on the Splitting of the Atom: The Environmental Legacy of Nuclear Weapons Production in the United States and What the Department of Energy is Doing About It.* January 1995; *Estimating the Cold War Mortgage: The 1995 Baseline Environmental Management Report,* Volume II, March 1995, Office of Environmental Management, U.S. DOE, U.S. Department of Commerce, Technology & Administration, National Technical Information Service, Springfield, VA, p. 1–2; General Accounting Office. 1995. *The Galvin Report and National Laboratories Need Clearer Missions and Better Management: A GAO Report to the Secretary of Energy.* Washington, D.C.: U.S. Government Printing Office.

4. Hearing Before the Committee on Energy and Natural Resources, United States Senate, 104th Congress, 1st Session, Waste Management and Cleanup Activities at the Hanford Nuclear Reservation. March 22, 1995, Senate Hearing 104–99, Washington, D.C.: U.S. Government Printing Office, 1995.

5. Ibid., p. 31.

6. Ibid., p. 33.

7. Letter from the Secretary, in *Closing the Circle on the Splitting of the Atom,* The U.S. Department of Energy Office of Environmental Management, January 1995, p. vii.

8. Ibid., p. xiii.

9. Testimony of Thomas Grumbly before the U.S. Senate Committee on Energy and Natural Resources, March 22, 1995, p. 9.

10. *Estimating the Cold War Mortgage: The 1995 Baseline Environmental Management Report,* Volume II, March 1995, Office of Environmental Management, U.S. DOE, U.S. Department of Commerce, Technology & Administration, National Technical Information Service, Springfield, VA, p. 1–2.

11. Ibid., p. I.

12. Ibid., p. I.

13. Ibid., p. iii.

14. Ibid., p. v.

15. Ibid., p. xvii.

16. Ibid., p. xvii.

17. Ibid., p. 702.

18. There are a number of matters on which it would have been instructive to have a fuller descriptive narrative. For example, technology development did not appear as a specific category of expenditure, even though it was explicitly counted on in the first volume as a key contributor to a successful program. Also, program management was listed as a $14 billion expenditure in two places, once as a separate category ($14.8 billion) and once as a part of the defense funding estimate ($14.6 billion). It is not clear from the text how these program management expense estimates relate to one another, whether they overlap, or who is doing the managing.

19. *Cleaning up the Department of Energy's Nuclear Weapons Complex,* Congress of the United States Congressional Budget Office, Washington, D.C., May 1994.

20. Hanford Nuclear Reservation Hearing before the Committee on Energy and Natural Resources of the United States Senate, 104th Congress, First Session on the Waste Management and Cleanup Activities at the Hanford Nuclear Reservation, March 22, 1995, S.Hrg. 104–99, Washington, D.C.: U.S. Government Printing Office, 1995.

21. Ibid., p. 1.

22. Ibid., p. 1.

23. Ibid., p. 6.

24. Ibid., p. 5.

25. Ibid., p. 9.

26. Ibid., p. 36.

27. Ibid., p. 57.

28. Ibid., p. 22.

29. A U.S. District Court judge dismissed the downwinders' class action suit in August 1998, but lawyers for the plaintiffs have said they will file an appeal to the 760-page ruling.

11

Still Tilting against the Environment: The Struggle over the Russian Governmental Response

John M. Whiteley

Fancy a brush with the nasty side of what is left of the Russian state? Then make sure you do it in a relatively civilised, easily accessible part of the country, backed by energetic foreign friends. That, more or less, is the lesson from the recent fortunes of Russia's most notable political prisoners.

The best-known, Alexander Nikitin, is no longer in jail. He was accused of stealing secrets about the navy's (shockingly careless) habits with nuclear waste, and passing them to foreigners. Luckily for him, he was working for an alert and able Norwegian-based environmental group, which has made him a cause célèbre. In October, a St. Petersburg court said the charges against him were inadequately prepared—quite an understatement, given that the information he collected was from public sources. Though not formally acquitted, Mr. Nikitin is now a free man.

—"Old Russian Habits Die Hard," *The Economist,* January 23–29, 1999

The struggle over Russian governmental response to mounting nuclear crises and to the nuclear environment in the Chelyabinsk region occurs in the context of the end of the Cold War nuclear arms race. It also occurs in the context of the legacy of Soviet approaches to environmental protection and of the historical relationship between the environmental community and governmental authority in the former Soviet Union (Whiteley and Gontmacher 1998).

A number of factors in present-day Russia, both institutions and circumstances, have coalesced to exacerbate the flawed nuclear environment that was inherited from the Soviet Union. This dubious inheritance lacked the financial resources to stabilize or restore the affected environment. The title of this chapter reflects the continuing antienvironmental tone of Russian government policy that has major elements in common with that of government policy in the Soviet era. There are also major differences in Russian government policy from that of the Soviet era, reflecting redefinitions of nuclear threat and the beginnings of international cooperation on the environmental consequences of nuclear weapons development.

The basic argument of this chapter, that important institutions of the government of Russia still tilt against the environment, is developed in seven sections. The first section briefly summarizes the status of environmental regulation under the Soviet regime. The next chronicles the deteriorating relationships between the environmental community and government authorities at the end of the Soviet era and the transition to the Russian Federation. The third reviews the antienvironmental themes in the first years following the dissolution of the former Soviet Union, and the efforts to establish a law on nuclear waste as examples of the response of governmental structures to environmental concerns in the new Russian Federation.

The fourth section examines the expansion of Mayak's role as a nuclear enterprise in support of nuclear security initiatives with the United States. The common denominator of the governmental responses to crises elsewhere in Russia, as explored in section five, is to initiate programs that will increase environmental contamination at Mayak. The fifth section introduces the proposed role for Mayak in solving the most urgent nuclear waste crises elsewhere in Russia. The sixth section presents the "business plan" for Mayak's future.

The seventh section returns to "tilting against the environment" as a recurrent theme in the Russian Federation, as well as a legacy from the Soviet era. This section illustrates the bias toward economic and nuclear development with minimal explicit consideration of broader environmental impacts. The effect of this development is to intensify the dependence of Russia's short- and medium-term nuclear future on external sources of funding.

Soviet Governmental Response to the Environment Prior to Glasnost and Perestroika[1]

The former Soviet Union had a policy of officially espousing environmental values. Article 11 of the 1977 version of the Soviet Constitution stated that the "land, its mineral wealth, the waters and the forests are the exclusive property of the state." Article 18 outlined the steps to protect and make rational use of land, mineral, water, and other resources "in the interests of the present and future generations." Article 67 indicated that citizens are obligated to protect nature and conserve its riches (Butler 1987).

By 1987, within this constitutional context, over 1,000 environmental laws had been enacted or environmental decrees issued (Environmental Law Institute 1987). Most of the laws, however, required further implementing legislation, and were themselves quite general. It was left to overlapping agencies and ministries to work out such details as permissible emission levels, environmental controls, and penalties. It was frequently the very groups generating environmental pollution that collected the data about potential violations, and had no obligation to publish it (Timoshenko 1988–89; Butler 1983; Green 1991).

Radioactive waste and other nuclear issues had a special status under Soviet law. Maloney-Dunn (1993) traced specific legal requirements for radioactive waste to three general documents: 1) the USSR State Committee on Standards (GOST)—public health and sanitary standards; 2) the 1969 USSR Fundamental Principles of Legislation on Public Health—prevention of environmental pollution was a specific duty of state agencies, enterprises, and organizations; and 3) the 1972 USSR Supreme Soviet decree "On Measures for the Further Improvement of Nature Conservation and the Rational Utilization of Natural Resources." Maloney-Dunn noted that this latter decree "entrusted the Council of Ministers to take measures to intensify conservation efforts and to prevent the discharge of effluents and other polluting products" (Maloney-Dunn 1993). The Council of Ministers responded subsequently with a decree of their own entitled "On the Intensification of Nature Conservation and the Improved Utilization of Natural Resources." The purpose of this decree, according to Butler, was to prevent further pollution and promote efficient utilization of natural resources (Butler 1987).

In fact, however, prior to glasnost and perestroika the entire governmental framework for protecting the environment was vague, general, and contradictory. Efforts to integrate environmental concerns into centrally planned industrial management were overlapping and fragmented. Further, counterproductive competitive relationships existed between the state agencies and ministries responsible for implementation of mechanisms for environmental protection (Yosie 1988).

Green summarized the situation as follows: "On paper, Soviet environmental protection legislation is a paragon of ecological safety. The emission standards for factories releasing waste into water, air or soil are among the strictest in the world. In some regions there are outright bans on the use of certain substances" (Green 1991). The analysis by Green pinpoints the reasons that Soviet approaches to environmental protection did not work, and identifies the vested interests whose actions led directly to pollution of the environment, and the incentive structure that compounded the problem.

First, during the Soviet era prior to glasnost, public input on environmental policy was essentially nonexistent. Economic development to advance the command economy was a principal criteria the Central Committee of the Communist party used to guide environmental protection policy making. Green (1990, 1) states: "Until recently, the Soviet Union based its environmental protection strategy on the premise that state ownership of natural resources would ensure their rational, benign utilization. This deluded notion that state institutions could regulate themselves effectively, coupled with the breakneck drive for industrialization, resulted in widespread ecological destruction." Environmental problems began in the initial planning stage. Extensive methods of economic growth (increasing inputs of land, energy, labor) on behalf of heavy industry were initiated with lax or nonexistent review.

Second, "centralized pluralism" describes a system where actual decisions were made in Moscow but a large number of government organizations participated in the process (Green 1990, 3). This did not work for several reasons. The economic structure contained implicit incentives to pollute. Planned output, not environmental protection, was the criteria by which managers and enterprises were judged. In addition, natural resources were available at essentially no cost. This circumstance provided no incentive to conserve. Enterprises and ministries also incurred basically no costs for breaking environmental regulations. Furthermore, prior to 1988, up to twenty-six separate state committees and branch ministries participated in the design and implementation of environmental regulations. The creation of the State Committee on Environmental Protection

(*Goskompriroda*) in January 1988 was intended to address this problem. But *Goskompriroda* proved to be ineffective in countering the established interests of industry and the military (Green 1990, 8; Pryde 1991). Finally, the weakness of Soviet environmental law in the feeble Soviet judicial system was coupled with negligent and inadequate enforcement practices.

Virtually all of the industrial development in the Chelyabinsk region occurred under the system characterized above. Its status as a closed military industrial region exacerbated the secrecy. But the problems that plagued the region reflected the broader environmental problems in the former USSR:

> In many ways the USSR's environmental problems embodied, in microcosm, the array of problems plaguing Soviet society as a whole. The economic structure essentially encouraged enterprises to degrade the environment just as it pushed development of heavy industry at the expense of high living standards. In determining environmental and other policies, the legislature and judiciary were at best symbolic counterweights to the dominance of the Council of Ministers. As a result, the public had no institutionalized means of translating its concerns into policy changes and virtually no legal protection from arbitrary government actions. Despite volumes of well-intentioned rhetoric, the leadership did not create regulatory mechanisms to correct these problems. In sum, the factors that led to economic stagnation and eroded the regime's legitimacy also caused widespread destruction of the natural environment. (Green 1991, 6)

Of particular relevance to Green's characterization of the USSR's environmental problems as a microcosm of the array of problems of Soviet society as a whole, is the notion of the legislature and judiciary as only "symbolic counterweights" to domination by the Council of Ministers, with their command economy and output goals. There was no effective counterweight within Soviet society to executive government domination, no actual separation of powers and system of checks and balances, and essentially no opportunity for citizen participation in policy formation.

Environmentalism and Government at the Transition to the Russian Republic

Before the demise of the former Soviet Union at the end of 1991, relations between the "greens" and governmental authorities had deteriorated. This resulted from a combination of factors. One factor was the power struggle between the structures of the Communist party that supported the Soviet Union on the one hand, and the emerging institutions of Russia on the other. An academic specialist on the Russian environmental movement, Oleg Yanitsky, has been highly critical of the government during this period of transition. Among other charges, he claimed there had been a disintegration of state authority, leading to the loss of policy making and respect for authority, and excesses of unregulated privatization. All this led, in his opinion, to a crisis of legitimacy (Yanitsky 1993, 1995).

Another prime factor in deteriorating relationships between environmentalists and the government were the twin economic realities of declining production and increasing inflation. These realities prompted unwise choices in public policy affecting both the environment and public health. For example, the government closed the Armenian nuclear station after a big earthquake in 1989. Though undamaged by the earthquake, the nuclear station is located near an earthquake fault, and people reasonably worried about the potential for another Chernobyl. Still, by 1993, the Armenian government, after the demise of the Soviet Union, which had been responsible for the plant, decided to open this nuclear station again. In the face of obvious (and largely unmitigated) safety concerns, the reasons for this decision were economic in nature: factories and homes needed the electric power, and other sources of energy were unavailable.

Similarly, the rise in unemployment in Russia forced unwise environmental choices. Even modest limits on environmental pollution could increase the risk of closure for a significant part of the industrial, transport, and agricultural sectors. In addition, many enterprises only avoided bankruptcy because of government subsidies and loans. Financial resources were simply not available to introduce pollution controls. Also, new environmental legislation was not being enforced by the executive branch of the government.

The judicial system was another factor. The judicial system was unresponsive to the environmental crisis. Environmental law had been one of the areas of the judicial system that historically received the least attention during the Soviet era. In addition, the usual sanctions available to the court were minor, consisting mostly of small fines. Finally there was simply a lack of information in the judiciary about applicable environmental law.

Response to Environmental Concerns in the New Russian Federation

There was a continuing antienvironmental theme in government policy in the Russian Federation. For example, a decree entitled "Questions of Nuclear Stations Construction in the Russian Federation" was released by the Ministry of Economy and the Ministry of Atomic Energy on December 24, 1992. This decree authorized the start of construction for some thirty new nuclear power plants. This was done without the completion of the prior environmental review stipulated in the 1992 Law on Environmental Protection. And the Ministry of Ecology had signed off on the decree!

In 1993, President Yeltsin's adviser on the environment and public health at the time, Alexei Yablokov (1993), noted what he considered to be a number of gross omissions in the activities of the Ministry of Ecology. The Ministry had been transformed into a bureaucratic monster with an excessive staff for what it actually did. In Moscow alone, there were 630 employees. In addition, it had failed to organize an ecology-natural resources conservation bloc in the government. Yablokov felt that government environmental activity would not be successful until the parliament could strongly influence the activities of the ministries. Financial support for the nature protection system also had declined. Although the previous parliament had demanded a separate section on "nature protection" in the state budget drafted by the government, it was absent in 1993. At the same time, *Goskompriroda* established pollution penalties for 1993 that were at a much lower level (in comparable prices) than was the situation in 1991.

Yablokov also believed that it was a blight on the Ministry of Ecology that the 1992 Law on Environmental Protection remained only in declaratory form more than a year after its enactment. The regulations necessary to implement successfully the law had yet to be promulgated. This was especially damaging in four areas relevant to regulation: circumstances where ecological expertise is required, statutes setting forth areas for environmental control and supervision, approaches to using the economy to further environmental protection, and the characteristics for identifying ecological disaster zones. Finally, Yablokov argued that the ecological expertise and commitment of the Russian Federation had been discredited. Important governmental decisions were made without preliminary environmental review and commentary. Perhaps worse, policy formulation was proceeding without entertaining environmental considerations.

Yablokov's assessments were harsh. There had been some progress by the Ministry of Ecology in the first years after the dissolution of the Soviet Union. For example, Russian state environmental experts examined over 55,000 projects in 1992: nearly 20 percent were rejected, about 40 percent were returned for alteration and further examination by experts, and 40 percent were adopted with commentary. Nevertheless, referring to the previous Parliament disbanded in October 1993, Yablokov stated that the Ministry of Ecology had not been able to influence a more ecological policy orientation by the Russian government.

Relevant to government policy toward Chelyabinsk, Yablokov cited two examples from reports of the former Parliament's Committee on Ecology and the Rational Use of Natural Resources. In the first example the government spent funds for other purposes that had been earmarked for the rehabilitation of the region contaminated by the 1957 Kyshtym nuclear waste explosion. In the second example the government elected to delay the whole program for rehabilitating the parts of the South Urals region adversely affected by nuclear development.

In 1992, Minister of Ecology Danilov-Danilyan was interviewed on television in Chelyabinsk, and addressed the problem of disposal of high-level radioactive waste.[2] In the beginning of the interview he expressed concern about the government plan for burying nuclear waste in containers in deep geological repositories (most likely in closed mines).

In another part of the interview he talked about promises that the U.S. Congress had made to help with developing storage facilities for plutonium and highly enriched uranium from dismantled nuclear warheads: "During recent negotiations with representatives of American environmental organizations we discussed possibilities for the construction of model controlled storehouses. One would be built at the beginning, then several more. Now in the American Congress there are debates about sending big sums of money for helping Russia solve these problems. Previously, the Congress promised several hundred million dollars in the forthcoming fiscal year, with an increase of 50 percent in the following year.

If this program is to be successful, it will mean great progress in making decisions about the ecological problems of the Chelyabinsk region."[3] The Cooperative Threat Reduction program that was under development by the U.S. Congress and the Clinton administration at the time of this interview did become a reality, and is described in the next section.

One response of the Ministry of Atomic Energy to the Chelyabinsk region was to talk about the need to establish public trust and to create an atmosphere of openness, to indicate the need to demonstrate the safe operation of nuclear power stations, and to plan to install modern equipment. The ministry also urged an energy policy for Russia that would increase the reliance on nuclear power from 11 percent to 22 percent for the generation of electric energy. As part of accomplishing this goal, construction would be completed at the South Urals Nuclear Power Station in the Chelyabinsk region (see chapter 3).

Nuclear Waste Legislation

Another example of the antienvironmental tone to government policy relevant to the Chelyabinsk region concerns the inconclusive debate on nuclear waste legislation. This debate began in the Russian Parliament in early 1992 and continued through the demise of that parliament in September 1993, a presidential decree was issued in April 1993, and in November 1994 there was an unsuccessful second reading of a draft law on nuclear waste in the Duma. A key policy debate in the development of the legislation centered around whether spent nuclear fuel would be classified as radioactive waste, or whether it should be in a separate category of material available for reprocessing and subsequent use in nuclear reactors (Perera 1992).

Early in 1992, the Ministry of Atomic Energy urged the Russian Parliament to classify spent nuclear fuel as acceptable for reprocessing. Such a classification would allow the ministry to import spent nuclear fuel from nuclear power reactors in Eastern Europe (and elsewhere), and potentially earn hard currency income. A consequence of such activity, of course, would be to increase the burden of radioactive waste at Mayak. The waste products from reprocessing are especially toxic to the environment. Environmentalists were united and vociferous in their opposition to this plan.

The sixth draft of the law entitled "On State Policy for Handling Nuclear Waste" was passed by the Committee on Ecology and Rational Use of Natural Resources in September of 1992. The Presidium of the Russian Parliament approved the bill for consideration at its next session.[4] Following the procedure in place at the time for the consideration of draft legislation, the deputy prime minister signed the bill. It was adopted at a first reading by the Russian Parliament on April 14, 1993.[5] A second (and final) reading was rescheduled for the fall 1993 session.

The Russian Parliament was abolished by President Yeltsin on September 21, 1993, and action on the legislation had not been completed when the Parliament ended. Recounting the dispute on pending legislation between the Ministry of Atomic Energy, some members of a parliament that still operated under a Soviet-era constitution, and the environmental community would not be particularly noteworthy (conflict is after all endemic to pending legislation and expected) except for the unexpected appearance of Presidential Decree Number 472 in April of 1993, five months before the Parliament was abolished, and while the legislation was still under active review.[6]

The wording of Presidential Decree No. 472 appeared to be relatively innocuous. It authorized the Russian Federation to fulfill intergovernmental agreements of the former Soviet Union. One portion of the cited agreements authorized Russia (read the Ministry of Atomic Energy) to deliver (read sell) nuclear fuel to countries in the former Soviet sphere of influence. A portion of the cited agreements authorized Russia (read the Ministry of Atomic Energy including the management of Mayak) to accept spent fuel from nuclear power stations abroad for reprocessing (for which the Ministry of Atomic Energy would presumably be paid in hard currency).

The decree directed that vitrified (glassified) waste from reprocessing spent fuel should *preferably* be returned to the country of the nuclear fuel's origin. This portion of the decree appears to mitigate negative long-term environmental consequences for Chelyabinsk. In reality, as issued, the directive would have little practical effect in safeguarding the environment. Most high-level liquid nuclear waste from reprocessing would remain at Mayak, and it is impractical to transport large quantities of high-level solid waste, even in vitrified form. The decree allowed Prime

Minister Chernomyrdin to sign a protocol with Hungary that envisaged that high-level waste would remain on Russian territory. The disposition of the large quantities of low-level waste that would be generated by reprocessing was not part of the public record.

A most significant portion of this presidential decree called for an expansion of reprocessing activities. This was in direct contrast with (and contradiction of) the intent of Article 18 of the draft Nuclear Waste Law, which called for limiting and phasing out reprocessing and achieving the "lowest possible generation of radioactive waste."

Already in effect at the time of Presidential Decree No. 472 was the 1992 Russian Law on Environmental Protection. This law had been passed by the Supreme Soviet on December 19, 1991, and signed by President Yeltsin in February of 1992.[7] Subsequent environmental acts, including those on nuclear waste, were to be consistent with the principles and framework of this fundamental law (Robinson 1992). A literal interpretation of the Russian Law on Environmental Protection is that while it forbids storage and burial of foreign nuclear waste, reprocessing in Russia is not expressly prohibited.

Members of the Russian environmental community did not have advance word of the development of Presidential Decree No. 472. There are arguably adverse environmental consequences that will result from the decree, and it certainly appears to undercut the intent of both the draft Nuclear Waste Law and the enacted 1992 Russian Law on Environmental Protection. The policy of the government with respect to nuclear waste and reprocessing was widely perceived as antienvironmental.

A new Russian Parliament consisting of a Duma and a Council of Federations was elected in December of 1993, and a new constitution for the Russian Federation was ratified. After the delay caused by the election and a protracted dispute over what quarters would replace the Russian White House as their home (it had been set on fire by tank attack ordered by the Yeltsin government in October 1993), the new Duma began a consideration in 1994 of the nuclear waste topic.

"On State Policy for Handling Nuclear Waste" was reintroduced in the new Duma as the basis for the pending legislation. It passed its first reading, and was scheduled for its second reading in the Duma on November 11, 1994. Whether or not to classify spent nuclear fuel from outside Russia as acceptable for reprocessing continued to be a highly contentious issue. Earlier in 1994, in de facto response to the issues addressed by Presidential Decree No. 472, the Duma had indicated that it would review, on a case-by-case basis, proposals to bring spent nuclear fuel of foreign origin into Russia for reprocessing. Subsequently, the Duma revoked a Hungarian reprocessing contract (Weslowsky 1994, 2).

Two of the public protagonists in this ongoing dispute between the government and the environmental community were the Russian Ministry for Atomic Energy and the Russian branch of the international organization Greenpeace. The position of the Ministry of Atomic Energy, as articulated by Vladislav Kotlov, was that reprocessing spent nuclear fuel of foreign origin is not unsafe: "Of course, to people who don't study these issues and aren't informed, it seems like a dangerous undertaking. But we have people who study these issues and know how to deal with it" (quoted in Weslowsky 1994, 2).

A counter position from Greenpeace contained three central arguments: 1) the Duma's previous action in revoking the Hungarian reprocessing contract was correct, in effect equating spent nuclear fuel with radioactive waste; 2) unprecedented environmental damage would be caused by extracting plutonium from spent nuclear fuel; and 3) further security and storage problems would result from increasing the stockpiles of plutonium and highly enriched uranium (Weslowsky 1994, 2).

In support of the environmental damage argument, Greenpeace claimed that 2,000 tons of low-level waste result from the reprocessing of one ton of spent nuclear fuel. Their preferred solution was to store spent nuclear fuel: "We hear more frequently reports about the leakage of radioactive materials, of the sale of nuclear material by bandits. The best thing to do would be to store the spent nuclear fuel and not reprocess it" (quoted in Weslowsky 1994, 2).

There was also a vigorous difference of opinion between the environmental community and the Ministry of Atomic Energy on the financial costs associated with the reprocessing program. The Ministry of Atomic Energy proposed to construct a new reprocessing plant (RT-2) in Krasnoyarsk at an estimated cost of 3.5 trillion rubles (about $1.3 billion at the November 1994 exchange rate). The position articulated by Greenpeace was that such an expenditure was unnecessary and would further exacerbate Russia's budget problems: "Russia, as always, finds itself in a bad position. That is, it is accepting completely unnecessary expenditures, which will result in most of the world's reprocessing being located here" (quoted in Weslowsky 1994, 2).

The counterposition from the Ministry of Atomic Energy was that only a portion of the cost (unspecified) of the reprocessing plant would be paid by the budget of the Russian Federation, and that the countries where the spent nuclear fuel originates would bear the cost of the reprocessing. As Kotlov from the Ministry of Atomic Energy said, "Part of the cost of RT-2 will be paid from the federal budget, but clients would pay for the reprocessing" (quoted in Weslowsky 1994, 2). This counterargument by the ministry was silent on the costs associated with mitigating environmental consequences of reprocessing and the long-term storage of the high-level and low-level radioactive wastes.

The legislation "On State Policy for the Handling of Radioactive Waste" was scheduled for its second of three required readings in the Duma on November 11, 1994. On November 10, 1994, Greenpeace hung a massive banner from a fifth floor balcony of the Moskva Hotel, located across the street from the new Duma building. The banner stated in both English and Russian "STOP PLUTONIUM PRODUCTION." For unreported reasons, the second reading vote scheduled in the Duma from November 11, 1994 did not occur. Subsequently, the nuclear waste legislation was referred back to committee for reconsideration. The representatives from Greenpeace who hung the banner were arrested by police; the reasons for the arrests were not reported.

Presidential Rejection and Chernomyrdin Decree No. 773

In 1995, legislative and executive branch consideration of the issues surrounding nuclear waste continued. On July 21, 1995, the Duma passed a law entitled "On Radiation Safety of the Population."[8] After the passage of ten days it was deemed approved by the Federation Council. President Yeltsin sent a letter to State Duma Chairman Ivan Rybkin claiming this law was contrary to the constitution, to Russian legislation, and to existing juridical practice, and chose not to sign it into law.

There were two reasons for this rejection and return for further consideration according to the presidential statement: The provisions of the law needed to be adjusted to meet the requirements set forth in the law of the Russian Federation "On State Secrets," and it needed to be adjusted to conform with "other legislative acts which are part of the system of the constitutional, civil, administrative and conservation legislation of the Russian Federation."[9] Yeltsin's reasons for the rejection of these two legislative bills are vague. These actions are, however, entirely consistent with the underlying thesis of this chapter, that there has been a continuous process under way in the Russian government of tilting against the environment.

On July 29, 1995, Prime Minister Viktor Chernomyrdin signed Decree of the Government of the Russian Federation No. 733 entitled "On Confirmation of the Procedure for Accepting Spent Nuclear Fuel from Foreign Nuclear Electric Power Plants for Subsequent Reprocessing at Russian Enterprises and for the Return of Radioactive Wastes and Materials Formed During Its Reprocessing."[10] Decree No. 733 is worth examining in detail in the context of President Yeltsin's contemporaneous rejection of the legislation on nuclear waste and radiation safety.

Decree No. 733 has three stated goals. The first is to "fulfill" Presidential Decree No. 389 of April 20, 1995, entitled "On Additional Measures for Increasing Monitoring of Compliance with Ecological Safety Requirements During the Reprocessing of Spent Nuclear Fuel." The second is to standardize "the acceptance of spent nuclear fuel from foreign nuclear power plants." The third is to ensure "ecological safety" during the course of the treatment (reprocessing) of spent fuel.

The origin of Decree No. 733 was the Ministry of Atomic Energy. Among the groups listed as approving the decree are the Ministry of Environment and Natural Resource Protection, the Ministry of Health Care and the Medical Industry, and the Russian Federal Nuclear and Radiation Safety Inspectorate (*Gosatomnadzor*).

It is useful to review the chain of events. By the start of 1995, it was very clear that the environmental community vehemently opposed importation of spent nuclear fuel for reprocessing. It was also obvious that the Duma was intent on enacting nuclear waste legislation, although it was not clear what its final position would be on the fuel reprocessing issue. Earlier in the process the Ministry of Atomic Energy had argued that they had an obligation to accept spent nuclear fuel from reactors in former Soviet bloc countries where they had helped build the power plants.

Effectively bypassing the structure of both the parliamentary and (presumably) the Chernomyrdin government, the Ministry of Atomic Energy had previously obtained the signature of President Yeltsin on two decrees relevant to reprocessing spent fuel. Decree No. 472 (April 21, 1993), reviewed earlier in this chapter, focused on the obligation claimed to import spent fuel from former Soviet bloc countries. The 1995 Presidential Decree No. 389 appeared narrower in scope: "On Additional Measures for Increasing Monitoring of Fulfillment of Ecological Safety Requirements During the Reprocessing of Spent Nuclear Fuel."

In contrast, the Chernomyrdin Decree No. 773 is much broader in scope. It authorized importation of spent nuclear fuel from plants abroad constructed with *technical assistance* (emphasis added) from the USSR, from *those built recently* (emphasis added) abroad, based on Russian Federation plans, and from a new open-ended category: "spent nuclear fuel from nuclear electric power plants built *according to the plans of other countries* (emphasis added) and accepted for reprocessing at the Gorno-Khimicheskiy Combine in the city of Zheleznogorsk, Krasnoyarsk Kray."[11]

Decree No. 733 went well beyond the scope of the two previous presidential decrees, as indicated by the italicized portions. Furthermore, the Ministry of Atomic Energy was essentially authorized to regulate itself on a host of critical matters including: the conditions for acceptance of spent nuclear fuel, determining what reprocessing products are not earmarked for further use in the Russian Federation (the rest must be returned), and how long and where nuclear materials may be stored (in apparent violation of the relevant provision of the 1992 Russian Law on Environmental Protection). There was also a provision that authorizes only "interested federal executive branch organs" to approve draft intergovernmental agreements with countries whose power plants were built to their own plans. Presumably, interested legislative and judicial branch organs are not authorized to review draft agreements.

As might be expected, there was an immediate negative reaction in the press and in the informed public. The *Moscow Times* (Dmitrieva) for September 13, 1995 carried a news story caption that read, "Activists

Decry Fallout From Decree on Waste." The reporter covering the story, Irina Dmitrieva, summarized the views of the environmental community: "a presidential decree providing for reprocessing spent foreign nuclear fuel in Russia will threaten the ecosystem and sharply increase the quantity of radioactive nuclear waste in the country." Marina Khotuleva of the Center for Independent Ecological Research was quoted as saying that "The decree breaks Russia's law on environmental protection and flouts the right to live in a healthy environment" (quoted in Dmitrieva 1995, 4).

In order for the reprocessing to occur, it would be necessary to complete the controversial RT-2 plant near Krasnoyarsk. In 1994, the cost for the partially constructed RT-2 plant was estimated at $1.3 billion (Weslowsky 1994). In a presentation at a news conference on September 12, 1995, Georgy Kaurov of the Ministry of Atomic Energy said that almost $2 million was required to resume construction of RT-2. He said the Ministry of Atomic Energy was considering international deals with Taiwan, North Korea, and Iran that would help complete the plant. It was reported that attempts had also been made to consummate deals with countries in Europe, though none have been successful so far.[12] The actual financial circumstances of the RT-2 plant, and the economics of reprocessing of spent nuclear fuel from abroad, are unclear. Chernomyrdin's Decree No. 773 authorized private firms to accomplish the task. Selection of the private firm would presumably be decided upon by officials of the Ministry of Atomic Energy, including a private firm they established themselves.

Another line of opposition to storing and reprocessing spent nuclear fuel from abroad was provided by Valery Menshikov, an adviser to President Yeltsin's Security Council. He indicated that Russia does not have sufficient storage space for the spent nuclear fuel generated by its own nuclear power plants: "The stores at all 29 of Russia's nuclear stations will be full by the year 2000" (quoted in Dmitrieva 1995, 4). He also said that the required environmental impact study had not been conducted.

The conflict in Russia over the importation for reprocessing of spent nuclear fuel is not going to be resolved in any straightforward manner. It has served to expose the manner in which the interests of powerful ministries are served by the government under the Constitution adopted on December 12, 1993. The flawed economy ironically serves as a major protector of the environment from new projects of dubious impact, not the established institutions of the government. Also serving as a protector of the environment are the institutions of a free press, a vital legacy from the glasnost era during the last years of the Soviet Union. And while the debate on reprocessing policy goes on, the Duma continues to review on a case-by-case basis, contracts for reprocessing spent nuclear fuel from foreign countries.

It is clear that there are different factions within the Russian government and broader Russian society that have fundamentally contrasting positions on nuclear safety and the effect of nuclear policy on the environment. The economy has prevented action in any consistent direction. Governmental actions (and inactions) simply defer addressing critical problems of nuclear safety and nuclear waste disposal.

Mayak's Role in Support of Government-to-Government Security Initiatives

There is under way a planned expansion of Mayak's role as a nuclear enterprise in support of cooperative national security initiatives with the United States. The rationale for Mayak's role is part of a fundamental paradigm shift in the meaning of nuclear threat. The form of this fundamental paradigm shift is manifested in a series of government-to-government initiatives known as the Cooperative Threat Reduction program (CTR) and the Materials Protection, Control, and Accounting program (MPC&A).

The rationale for these government-to-government programs is a redefinition of nuclear threat (and that of other materials for weapons of mass destruction). The principal actors in this fundamental paradigm shift are the institutions of Russian and American governments that control nuclear weapons and nuclear materials: the Russian Ministry of Defense, the Russian Ministry of Atomic Energy, the United States Department of Defense, and the United States Department of Energy.

The unprecedented cooperation included participation at the highest levels of the governments of the former Soviet Union and the Russian Federation, and of the United States. As part of developing H.R. 3807

(Public Law 102–228), which was agreed to on November 27, 1991, the Congress of the United States stated three formal findings: 1) Soviet President Gorbachev had requested Western help in dismantling nuclear weapons; 2) that the profound changes under way in the Soviet Union posed multiple threats to nuclear safety and stability, involving the disposition of weapons, the possible theft or seizure of nuclear materials, and the proliferation of weapons or knowledge outside of the territory of the Soviet Union and any successor entities; and 3) it was in the national security interests of the United States to facilitate the destruction of nuclear and other weapons in the Soviet Union and to assist in the prevention of weapons proliferation.

H.R. 3807 is formally called the "Soviet Nuclear Threat Reduction Act of 1991," popularly known as the Nunn-Lugar Act, after its principal legislative sponsors, Senators Sam Nunn of Georgia and Richard Lugar of Indiana. Both Presidents Boris Yeltsin and Bill Clinton have been active participants in advancing the work of the CTR program, as have Prime Ministers Viktor Chernomyrdin, Sergei Kiriyenko, and Yevgeny Primakov and Vice President Al Gore through regular meetings of the Gore-Chernomyrdin (and Gore-Kiriyenko and Gore-Primakov) commissions.

An extended rationale for the paradigm shift in the meaning of nuclear threat is provided by John P. Holdren (1996). Among other significant scientific service, Professor Holdren has been chair of the Committee on International Security and Arms Control of the National Academy of Sciences, and a member of the President's Committee of Advisers on Science and Technology. The heart of his argument is that "nothing could be more central to U.S. security than ensuring that nuclear weapons and the materials needed to make them do not fall into the hands of radical states or terrorist groups" (Holdren 1996, 14).

Holdren (1996) identified the most pressing concerns of this new security paradigm: 1) improving the security and accountability of nuclear materials; 2) combating nuclear smuggling; 3) increasing transparency in the management of weapons-usable nuclear materials; 4) halting or minimizing production of these materials; and 5) carrying out disposition procedures to reduce the risks from excess fissile materials by making them far more difficult to use in weapons (adapted from Holdren 1996, 14). A central point in Holdren's argument for reducing the nuclear threat in the former Soviet Union is that neither Soviet-era security systems nor storage systems were designed for the new circumstances confronting the Russian Federation.

Holdren (1996) assigned the "first priority" to establishing effective material protection, control, and accounting. In the case of Mayak, it was selected to be the location for a storage facility for plutonium and highly enriched uranium from dismantled weapons. The importance of this facility, partially funded by Nunn-Lugar monies, is to be found in its "greatly improved security and accounting compared to the locations where nuclear materials are currently being stored, and, in exchange for its assistance, the United States will gain transparency measures it has sought" (Holdren 1996, 15). Construction of appropriate facilities at Mayak began in 1994. In 1996, Holdren commented on the many delays that have plagued this effort (Holdren 1996, 15).

Cooperative Threat Reduction Program

The spirit behind the Cooperative Threat Reduction program is reflected in Senator Sam Nunn's (1996, 41) comments that: "Possession of nuclear, chemical, or biological weapons by rogue nations or terrorist groups could pose a clear and present danger to our society. . . . Although the risk of nuclear war is vastly reduced and the overall outlook for our security is greatly improved, the risk of chemical, biological, or some form of nuclear terrorism has increased."

Senator Richard Lugar (Lugar 1997, 6) expanded on the theme that prevention at home begins with increased security in the former Soviet Union and elsewhere: "If the United States is to have any choice of stopping the detonation of a weapon of mass destruction on our own soil, prevention must start at the source—the weapons and material depots and research institutes in the former Soviet Union and elsewhere."

The scope of the CTR program is indeed broad. One publication from the Department of Defense (DOD 1997) described the January 1996 meeting in the Ukraine of the defense ministers of the Ukraine and Russia and the U.S. secretary of defense as an event of significance: the destruction of a silo that formerly housed a strategic nuclear missile targeted at the United States. Six months later the three defense ministers met again at the former site of the missile silo to observe its transformation into a commercial sunflower field.

Robert Boudreau, chief of the Defense Cooperation Programs Office at the United States Embassy in Moscow, cited a wide range of general accomplishments with the CTR Program: working toward agreements for chemical-weapons elimination, beginning a dialogue on biological weapons, providing computers and software to help account for nuclear weapons, increasing the rate of elimination for SS-18 missiles, and supporting and tracking the reductions agreed to in arms control treaties (Boudreau, 1998).

Boudreau went on to elaborate a number of specific, tangible accomplishments of the CTR program: "In Russia alone, 80 submarine launch tubes have been destroyed, 50 missile silos and 181 missiles have been eliminated, and 27 bombers have been cut to pieces. With more support already provided or agreed to, approximately 690 additional missiles, 10 additional bombers, many more submarine launch tubes, more missile silos and additional rail- and land-based transporters for mobile missiles will be eliminated. Threat-reduction funding also made possible the removal of more than 3,800 deployed strategic warheads from Russia's weapons-delivery systems and assisted in the elimination of 214 missile silos" (Boudreau 1998, 31).

There are a number of indexes of the significance of the CTR program beyond these tangible accomplishments. First and foremost is the depth and consistency of the commitments, and the centrality of the participating Russian and American organizations to the national security of their respective countries. The depth of the commitment is reflected in table 11.1, which displays the categories of CTR program funding with Russia from 1992 until 1997. One fundamental meaning conveyed by the table is the centrality of national security in the core categories included within the CTR program. Another fundamental meaning is the inclusion of Mayak in important program areas, reflecting its prominent role in the production of nuclear materials and its central contribution to Russia's national security. Mayak is formally involved in aspects of the major categories of fissile material storage ($238.5 million), nuclear weapons protection, control, and accounting ($114.3 million), and the additional programs portion directed to the problem of arctic nuclear waste ($129.5 million). The depth of the CTR program within the Russian national security establishment is reflected in the participation by the Ministry of

Table 11.1
Cooperative Threat Reduction Program with Russia, 1992–1997

Program	Partners	Current value (in millions of dollars)
Strategic offensive arms elimination	Department of Defense Ministry of Defense Industries	$295.8
Fissile material storage	Department of Defense *MinAtom*	$238.5
Nuclear weapons protection, control, and accounting	Department of Defense MOD (12th Main Directorate)	$114.3
Chemical weapons destruction	Department of Defense President's Committee, MOD (15th Main Directorate)	$136.5
Reactor core conversion	Department of Defense, Department of Energy, NRC *MinAtom, Gosatomnadzor*	$10.0
Defense conversion/nonproliferation • "Fast Four" projects • R&D foundation • Defense Enterprise Fund	Department of Defense, Dept. of State Defense Enterprise Fund PM's Department of Defense-Industries MOD, Min. of Economics, Min. of Science & Technology	$58.0
Defense and military contacts	Department of Defense MOD	$14.5
Additional programs • Emergency response • Export control • International Science and Technology Center • CW production facility dismantlement • Arctic nuclear waste • Material control and accounting	Dept. of Defense, Dept. of Energy, Dept. of State, Dept. of Commerce MOD, MinAtom, Min. of Economics	$129.5
		Total: $997.1

Defense, the Ministry of Atomic Energy, and the prime minister's Department of Defense Industries.

Second, the consistency of commitment by the United States is reflected in the steady growth of the financing for the CTR program, as presented in figure 11.1. There are numerous compelling and competing pressures in the formulation of the U.S. national security budget. For a national security program such as the CTR to experience the continual growth in funding reflected in the figure normally means that a bipartisan consensus has coalesced around the importance of the overall mission. That has occurred with the CTR program.

Materials Protection, Control, and Accounting Program

The mission of the Materials Protection, Control, and Accounting (MPC&A) program is focused on the need for improved nuclear material security against both insider and outsider threats. Following the passage of Nunn-Lugar legislation in November 1991, the first MPC&A discus-

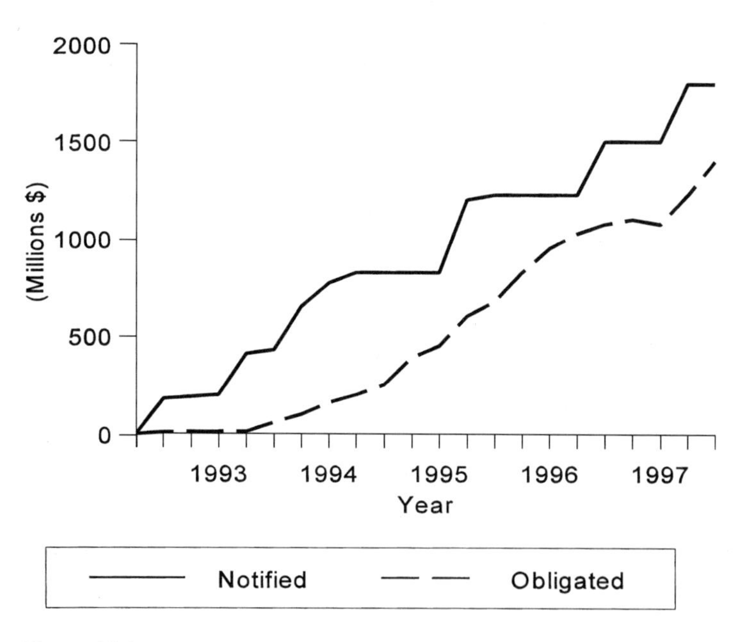

Figure 11.1
Growth in cooperative threat reduction program funding, 1992–1997
Source: Department of Defense (1997, p. 13).

sions were held in March 1992. In 1995, responsibility for the MPC&A effort was shifted by President Clinton from the Department of Defense to the Department of Energy.

DOE (1998) reports that only four months after Presidents Bush and Yeltsin signed an agreement in June 1992 authorizing expenditure of Nunn-Lugar funds, there was an authenticated theft of highly enriched uranium from Luch, Russia. In November 1993, there were thefts of highly enriched uranium from the Russian Navy. In May, June, August, and December of 1994, authorities in Western Europe seized illegally obtained plutonium and highly enriched uranium.

Although the MPC&A program had already been initiated when these thefts occurred, and the stolen nuclear materials were recovered, these events underscored the significance of the problem of nuclear material security. An early milestone of accomplishment was the demonstration in January of 1995 of pilot MPC&A systems at the Kurchatov Institute and the Arzamas-16 nuclear laboratory. All three of the elements of a MPC&A system are present at Mayak: physical protection (multiple barriers to entry, motion detectors, etc.), material accounting (measurements of the types and quantities of nuclear materials in a given area), and material control (controlled egress from storage sites, secure containers and identification codes, etc.).

The MPC&A program since September of 1995 has operated under the authority of presidential decision directive on "U.S. Policy on Improving Nuclear Material Security in Russia and the Other Newly Independent States" (PDD/NSC-41). This presidential decision directive established as one of the nation's top security objectives improving the security of nuclear materials in Russia, in the Newly Independent States formed in the dissolution of the Soviet Union, and in the Baltic states. In October 1995, Vice President Gore and Prime Minister Chernomyrdin made MPC&A improvements "a top priority for both governments" (DOE 1998, 7). Through the year 2002, the budget for MPC&A cooperation is estimated at $800 million.

Mayak's participation in MPC&A cooperation is scheduled from 1996 through 2000. It is not at all clear how this program at Mayak can end by the year 2000 if its achievements are to be fully realized and permanently maintained. In addition, the Fissile Materials Storage facility associated with the Cooperative Threat Reduction program is an ongoing initiative with strong MPC&A elements. The bottom line is that the Russian-American partnership involving nuclear materials security at Mayak, if it is to attain its fundamental program goals and sustain those accomplishments, will extend well into the twenty-first century. As for many of the U.S. nuclear facilities, stockpile stewardship will be a continuing responsibility and source of funding for Mayak during the post–Cold War era.

Implications for Mayak's Future of the CTR and MPC&A Programs

The record of accomplishment and potential of the Cooperative Threat Reduction program and the MPC&A program together contribute a dimension to Mayak's future that would have been unthinkable a decade ago. The extraordinary potential of these government-to-government agreements for contributing to a safer nuclear world have fundamentally redefined elements of Mayak's future. They mask, however, continued struggle within Russian society over elements of that future.

Mayak's Role in Solving Other Russian Nuclear Waste Crises

The Russian government has two major nuclear waste crises that were inherited from the former Soviet Union—without either the physical infrastructure or the fiscal resources with which to solve them. The first crisis is the legacy of the Soviet Nuclear Navy and Soviet icebreaker fleets. The Soviet Union failed to make provision for environmentally sound methods of disposal of its decommissioned nuclear fleet, especially submarines. Neither the Soviet Union nor the Russian Federation has complied with international treaties prohibiting the dumping of nuclear waste at sea.

The second crisis is that of the accumulation of spent nuclear fuel across Russia from both commercial nuclear power plants and from the propulsion reactors of nuclear submarines and icebreakers. A role has been proposed for Mayak in addressing these crises, which will only increase its burden of nuclear pollution.

Another factor relevant to considering Mayak's role in Russia's nuclear waste crises is the status of spent nuclear fuel in a country that utilizes a "closed fuel cycle." As with reactors in Japan and the United Kingdom, in a closed fuel cycle, uranium and plutonium from spent nuclear fuel are accorded by policy choice the status of a resource to be reprocessed and reused. A consequence of employing a closed fuel cycle is that the necessary chemical reprocessing of fuel generates an increase in the amount of radioactive waste, with much of it in liquid form (see chapter 3). The RT-1 chemical separation facility at Mayak is a principal location for nuclear fuel reprocessing in Russia.

The Russian Navy's Nuclear Waste Problem

Eaton (1995) summarized the situation today for the Russian Navy as "being swallowed by nuclear waste." Operating ships produce each year new liquid waste in the amount of 20,000 cubic meters, and solid waste in the amount of 6,000 tons. In the Russian Pacific Fleet, only one-half of the thirty-five out-of-service submarines have had their nuclear fuel removed; in the Russian Northern Fleet, only one-quarter of the forty-five out-of-service submarines has had their nuclear fuel removed. The total of four storage facilities available to the Northern and Pacific fleets are essentially full; and the Northern Fleet, the Pacific Fleet, and the Murmansk Shipping Company (icebreakers) have together more than 30,000 spent fuel assemblies in temporary storage (adapted from Eaton 1995, 290–291). According to Vraalsen (1998), there are now one hundred submarines in the Northern Fleet in need of decommissioning.

Since the inception of the Soviet nuclear fleet, Soviet and now Russian nuclear waste has been dumped at sea contravening the London Convention on the Prevention of Marine Pollution by Dumping of Wastes and Other Matters (1972); the Convention for the Protection of the Marine Environment of the Baltic Sea (1974); and the Convention on the Protection of the Black Sea Against Pollution (1993). In addition, Russia has not fully complied with the International Atomic Energy Agency requirement that it be notified of nuclear dumping. Radioactive materials have been dumped into the Seas of Japan and Okhotsk, and into the Kara and Barents seas.

The depth of the nuclear waste disposal problem of the Russian Navy is clearly illustrated by a series of events in October 1993. On October 12, 1993, in response to concerns voiced by the Japanese government about dumping nuclear waste at sea, Yeltsin signed a declaration jointly with Japanese Prime Minister Morihiro Hosokawa expressing grave concern about the ocean dumping of radioactive wastes (Joint Russian-Japanese Declaration 1993). At the time President Yeltsin was signing the joint declaration, Tanker TNT-27 of the Russian Navy was en route to a location in the Sea of Japan 120 miles southeast of Vladivostock and 341 miles west of the Japanese island of Hokkaido. The tanker was being monitored by the Pegasus, a ship belonging to the environmental organization Greenpeace (Sanger 1993). It took the tanker over twelve hours to dump its cargo of 237,000 gallons (900 tons) of low-level waste composed of coolant and cleansing fluid that had been used in servicing nuclear submarines (Watanabe and Boudreaux 1993). While Russia had served notice that it would continue to dump nuclear waste at sea, the timing coinciding with Yeltsin's visit to Japan and the location in fishing grounds resulted in significant protests. A senior Japanese Foreign Ministry official was quoted by Johnson as saying that "the Japanese people were shocked and outraged by this action and particularly so because it took place only days after President Yeltsin's visit" (Johnson 1993).

The actual dumping operation had received approval from the Russian Ministry of Environmental Protection. The reason for the dumping operation was that both tankers (TNT-27 and TNT-5) were full, and there was fear that one of the vessels might spill its cargo into a Russian harbor where its concentration would be a danger to Russian citizens (Lavine 1996, 422–23).[13]

Handler (1993) noted that a direct relationship has been developed between the Russian Navy's problems of nuclear waste disposal and Mayak's role in Russia's nuclear future. The logistic and financial difficulties in accomplishing this role, however, appear insurmountable. ITAR-TASS news agency reported the arrival at Mayak on August 5, 1996 of the first shipment of spent nuclear fuel from decommissioned Pacific Fleet nuclear submarines. Specialists from Mayak were quoted as saying that it would take up to ten years to completely remove the spent fuel from Pacific Fleet submarines and transport it to Mayak even if funding were available.

The lack of adequate financing has negatively impacted the transportation of nuclear fuel to Mayak. The Pacific Fleet only has thirty-two of the new TUK-18 containers that are necessary to transport spent or decommissioned fuel safely,[14] and that is not enough for the task. A further difficulty is that use of the new 40-ton TUK-18 containers requires the upgrading of service roads and the acquisition of new, mechanized handling equipment, including new, powerful hoisting cranes.

Another problem is that in response to its own budget dilemmas, Mayak is insisting on full payment in advance from the Russian Navy for each train load. In May 1995, Mayak required that each train, made up of four cars carrying a total of 12 TUK-18s, would entail a payment of up to $1.2 million.[15] The Russian Navy simply does not have the funds to support transport and reprocessing of its spent or decommissioned nuclear fuel.

An additional indication of government response to problems of the nuclear environment is found in another area involving the Russian Navy, the Kola Peninsula in Russia's northern territory located near Norway and Finland. A Norwegian environmental group called the Bellona Foundation undertook a study of military nuclear waste in that region. The foundation called the situation on the Kola Peninsula "a Chernobyl in slow motion" (*Energy Economist,* May 1996, 13). Their analysis attributed the origins of the nuclear waste crisis to the time pressures of the Cold War arms race and a failure to anticipate the magnitude of the problem.[16]

One response of the Russian security police to the release of the Bellona Foundation report was to arrest and jail Alexander Nikitin, a retired captain in the Northern Fleet who had served as a technical consultant to the Norwegian group. He was charged with treason and espionage for releasing confidential information on nuclear submarine accidents. The response of the Bellona Foundation to Nikitin's arrest was to assert that all of the information in their report was publicly available in Russia. Bellona's managing director, Frederick Hauge, cited the Russian Constitution as the source of legitimacy for their work: "According to the Russian Constitution, it is illegal to keep secret accidental situations that can affect another country, then there is no way to keep them secret" (quoted in *Energy Economist,* May 1996).

A representative of the Bellona Foundation stated the concern that there is an effort by Russian authorities to stifle freedom of expression:

"I think they want to make an example of Bellona because if Alexander is jailed and the courts find him guilty, that will mean very few other people inside Russia who are willing to work with this kind of problem" (quoted in the *Energy Economist,* May 1996, 14). In a concluding statement to their coverage of the Kola Peninsula, the *Energy Economist* May 1996 stated: ". . . by arresting Nikitin and resisting international efforts to tackle the nuclear waste issue, Russia is in danger of being seen to be flouting important environmental and human rights issues—a position which may result in Russia jeopardizing relations with funding organizations like the European Union, which is pushing for increased environmental awareness and nuclear safety throughout the former Soviet Union" (*Energy Economist* May 1996, 15).

Alexander Nikitin was arrested by the FSB on February 6, 1996, and held in jail until December 14, 1996, at which time he was released into confinement in the city of St. Petersburg. Since his release into city arrest, perspectives on the case have continued to polarize. As stated by the Bellona Foundation (Kudrik 1997), the case is far more than a false criminal complaint against an innocent citizen. It is about the "peoples' right to have access to information on the environmental situation in the new Russia, a right which has been relentlessly suppressed by various state bodies striving to keep public attention away from the hazardous mess that is the military nuclear complex, cloaking the embarrassing information in secrecy and justifying their actions by proud phrases about 'guarding the national security.' "

According to the Bellona Foundation, one of the fundamental flaws with the FSB case against Nikitin is that the definition of state secrets is unclear. For example, military experts disagree over whether various portions of Nikitin's North Fleet report include classified secrets. The FSB has used the threat of violating state secrets as a method to intimidate environmentalists. In addition, in October of 1997, President Yeltsin signed a state secrets law that established "all information on military bases including shipyards, labor conditions, and radioactive waste as secrets" (Ivanov and Perera 1997).[17] All this works to constrain the information and legal rights guaranteed to Russians under the Constitution, which is a special impediment to environmental groups that are attempting to challenge activities of the military nuclear complex.

The St. Petersburg court that heard the Nikitin case rejected it as too weak and sent it back to the prosecutors for more investigation (there is no provison in Russian law that authorizes an exoneration). Both the prosecution and the defense appealed the case to the Russian Supreme Court, which upheld the lower court ruling that the prosecutors' case was too weak, but concurred that the case should be returned to the prosecutors for more investigation.

Under a headline of "Court Puts Nikitin in Purgatory," the *Moscow Times* for February 5, 1999, said in part: "In a move that probably damns environmentalist Alexander Nikitin to many more years of living in limbo, the Russian Supreme Court refused Thursday to dismiss treason and espionage charges against him."

In another high-profile case, charges of treason have been leveled against Captain 2nd Rank (Commander) Grigory Pasko, senior editor of the Pacific Fleet's *Boyevaya Vakhta* (Combat Vigil) newspaper. On November 23, 1997, Captain Pasko left Vladivostok for Tokyo, Japan, for a visit under the sponsorship of the Mayor's Office of Vladivostok. On his way out of the country, customs agents confiscated several documents but allowed him to leave the country. Upon Captain Pasko's return to Russia on November 28, 1997, he was arrested and charged with "high treason," and accused of spying for Japan. In October 1993, on behalf of Japanese NHK TV, Captain Pasko had participated in the videotaping of a Russian tanker dumping into the Sea of Japan 900 tons of low-level nuclear waste from servicing dismantled or decommissioned nuclear powered submarines (see Inter Press Services 1998; Wadham 1997).

Pasko's lawyers claimed that the arrest was the Russian Navy's way of retaliating for his exposure of the environmental damage caused by the Pacific Fleet. General Viktor Kondratov, head of the local branch of the FSB in the Russian Far East, deemed that Pasko's arrest was not at all related to his environmental investigations: "It (the arrest) deals with Russia's state security secrets" (Inter Press Services 1998). In response, Vladimir Spiridonov, a fellow journalist with Pasko at *Boyevaya Vakhta*, said that the government's position of defining Pasko's arrest as based on treason for passing state secrets limits the extent to which defense lawyers can accuse the government of "suppressing press organs" (quoted in Wadham 1997).

Members of the nuclear environmental community surrounding Mayak have expressed parallel concerns to those noted above: that one response of the Russian government has been to take actions that flout both human rights and environmental rights issues in the name of state secrets. Indeed, since the mid-1990s, the successor organization to the KGB, the Federal Security Service (FSB), has continued to assert its role in affairs related to Mayak. Local environmentalists feel this as a constraint on their futher activities and especially those that involve foreign contacts.

Mayak's Business Plan Circa 1999

The new business plan for Mayak circa 1999 is based on fundamental changes in cooperative security policy between the Russian Federation and the United States. The principals in this new business plan are the Russian Ministry of Defense, the Russian Ministry of Atomic Energy, the United States Department of Defense, and the United States Department of Energy. The focus of the new business plan for Mayak is full participation in the government-to-government initiatives for the safe and secure storage of nuclear materials, their transportation, and the dismantlement of weapons of mass destruction.

The reality driving the new business plan for Mayak is that both the United States and Russia possessed excess nuclear and chemical weapons at the end of the Cold War. Russia did not inherit from the former Soviet Union the financial resources necessary to reduce excess weapons and to come into compliance with its treaty commitments, and was not likely to have the resources necessary to make reductions in the immediate next decades.

Two other circumstances that are related to its nuclear future have contributed to the new business plan for Mayak. The first is the presence at Mayak and elsewhere in the Russian Federation of vast stockpiles of weapons-grade plutonium and highly enriched uranium, which has been utilized in nuclear weapons. It is simply beyond the capacity of terrorist organizations and rogue nation states to build the facilities and industrial infrastructure to produce plutonium and highly enriched uranium. Theft or diversion of nuclear materials, however, is not beyond the capacity of either group. In the former Soviet Union, nuclear materials were protected by a system, variously characterized as "guards, gates, guns, and gulags." At the end of the Soviet Union there were insufficient fiscal resources to maintain the physical infrastructure of this approach to nuclear materials protection. Therefore, the United States and Russia have undertaken a cooperative program to improve nuclear material protection, control, and accounting. Furthermore, with more open borders, there were a variety of new and increased threats to the security of nuclear materials.

The second changed circumstance is highly ironic: as mentioned earlier, there are separately motivated proposals by the International Atomic Energy Agency and the Norwegian government to expand Mayak's role as a repository of nuclear waste. Given its history of nuclear waste tragedies and accidents, a business future for Mayak with an even greater nuclear waste burden reflects the significance of imminent nuclear waste dangers elsewhere in Russia that demand urgent solutions. Mayak's business future is integrally related to reducing imminent nuclear waste dangers elsewhere in Russia or to the international community, as the following list shows.

1. As discussed earlier in this chapter, Mayak has been selected under the Cooperative Threat Reduction Program as the location for a Fissile Material Storage Facility. The role of the United States Department of Defense is to provide the design, equipment, training, materials, and construction support that are associated with the building of a storage facility for the plutonium and highly enriched uranium from 12,500 dismantled nuclear warheads. A U.S. company, the Bechtel Corporation, was awarded the contract for design and construction. A nuclear weapons laboratory from Russia, Arzamus-16, was awarded the contract for design of a new system for nuclear materials protection, control, and accounting.

The project is proceeding at a pace that reflects the magnitude of this unprecedented undertaking. The concrete slab for the foundation was completed in October 1996. The fabrication of the roof was ongoing in 1998. Personnel from Arzamus-16 demonstrated prototype systems for nuclear materials protection, control, and accounting in July and December of 1997. Final equipment design and procurement was to be completed in fiscal year 1998, with construction completed in fiscal year 2001. Within the U.S. government, responsibility for this initiative at Mayak rests in the Department of Defense. In 1998 dollars, the value of fissile materials storage initiatives between the Russian Ministry of Atomic Energy and the United States Department of Defense is $238.5 million from 1992 through 1997.

2. Mayak is the location for an associated project within the broader initiative in Fissile Materials Storage. It is the destination for the actual containers that will store nuclear materials from dismantled nuclear warheads. Sandia National Laboratory in the United States developed the design for the fissile materials containers. A commercial firm from the United States, Scientific Ecology Group, Inc., of Carlsbad, New Mexico, received the contract to produce the first 24,000 containers. The U.S. Department of Defense exercised an option in its contract with Scientific Ecology Group, Inc., for an additional 8,400 containers.

3. Mayak is one of the locations for a government-to-government program to reduce the threat of nuclear proliferation and nuclear terrorism by improving the security of weapons-usable nuclear material in forms other than nuclear warheads. As of 1998, there are fifty-three sites within Russia, other Newly Independent States, and the Baltic Republics with agreements for cooperation in the area of nuclear materials protection control, and accounting. Of those fifty-three sites, twenty-seven have actually upgraded their security systems. The scope of this undertaking is reflected in the sheer magnitude of the amount of weapons-grade fissile materials produced in the Soviet era: 1,350 metric tons of plutonium.

According to a 1998 report from the Department of Energy, 700 metric tons is in the form necessary for nuclear bombs. The 650 metric tons in the form of metals, oxides, solutions, and scrap is stored in fifty sites across Russia, the Newly Independent States, and the Baltic countries. This portion contains more than enough nuclear materials for 40,000 nuclear bombs. The 700 metric tons of plutonium in nuclear bombs are under the control of the Russian Ministry of Defense, and are comparably less of a security risk.

Among the accomplishments under this initiative in 1997 at Mayak were: 1) the completion of interior sensor upgrades in buildings that store plutonium oxide; 2) the finalization of the design phase of a nuclear materials protection, control, and accounting system for the RT-1 Plutonium Storage Facility; and 3) the installation of four sophisticated nuclear materials portal monitors at pedestrian entry and exit points.

In 1995, the United States Department of Energy created a Russia/NIS Nuclear Material Security Task Force within the DOE Office of Arms Control and Nonproliferation. In coordination with the DOE National Laboratories the task force administers this elaborate initiative. Mayak is specifically included in the current program plan from 1996 through 2000.

Among the types of upgrades to existing Mayak systems were physical protections such as interior and exterior sensors and improved barriers and badges; improvements in responses to alarms and computer upgrades to process data from sensors; installation of nuclear material detectors at vehicle and pedestrian portals; barcode and other computerized systems to track nuclear materials inventory and tamper-indicating devices; structural improvements and perimeter clearing to enhance physical protection; and nondestructive assay measurements and such physical inventory control techniques as computerized material accounting systems.

From the United States government, the magnitude of the budget commitment for this general element of nuclear materials protection, control, and accounting through the year 2002 is estimated at $800 million. Mayak's central role within Russia's plutonium control and accounting process will undoubtedly extend its participation beyond the current 1996–2000 schedule.

4. The International Atomic Energy Agency has recommended "strong and concerted action" by the Russian Federation and the international community to address the radioactive waste situation in Northwest Russia. Among those projects "having the highest immediate priority requiring accelerated action" (Armbruster 1998) is the construction and commissioning of interim storage for spent nuclear fuel at Mayak.

Highly radioactive spent nuclear fuel will be an environmental hazard for thousands of years. Whatever method is employed for interim storage (for example, dry cask technology), the underlying proposal from the International Atomic Energy Agency is to use Mayak for a new purpose: interim storage of spent nuclear fuel from military and commercial sources.

5. The Norwegian government and the government of the Russian Federation entered into an agreement on May 26, 1998, under which Norway will provide the equivalent of $30 million over three to four years on actions to promote nuclear safety (Vraalsen, 1998). The agreement to promote nuclear safety removed two obstacles that had blocked previous attempts at cooperation: "1) a lack of exemption from taxes, duties, and fees on technical assistance grants; and 2) a lack of indemnification from lawsuits that might arise following a nuclear incident or damage to property owned by the Russian Federation" (Vraalsen, 1998, 1). These two obstacles have hindered a number of other international nuclear safety initiatives that had a focus on improving nuclear reactor safety.

The Norwegian pledge of $30 million to improve nuclear safety focuses on the dismantling of retired nuclear submarines. It creates a joint Norwegian-Russian Commission that will approve joint projects and coordinate and control the implementation of the agreement. Noting that there are well in excess of one hundred retired Navy submarines awaiting demolition in Northwest Russia, Ambassador Vraalsen indicated that "the safe handling and ultimate disposal of the large amounts of resulting radioactive waste and spent nuclear fuel represent formidable challenges."

The joint Norwegian-Russian Commission is too recent in its creation (May 26, 1998) to understand what role will evolve eventually for Mayak. There are numerous challenges along the road to solving the legacies of the Soviet Nuclear Navy, including the call by the International Atomic Energy Agency for the design and construction of a facility for defueling decommissioned nuclear submarines.

Evolving for Mayak, as a functioning nuclear enterprise on a major national railway line, is a key role in solving Russia's problems of nuclear materials from dismantled nuclear weapons, of nuclear materials that could be fabricated into nuclear weapons, of realization of aspects of the earlier business plan circa 1993 reviewed in chapter 3 of this volume, and of at least interim storage of spent nuclear fuel from military and commercial sources.

Still Tilting against the Nuclear Environment: The Future of Russian Government Policy

The future of Russian government policy on the nuclear environment is uncertain. There are a number of indicators that coalesce into a general assessment that in the nuclear area the government will continue to tilt against the environment in key policy decisions. At the same time, reflecting the profound changes in Russian society today, there are nuclear security initiatives that would have been unthinkable in Soviet times. Some of these initiatives are highly positive, such as in the government-to-government Cooperative Threat Reduction program, the Materials Protection, Control, and Accounting program, and the new Norway-Russian initiative on the nuclear waste crisis in and surrounding the Kola Peninsula and the Kara Sea.

In this concluding section, a number of situations will be introduced that reflect the countervailing pressures present in modern Russia that exert conflicting pressures on the nuclear future and the nuclear environment. One pressure is the bias toward economic and nuclear development with minimal explicit consideration of broader environmental impacts. Examples that will be reviewed are the contemporaneous decisions to reveal the nuclear waste tragedy from the injection into deep wells of 3 billion curies of radionuclides while deciding in the same time frame to double nuclear power without designing a solution to the nuclear waste crisis. The related conflict over the construction of the South Urals Nuclear Power Station has been highly contentious. The project has been in suspension because of economic factors, not government policy decisions.

The division over what constitutes state secrets and what constitutes constitutionally guaranteed access to environmental security information is an ongoing conflict between the Federal Security Service on the one side and the environmental community and the free press on the other side. The current form that the conflict has taken is over criminal charges against Alexander Nikitin and Grigory Pasko. The effectiveness of the response of the Bellona Foundation to the arrest of their colleague Nikitin reflects the profound changes in Russia. While the representatives of the Bellona Foundation are banned from Russia, the ideas they represent are part of the debate both within Russia and in the broader international community. In one of Russia's plutonium cities, Tomsk-7, another situation is unfolding that is pitting an environmental lawyer against government authorities. The local Tomsk court system is the current location of the Tomsk-7 conflict.

In the nuclear arena, some aspects of the continuing funding crisis within the Russian Federation have been ameliorated in part by the financial consequences of the paradigm shift in the definition of nuclear security resulting from CTR and the MPC&A programs. There is another program, which, while experiencing difficulties in implementation, represents yet another example of the paradigm shift in conceptions of security and international cooperation: The multiyear U.S.-Russian agreement for American purchase of highly enriched uranium is valued at $12 billion. The fact remains that, despite such positive initiatives, adequate funding for a safer nuclear environment is an essentially insurmountable obstacle for the Russian Federation.

A final indicator of the irreversible changes in Russian society that impact the nuclear environment is the introduction by the mayor of

Moscow of a proposal that the nuclear reactors of the Kurchatov Institute, once an integral part of the core of the national security scientific structure of the Soviet Union, move out of Moscow. Mayor Luzhkov may be motivated by the prospect of securing more land for development within the city limits of Moscow, and 44 hectares of prime urban land in Russia's major city is a very valuable commodity. The proposal of a "nuclear-free city" implies a motivation beyond concern with the increase in interim storage of nuclear waste in Moscow. Surprisingly, the head of the Kurchatov Institute subsequently endorsed a variation of the plan. That such a proposal was floated in the prelude to a national election and endorsed by the institute reflects a new climate of debate and discussion, one that would not have occurred in the former Soviet Union.

Bias Toward Economic and Nuclear Development

Many members of the environmental community within Chelyabinsk, and scientists outside the area, have expressed grave reservations about the South Urals Nuclear Power Station. But the most likely reason that construction has not resumed is the lack of necessary financial resources. In this regard the status of the South Urals Nuclear Power Station as an endorsed project is representative of government response to both nuclear and nonnuclear environmental problems in current-day Russia. There is a general antienvironmental tone to government policy and a willingness to subordinate environmental considerations to economic development.

The primary advocate within the Russian government for expanding nuclear power and reprocessing spent nuclear fuel from foreign countries has been the Ministry of Atomic Energy. This same group is responsible for the latest nuclear waste catastrophe, the roots of which go back to the Kyshtym nuclear waste explosion of 1957. An obvious lesson of the Kyshtym explosion was that storage of high-level nuclear waste requires a significant financial investment in waste storage tanks that must be vigilantly monitored for leakage or for the buildup of explosive gasses. In the United States, for example, at least sixty-seven of the single-shell steel tanks at the Hanford Reservation have leaked extensively into the surrounding environment (see chapter 2). In the absence of careful records about the chemical composition of the nuclear waste in U.S. storage tanks, it costs the Department of Energy at least $1 million a tank to take a core sample and from that determine a tank's contents.

Rather than confront the investment of financial resources and allocation of ongoing maintenance required by storage of nuclear waste in tanks, Soviet officials decided after the Kyshtym disaster to inject high-level nuclear waste into deep wells in what, at the time, were thought to be stable and impervious geological formations (under layers of shale and clay). All three injection sites have experienced unexpected migration of radionuclides into the surrounding groundwater. All three sites are in harsh and inhospitable climates. There is no feasible process for reversing this Soviet-era nuclear waste injection of 3 billion curies. This decision to utilize deep-well injection of uncontainerized waste was made before the method could be subjected to scientific study and environmental review. Henry W. Kendall, a recipient of the Nobel Prize in Physics, characterized this injection as "Far and away, . . . the largest and most careless nuclear practice that the human race has ever experienced. . . . It's just an enormous scale of irresponsibility" (quoted in Broad 1994, A8).

Knowing that this nuclear waste catastrophe was unfolding, however, the Ministry of Atomic Energy has continued to proceed with plans to double the capacity for the generation of electricity from nuclear power plants. In parallel, the ministry proceeded to lobby successfully for Presidential Decree No. 472 and Government Decree No. 773, which together have served as authorization for continued reprocessing of imported spent nuclear fuel. The legacy of excessive secrecy from the Soviet era has kept most Russians from learning about either the extent of the dumping-by-injection, or of the subsequent migration of radionuclides.

The point is that the Russian government has advocated major nuclear power expansion and the generation of large amounts of new nuclear waste at the very time it knew that another, perhaps irreversible, nuclear waste catastrophe had occurred with the deep-well waste sites. A long-term solution to nuclear waste storage was neither in place, nor included in the government planning process for the new undertakings. By definition, this is the government, through one of its ministries, again tilting against the environment in the pursuit of other goals. In the foreseeable future, it does not appear likely that the judicial branch of the

Russian government will serve as a counterbalance to government sanctioned assaults on the environment. Not only has the judicial system yet to become a positive force in environmental law, it does not have a basis in law to employ effective remedies against environmental transgressions.

It is too early to tell what role the Duma will take in environmental protection. Under the new constitution, it will take a two-thirds vote to successfully force a new policy on the government. In the meantime, it is ironic that the dire state of the economy itself is the major bulwark against an acceleration of the decline of the nuclear environment in Russia. In the current economic situation, even when there is better access to more reliable information, the great majority of the population does not seek active participation in political life affecting the environment.

Two dynamics are at work that affect participation in the environmental movement. The first is that the "greens" can expect only the passive support of the population in the context of a flawed economy requiring extraordinary efforts simply to provide basic necessities of life for individuals and families. The second is that environmentally adverse decisions by authorities do not meet with sufficient resistance from the population to cause their reversal. Therefore, there is a strong tendency for governmental authorities to revert to Soviet-era decision making, giving primacy to production and economic values instead of ecological ones.

Contentious Demarcation between State Secrets and Citizen Access

As was discussed in chapter 3, prior to glasnost and perestroika in 1985, attempts to articulate problems of the nuclear environment in the former Soviet Union were suppressed by organs of the state, particularly by the KGB. In recent years, environmental activists have claimed that the successor organization to the KGB, the FSB, is attempting to suppress individuals who are bringing attention to the most acute problems of the nuclear environment. The evidence offered for this assertion are the treason and espionage charges filed against Alexander Nikitin and Grigory Pasko, and their jailing while those charges are under investigation.

Another situation is unfolding in Tomsk-7, where Konstantin Lebedev, an environmental lawyer in Tomsk, charged in June of 1998 that "we have more than 22 Chernobyls under our feet" (quoted in Shulyakovskaya 1998). Lebedev was referring to the consequences of the injection by authorities at Tomsk-7 of from 500 million to 1.2 billion curies of radioactivity into local groundwater. Lebedev filed suit in August of 1997 to force the revocation of Tomsk-7's dumping permit. The background of the dispute is that Tomsk regional authorities in the Soviet era gave permission for the dumping. In 1996, the Tomsk Regional Administration issued permits to continue the practice. Valery Konyashkin, head of the Department of Radiological and Environmental Safety at the Tomsk Regional Administration, said "If we didn't give them the license, they would continue dumping anyway" (quoted in Shulyakovskaya 1998).

There are numerous elements to the dispute. Tomsk-7's permit allows it to dump radioactive waste water from cooling the nuclear reactors within the complex, and from processing nuclear materials, into two water-bearing layers (aquifers), which are 280 to 400 meters below the surface. The city of Tomsk, fifteen kilometers away, has artesian wells for drinking water, which are about 150 meters deep. The director of Tomsk-7, Gennady Khandorin, claims that a water-resistant layer of clay keeps water from the top and bottom layers from mixing (quoted in Shulyakovskaya 1998).

The facts of the contaminated groundwater's migration are disputed. Konyashkin said that recent tests indicated that traces of chemicals from the bottom aquifers where the nuclear waste was injected indicate that the waste could theoretically travel between layers. Shulyakovskaya (1998) reported that studies of village wells seven to fourteen kilometers away found contamination by radionuclides. Dr. Andrei Rybalchenko of the Russian Institute for Industrial Design has maintained that Lebedev's claims of contamination of the land or drinking water were groundless.[18] Rybalchenko was quoted as saying that the deep-well injection has been taking place for thirty-eight years, and that inspections by both geologists and environmentalists have found "no contamination of the region's land or drinking water."

Lebedev turned to the Tomsk District Court to settle the dispute. Officials at the court first refused to accept Lebedev's lawsuit on the grounds that it covered matters that were state secrets. Lebedev also encountered difficulties in obtaining information: "the regional administration refused to show us the license for underground disposal, saying it was the intellectual property of the administration."[19] A final illustration of the profound differences in perceptions comes from Lebedev's presentation of the attitude of the Tomsk Regional Administration: "They don't want to know the law . . . They are dependent on the nuclear complex. They do everything backwards. I say, let's get the environmental impact study and then issue the license. But they say, well, let's give them the license and see what happens."[20]

The struggle between environmental lawyer Lebedev over nuclear waste disposal practices at Tomsk illustrates the differences that exist between some members of the Russian nuclear expert community and some people in the Russian environmental community. It also illustrates some of the factors at work within the Tomsk Regional Administration and the Tomsk District Court.

The responses of governmental entities to environmental lawyer Lebedev and his lawsuit have ranged from dismissal of his claims as groundless to reluctant acceptance of the lawsuit itself. It is not at all clear that the court system will play a viable role in the foreseeable future in adjudicating environmental claims where the subject is a dispute about the nuclear environment. While the illustration offered in support of this generalization is at Tomsk and not Mayak, the earlier and continuing history of unresolved controversy over the South Urals Nuclear Power Station at Mayak contains many parallels.

An additional factor operating at Tomsk-7 that is relevant to Mayak is the role of the nuclear facility as an employer in a difficult economy and the political influence it derives from that role. According to Nikolai Ilyinskikh, a geneticist at the Siberian Medical University: "An entire city of 120,000 is built around the complex, which is based on a process that generates nuclear waste. . . . They don't know how else to make money. Besides, the complex is our region's only cash cow. Even the governor says that when he is short of money, he turns to Khandorin for help" (quoted in Shulyakovskaya 1998).

Continuing Funding of the Russian Federation

The comparatively large amounts of funding available to Mayak from the Cooperative Threat Reduction program and the Materials Protection, Control and Accounting program, comes from the United States government. This contrasts sharply with the funding available in current circumstances from the Russian government itself. During the summer of 1998, Siberian coal miners blocked railroads in Eastern Russia, which crippled several stretches of Russia's railway system. As the summer progressed, and the Russian government continued to fail to pay wage arrears, some miners moved their protest to the Russian White House in Moscow.

Polikarpov (1998) reported the unexpected circumstance that nuclear scientists chose to join the miners at the White House to protest against the poor financing of their enterprises. The trade union of the Russian Academy of Sciences planned to have 4,000 in attendance at a White House rally, including nuclear scientists from the Academy of Sciences research centers at Serpukhov and Troisk, who marched in from ninety-eight kilometers and forty kilometers away, respectively.

Polikarpov reported that both Serpukhov and Troisk have huge debts to their energy suppliers. The nuclear scientists are asking that salaries be raised. Earlier in the year the Russian government had agreed to budget the allocation for science at 4 percent of the total available for allocation in 1999, but this pledge was abandoned when faced with the ongoing economic crisis. The 2 percent budgeted for 1998 has been reduced subsequently by 27 percent, and then disbursed irregularly. In July 1997, workers at the Smolensk nuclear power station marched to Moscow on foot. Their protest, according to Polikarpov, concerned the failure by the government to guarantee regular funding for nuclear power stations: "The nuclear sector had received only 30 percent of the money allocated by the 1997 budget."

Vice Premier Viktor Christenko ordered that the Finance Ministry pay the energy debts of the Serpukhov and Troisk nuclear research centers, but also said that the government does not have the resources to increase financing this year. The continuing fiscal crisis affecting almost all elements of Russian government was reflected in an article by John Helmer (1998): "What government in the world could hope to attract foreign currency flows and investment, if, according to its independent state auditor, it refuses to pay one-third of its work force; cheats its pensioners with payment formulas and delays that are contrary to law; orders electricity and water supplies cut off from towns that have tax arrears, and misdirects about one-third of the annual budget."

For the fiscal stability of Mayak and other institutions in Russia that affect the nuclear environment and nuclear safety, there are clear structural impediments to their receiving dependable revenue from the Russian government, no matter what priority is assigned. The devaluation of the ruble in late summer of 1998, and the default on government debt obligations and associated financial instability, only exacerbate the underlying economic flaws hindering a safer nuclear future for Russia.

The Nuclear Environment in Russia's Future

Today Russia is again linked to the United States in a nuclear arms race. This time, however, the race is to protect the world from the environmental and political legacy of nuclear weapons. Even in the United States, the costs of environmental remediation severely strain Department of Energy resources (see chapter 10). Russia faces an even greater task to redress past environmental damages, and, frankly, the nation lacks the resources to accomplish this task even if there were proven technology.

In addition, nuclear weapons still exist, and their existence represent a military resource but also a military threat to both nations. The U.S. government is developing a program of stockpile stewardship to maintain the core resources of the nuclear weapons complex, such as the expertise that exists at Livermore and Los Alamos, while ensuring the continued maintenance and security of the present nuclear weapons arsenal. An even greater challenge exists for Russia. The monitoring of the nuclear stockpile falls dramatically short of what might be considered minimal security standards. The Russian government is unable to support the research scientists and technical expertise of the former weapons facilities, nor can these facilities easily be converted to nonnuclear applications. The threat of nuclear proliferation, either involving nuclear material or nuclear weapons expertise—represents a new global threat from the Cold War arms race.

Russia is also unprepared to meet this challenge. There are insufficient resources for the task of safeguarding Russia's remaining nuclear weapons arsenal. For example, in summer of 1998 the governor of the Krasnoyarsk region (former General Alexander Lebed, a likely presidential candidate in the 2000 elections) threatened to assume control of the nuclear forces in the region because military officers had not been paid in months and security for nuclear weapons and material was insufficient. The antinuclear stance of Moscow's mayor (another likely presidential candidate, discussed earlier) is another indicator of the political posturing in the current situation, something that would have been inconceivable in the Soviet era.

To address this challenge, the U.S. and Russian governments have established a joint commission on nuclear security issues. In July 1998, Vice President Gore and Prime Minister Kiriyenko reached a series of bilateral agreements. These agreements will provide for the retraining of nuclear scientists and bilateral cooperation to control the plutonium removed from dismantled nuclear weapons. These programs will largely be funded from U.S. government sources. Ironically, the resolution of the Cold War has forced both nations to cooperate to resolve the legacies of the past.

The struggle over the future of the nuclear experiment in Russia is intense. It is being fought over the choice of values that will influence policy decisions, over which institutions of government and society will have dominant roles in decision making, and even over whether (and how) the rule of law will be applied and disputes adjudicated.

There is also a specter that intrudes into many debates—the image of another Chernobyl. In the disputes over the safety of the decommissioned submarines around the Kola Peninsula or the deep-well injection of nuclear waste at Tomsk, the threat of another Chernobyl is an explicitly used metaphor for the severity of the threat to health and the environment. The wisdom and judgment of nuclear elites is also constantly under attack over the risk of catastrophe in the nuclear age, when safety provisions have already proven inadequate. Furthermore, nuclear dangers inherited from the Soviet Union have not been accompanied either by the financial resources or the technology necessary to reduce the proven risks.

The consequences of past Soviet-era decisions about the disposal of nuclear waste and the failure to develop adequate plans for decommissioning nuclear submarines are sobering reminders of how great the known dangers are. The perceived lack of credibility of the nuclear elites' judgment intensifies public concerns about the lurking presence of unknown nuclear risks. This is compounded by the nuclear elites' record of deliberately covering up past threats to human health in order to advance their nuclear agenda.

The institutions of the Soviet or Russian governments have not been the most influential in alerting the country to its nuclear peril. Rather, Russian citizens working through nongovernmental organizations, both foreign and domestic, have signaled the peril. Individuals who exercised personal responsibility by alerting their fellow citizens to information on these environmental matters have done so in a context where access to environmental security information is constitutionally guaranteed, but impeded by claims of violation of state secrets.

Executive branch domination of the nuclear environmental agenda continues. Neither the legislative nor the judicial branches of government have acted as real or symbolic counterweights to the executive. In 1993, Alexei Yablokov, then President Yeltsin's primary adviser on the environment and public health, stated that environmental preservation by the Russian government would not be successful until the Parliament could strongly influence the activities of the ministries. As was evidenced by Presidential Decree 472 and Prime Minister Chernomyrdin's Decree 773, the ministries can sometimes still get what they want, by secret or at least nonpublic dealings with the executive branch and the circumvention of the legislative branch.

The flawed Russian economy remains another major obstacle to new nuclear development, but it is also an obstacle to improved nuclear safety and environmental restoration. Representatives of the international community have become important sources of funding for the Russian nuclear elites dealing with nuclear weapons development programs and with programs to safeguard nuclear materials. Support comes from the U.S. CTR and MPC&A programs, the U.S. Nuclear Regulatory Commission, the Norwegian government, the International Atomic Energy Agency, and other sources. The allocations available to the nuclear elites exceed by many orders of magnitude the resources available to the U.S. Agency for International Development for Russian programs. Despite USAID's many successes in helping to build the institutions of a civil society in Russia, these efforts are severely underfunded. There is not the level of bipartisan support in the U.S. Congress for sufficient technical assistance to the NGO environmental sector in Russia. Until there is a more effective voice on behalf of a safer nuclear world and environmental restoration, Russian government policy will continue to tilt against the environment.

Notes

1. See Whiteley and Gontmacher (1998) for a more extensive discussion of environmental and nuclear weapons policy during the Soviet era.

2. *Ekologicheskiy Vestnik,* "Yuzhniy Ural," N4–5, 1992.

3. Ibid.

4. Session of the presidium of the Russian Supreme Soviet opens. BBC Summary World Broadcasts, September 9, 1992.

5. Law on Radioactive Waste, *CIS Environmental Watch* (1993): 83.

6. Decree 472 is entitled "On Fulfillment by the Russian Federation of Intergovernmental Agreements on Cooperation in the Construction of Nuclear Electric Power Plants Abroad."

7. RSFSR Law on Environmental Protection, *Rossiyskaya Gazeta,* March 3, 1992. Translated by Foreign Broadcast Information Service (FBIS) JPRS-TEN-92-007, April 15, 1992, RSFSR Law No. 2060–1.

8. FBIS-SOV-95-166, August 28, 1995, p. 24.

9. FBIS-SOV-95-166, August 28, 1995, p. 24.

10. FBIS-TEN-95-007-L, August 24, 1995, pp. 3–4.

11. FBIS-TEN-95-007-L, August 24, 1995, p. 4.

12. Lydia Popova, Personal Communication, December 19, 1995.

13. In a context of having no facilities to process the low-level waste from the Pacific Fleet, Russia announced that an additional load of it would be dumped into the Sea of Japan between October 20 and November 15, 1993. In response to protests about this contemplated action, Prime Minister Viktor Chernomyrdin delayed the second dumping, but the Agence France (October 21) reported that other Russian officials indicated that the suspension was only temporary. Subsequently, after negotiating a transfer (diversion) of funds intended for the dismantlement of nuclear weapons for use on the nuclear waste problem, Viktor Mikhailov, then Minister of Atomic Energy, pledged that dumping would cease in the Sea of Japan (Usui 1993a; 1993b). Sun (1993) also reported speculation that since Minister Mikhailov did not pledge to stop dumping at sea, the location would merely switch to the Arctic Ocean or the Pacific Ocean.

14. British Broadcasting Corporation. 1996. Chelyabinsk plant receives used nuclear fuel from Pacific Fleet for recycling. In British Broadcasting Corporation Summary of World Broadcasts, August 5, ITAR-TASS (Record Number: 00805*1996 0805*00262).

15. Interpress Service. 1996. The removal of nuclear fuel from Pacific bases, Global NewsBank, July 19. (Record Number: 01084*1996 0719*00014).

16. The coauthor of the Bellona Foundation report, Thomas Nilson, was quoted as saying, "After we released our report at the G-7 summit in Moscow, the Russian security police have given an order that our report is not to be owned or distributed within Russia and they are trying to close down even more information which was open one or two years ago" (quoted in *Energy Economist* 1996).

17. The Bellona Foundation further argued that Nikitin was being charged under a decree that was enacted eleven months after the alleged activity was declared criminal by Decree 055:96 of the Ministry of Defense. In addition, the decree itself is secret. "Procedural violations in the Nikitin case." Bellona Foundation website: www.bellona.no/e/fakta/fakta84.htm, October 9, 1997.

18. The *Moscow Tribune,* Nuclear experts deny contamination, June 27, 1998.

19. Ibid.

20. Ibid.

VI

Conclusion

12

Conclusion: Legacies of the Nuclear Age

Russell J. Dalton, Paula Garb, Nicholas P. Lovrich, John C. Pierce, and John M. Whiteley

Plus ça change . . .

In studying Hanford and Mayak we have had the dubious distinction of examining two of the most polluted places on Earth. Hanford, the primary site of weapons-grade plutonium production in the United States, holds nearly two-thirds of the nuclear waste from the entire U.S. nuclear weapons production process. Over its lifetime, Mayak has released more than twenty times the radiation of Chernobyl, and encompasses some of the most radioactive land on this planet.

There are many important questions to be asked in connection with these two facilities. How could such environmental damage occur in the first place? The four decades of the Cold War arms race created an environmental legacy that will last for centuries. The earth has lost a part of itself that it will never regain; areas within the Hanford and Mayak facilities will not be habitable for hundreds or thousands of years. Along with the loss of territory has come a loss of human lives and health; deeply resentful feelings of disaffection and uncertainty hang over both regions. In the United States, the scale and lengthy duration of official governmental deception adds to the problem of declining public trust in political institutions (Nye, Zelikow, and King 1997). Especially in the Russian case, the magnitude of environmental and human damage is almost unfathomable—and the governmental response is even more uncertain.

An equally important question concerns what can be done to redress these past problems. Just as the development of nuclear weapons expanded the boundaries of human scientific and technical knowledge, correcting the environmental consequences of plutonium production will require comparable intellectual efforts to develop environmental remediation technologies—as well as an incredible amount from the public treasury (Martinez and Byrne 1996). Both the United States and Russia face an environmental cleanup challenge of unprecedented scale and cost. By many accounts, the cost of cleanup will exceed the total costs associated with producing the nuclear weapons in the first place.

Our research project focused on one crucial aspect of the Hanford/Mayak situation: How have the citizens and governments in both nations responded to these environmental challenges? For forty years nuclear weapons production created mounting environmental problems, and the effects of early miscalculations (or mistakes) in decisions about waste storage and atmospheric release policies were compounded with the passage of time. Despite ominous warning signs, such as the 1957 explosion at Mayak and the 1949 Green Run at Hanford, it is clear in retrospect that not nearly enough was done to protect the environment or the residents of the affected regions. The public accounting of environmental degradation only entered the political agenda when citizens in both countries began to question plant officials and government authorities, and to press for an open public discussion of the topic. The government in both cases responded in reaction to public pressure, but often the response was insufficient or illusory.

A promising beginning to solving the environmental problems at Hanford and Mayak, and the world's environmental problems more generally, is the frank and honest recognition that a substantial problem exists. We tracked the process of issue politicization and mobilization against the Hanford and Mayak facilities in considerable detail. We studied how the issue entered the agenda of politics, how it was framed in terms of the larger political debate, and how (or whether) citizen concern was translated into appropriate governmental action.

Comparing citizen action toward the facilities at Hanford and at Mayak offers the opportunity for unique comparisons and contrasts. The root nature of the environmental issue was similar at both sites. The government managers of the Hanford and Mayak facilities were primarily concerned with producing nuclear weapons to ensure their nation's defense; even if environmental concerns were important in their decision making, they were at best secondary considerations. Both facilities were similarly shrouded behind a veil of secrecy about their respective operations and about the environmental consequences of their activities. As the Cold War was ending, a partial lifting of strict secrecy provisions occurred in the late 1980s, which facilitated the politicization and mobilization of the residents in the surrounding areas. In addition, both nations now had an interest in playing up the benefits of rapprochement, which included telling their citizens how costly the nuclear arms race had been in financial and environmental terms. In both nations, however, a constructive government response to environmental devastation only resulted from the pressure of citizens and environmental groups to confront the legacy of nuclear weapons production in an open public forum.

The contrasts are also significant. Although both facilities scarred their immediate and surrounding environments, the damage wrought by the Mayak facility was much more severe. Hanford's safety and environmental history appears flawed by contemporary standards, but the problems created in the American Northwest fade in comparison to those that Mayak wrought upon its surroundings. From the selection of the site for these facilities, to their construction, to their on-site management, to their production standards and present mode of operation there have been clear differences in the environmental records of the two plutonium production facilities.

The nature and context for citizen action against the facilities were also very different. Political action at Hanford occurred in an established democratic system, built upon a legal foundation of strong individual civil and property rights, and with significant organizational and political resources for the local environmental groups to draw on. Environmental action against Mayak occurred in the midst of revolutionary changes in the social and political orders, with only nascent forms of democratic rights, civil liberties, and open governmental procedures in place, and with no local history of environmental activism.

Our goal was to use the two critical case studies of Hanford and Mayak to draw larger lessons about the nature of environmentalism and democracy in the two nations. This chapter discusses the broad implications of our research in three particular areas. First, the Hanford and Mayak experiences provide insights into the process of mobilizing citizen protest. By comparing how citizen groups formed and acted in these two cases, we learned a great deal about the general process of political mobilization. Second, our study illuminated how diverse elements of the citizenry think and act with respect to environmental matters. Prior public opinion research on environmental issues is almost exclusively based on the experience of Western democracies. We had the unusual opportunity to compare the attitudes of Americans and Russians on salient and similar environmental problems; this empirical foundation provides a better understanding of the nature of environmentalism as a unified paradigm. Third, the governmental response to citizen concerns about both facilities allowed us to compare how the policy process differs across these two nations. Simply put, are there significant differences in how the respective political systems performed in these two nations—and if so, why? The rest of this chapter is devoted to answering these questions.

The Mobilization of Protest

We begin our review by focusing on the environmental groups that mobilized citizen protest against the managers and government agencies responsible for the Hanford and Mayak facilities. Although many factors accounted for the politicization of public concerns in the late 1980s and early 1990s, sustained attention to the facilities' problems, and the government's eventual response, would not have occurred without citizen groups focusing attention on these problems and developing concrete political campaigns.

Citizen environmental protest around Hanford and Mayak offers a unique opportunity to compare the development of citizen action in these very different sociopolitical settings (see chapters 5 and 6). Both movements faced many of the same hurdles: to accumulate reliable information on the facilities despite the veil of secrecy, to disseminate this information to the public, to mobilize public opposition to the facilities, and finally to translate these popular sentiments into policy change. While the tasks were similar, the context for these actions of policy advocacy was widely different between the two nations. Consequently, a comparison of both movements illustrates which aspects of citizen mobilization are so essential that they transcend the specifics of the Russian and American situations.

Resource Mobilization

One important area of comparison involves the resources available to both movements. The literature on social movements stresses the point that political action begins with the mobilization of resources to support a "cause" (Zald and McCarthy 1987; Jenkins 1983). Both the Hanford and Mayak groups faced this challenge, and attempted to solve it in their own ways. The Hanford protest groups tapped into a variety of funding options. Many of the groups sustained themselves with membership dues; others were funded by donations; several received government grants for evaluation studies and policy analyses; private foundations also contributed to the Hanford groups. The Hanford groups also included branches of several regional and national associations—such as Physicians for Social Responsibility, the Sierra Club, and the Association of Atomic Scientists—so that funds, and more important, expertise and strategic advice were available from a broad network. The Yakama, Umatilla, and Nez Percé tribal governments have their own financial base. They often received grants from U.S. government agencies to support environmental protection activities, which led to conflict with the secretive actions of the DOE officials who managed the Hanford complex.

Another important resource for social movements is information, especially when the issue concerns complex scientific issues of plutonium production and the pollution caused by the disposal of plutonium production byproducts. After 1986, Hanford groups had access to a wealth of scientific and technical information. They used this information to combat and refute the environmental claims made by plant managers or government officials. This information most often came from public sources, other antinuclear protest organizations, and the government itself (particularly state and local agencies). The forced release of Hanford documents through citizen petitions under the Freedom of Information Act was a major resource for the Hanford protestors. This release was followed by a stream of other documents and government-funded studies, as John Whiteley described in chapter 2. Furthermore, Hanford environmentalists had access to independent scientific experts who reviewed and challenged government studies. Access to such information was important in developing a convincing public case against Hanford.

In contrast, the problems of resource mobilization were much more severe in Chelyabinsk. First, the groups protesting Mayak suffered because they were in a less affluent society, where financial resources were far scarcer (especially after the collapse of the Soviet-era economy). For instance, the Nuclear Safety Movement (NSM) in Chelyabinsk eventually developed a small dues-paying membership, but the ability of Russians to make a financial contribution to a political movement was severely limited in the early 1990s. Moreover, the society lacked the tradition of autonomous, dues-paying citizen movements that were common among Western democracies. These difficulties were further compounded because the Russian economy at the time initially lacked the financial instruments to collect donations by mail.[1]

Faced by these problems, the environmental groups protesting Mayak found financial support where they could. When the reform tide was at its apex and the regional government supported the environmentalists, the Nuclear Safety Movement received office space and administrative support from the local government. Later, when official support was withdrawn, environmental groups turned to their members and to contributions-in-kind from supporters. At times, foreign support—provided by Western foundations, environmental groups, and academic sources—were of critical importance in sustaining the Nuclear Safety Movement in Chelyabinsk and the Socio-Ecological Union (SEU) in Moscow.

Second, Mayak groups had limited access to the type of information resources that were normally available in an open, pluralist society. The Russian government and Mayak officials had a virtual monopoly of information about the facilities and their environmental impact. Gorbachev's policies of glasnost and perestroika provided some access to this information, and the privilege and ability to ask for more. Some of this information was provided by local residents who were personally concerned about the facility; and there were a few reformers in local and regional governments who assembled information on the facility. Again, however, it was necessary for outside actors—the Socio-Ecological Union (Russia), the Natural Resource Defense Council (United States), and Western environmentalists—to assist the Mayak groups in acquiring up-to-date information on nuclear issues, access to independent scientific expertise, and even medical supplies to deal with the health effects of Mayak.

In summary, both movements faced the same basic needs to marshal the resources necessary to organize an opposition to the official coverup of environmental and public health problems at the Hanford and Mayak facilities. There were significant differences, however, in their ability to mobilize such resources. The Hanford groups operated in a relatively rich resource environment, while the Mayak groups were constantly struggling to survive, let alone succeed in their policy advocacy. At most, the Mayak groups included a few thousand formal members in a region with a population in excess of 3.6 million. It is more difficult to determine the precise number of Hanford activists because the level of engagement and types of groups that became involved are so varied. It is clear, however, that the formal membership of the groups involved at Hanford was much greater than at Chelyabinsk, and the network of citizens and activists receiving newsletters and regular mailings from Hanford groups was at least several times larger. HEAL and the other Hanford groups were better able to mobilize the resources required to mount a publicity campaign, collect research data, and present testimony at government hearings or public discussions; the Mayak groups were unable to sustain the same level of activity.

Since the time of our study these contrasts have endured, and even increased. The environmental groups criticizing conditions at Hanford continue to attract substantial public attention. DOE cleanup efforts have shifted the focus of public policy from weapons production to environmental restoration and stabilization. National environmental laws now apply to the nuclear weapons facilities under the auspices of the Tri-Party Agreement between the DOE, EPA, and the state governments. The intense mobilization phase for the Hanford groups has passed, in part because they have been successful in gaining recognition as legitimate stakeholders and in stimulating changes in public policy. In contrast, the Mayak groups continue to struggle with an insufficient resource base and against an ingrained culture of secrecy. However, despite limited resources and a vigilant local security force (successors of the KGB),

Chelyabinsk environmentalists maintain a semblance of educational and legislative campaigns.

Alliance Networks

The ability of a citizen movement to challenge the government is partially dependent on its own abilities, and partially dependent on its capacity to attract allies to its cause (Diani 1995; Klandermans 1990). Emerging social movements can enhance their resources and opportunities by drawing support from existing interest groups and established political actors. Similarly, opposition from important political actors can create a major impediment to success.

One important difference between the two movements was their internal organizational diversity; another related to the diversity of elite-challenging actors. One of the best allies of an environmental group is another environmental group. The number and variety of groups active on the Hanford issue was much greater than those protesting against Mayak. In the American context there is a tendency for each group to specialize in its particular interests, and consequently in its oppositional activities. HEAL holds a leading position in the movement, and is active on a wide range of Hanford-related issues. The Hanford Downwinders' Health Concerns (HDHC) fills a special niche in the environmental health movement, focusing on public health issues for downwind residents; it is also pursuing litigation against Hanford. The Government Accountability Project focuses on oversight issues and serves as a protector of whistleblowers who revealed information about Hanford. Columbia River United's emphasis on water quality issues gives it standing on numerous government advisory and review committees. The tribal governments of the Yakama, Umatilla, and Nez Perce reservations represent another set of special actors with distinct environmental concerns and locally based culture. The tribes' legal status as sovereign nations creates a special government-to-government relationship between tribal officials and the federal government. This gives the tribes significant leverage in dealing with U.S. government agencies; the U.S. government must both exercise a trustee duty toward the tribes and must negotiate with them as a legally equal partner.

In comparison, the movement against Mayak was narrower in organizational diversity. The Nuclear Safety Movement is the most prominent group, and it developed ties with the Democratic Green Party and the Socio-Ecological Union based in Moscow. Another major group, the Kyshtym 57 Foundation, has a very narrow mandate, closer to the HDHC at Hanford. The Association of Greens is a conservative environmental group in the Chelyabinsk region. There are only loose ties among the various Mayak-related groups. Moreover, on occasion the various elements of the movement seemed to function more as rivals than as separate elements of a broad movement. The NSM was more likely to question the authorities at Mayak and to raise fundamental issues about the facilities; some environmentalists even claimed the Association of Greens was a Communist front organization.

There were some tensions among Hanford groups, though the magnitude of these differences was much smaller. Two Hanford groups—the Hanford Downwinders' Coalition and the Hanford Downwinders' Health Concerns—disagreed over the use of U.S. DOE funds to establish the Hanford Health Information Network. The former pursued litigation against the DOE and its contractors, and the latter worked with the DOE in setting up a system for documenting the adverse health effects resulting from Hanford releases. However, this division was relatively minor when compared with the fundamental divisions plaguing the Mayak groups. The HDC and the HDHC still work closely together and with other Hanford groups despite this one experience. In short, there was a clear division and rivalry among the Mayak-related groups that did not exist among Hanford groups.

We found that the citizen groups protesting Hanford drew upon an extensive support network outside of other local groups. Hanford groups could turn to national organizations, such as the Sierra Club, as sources of resources and technical expertise. In addition, there already existed an informal network of environmentalists in Washington State that the Hanford environmental groups drew upon (Rogers 1996). Another kcy ally of the Hanford groups was the regional news media. Karen Dorn Steele's meeting with the early HEAL activists, her engaging reporting of their campaign in the Spokane *Spokesman-Review,* and her exposés on the Hanford facilities were crucial events in developing broad public awareness and significant public support for these citizen groups. In short, the Hanford protests became integrated into a preexisting dense organizational and personal network of environmental action in Washington State, and could develop rather effectively in this political climate.

The groups protesting Mayak existed in a much different institutional context. The greatest contrast was the lack of a preexisting network of social movements and citizen interest groups. The Nuclear Safety Movement and the other Mayak groups were creating a new organizational form as Russia was democratizing. There were few other historical or contemporary models to draw upon for guidance, and few other groups that could provide resources, advice, scientific information, and political access for the Mayak groups. Furthermore, prior to 1992 the Chelyabinsk region was closed to foreigners, and the residents of the region faced restrictions on their foreign travel. These groups can be characterized as bootstrap organizations, creating their own resources, devising their own strategies, and developing their own political styles. Often scientific and technical expertise came from international sources. For example, the Nuclear Safety Movement held an international conference in Chelyabinsk in May 1992, which brought a wide range of Western environmental activists to Chelyabinsk. In 1993 and 1994, leaders of the Nuclear Safety Movement traveled throughout the United States, visiting with American environmental groups and learning about the organization and operation of citizen groups.[2]

Another distinctive feature of the Mayak protests was the relative weakness of potential allies for an environmental movement. Some regional media in Chelyabinsk assisted the movement, but the degree of their support was restricted. Media coverage of Mayak's environmental legacy in 1989 and 1990 sensitized the region's population to the magnitude of the environmental problems that resulted from the weapons production process. This contributed to the environmentalists' successful 1991 referendum against developing a new nuclear power plant at Mayak. On an individual level, specific reporters and journalists have remained sympathetic to the environmental movement. However, the authorities who control the Mayak complex still wield substantial influence that effectively limits critical reporting both in the press and on television, especially in rural areas of the oblast. After a period of relatively lax controls in the late 1980s and early 1990s, Mayak officials are exercising more influence on press coverage in recent years.

Besides having only a limited number of allies, the Mayak environmental groups also faced outright opposition from other political forces in Chelyabinsk. Before the breakup of the Soviet Union, the local Communist party organization strongly opposed the activities of the Nuclear Safety Movement and other environmental groups. They opposed the 1991 referendum preventing the construction of the Southern Urals Power Station at Mayak. Local party and government officials and officials from Mayak joined in these criticisms.[3] They charged that the environmentalists' allegations of environmental damage were greatly exaggerated, and they marshaled the government's resources to support their claims. Reflecting Mayak's stature in the region, its director was elected as the local representative to the Duma in the 1994 elections. When environmentalists conducted studies of radiation levels, Mayak officials impeded their efforts and criticized the results. When environmentalists cited the ill health of Muslyumovo residents as evidence of the facility's environmental consequences, Mayak officials claimed that their illness was due to diet and lifestyle factors. Mayak officials and local security forces have also tried to discourage local activists from working with Western environmentalists by accusing some Westerners of ties to foreign intelligence services. The scare tactics had limited effects on the Chelyabinsk activists, but it discouraged participation by some Westerners.

Officials at Hanford did not welcome the protests by HEAL and other environmental groups. But the nature and degree of response by the American government was much more modest than in Russia. Similarly, DOE officials were often openly critical of environmentalists' claims, but the degree of opposition was moderated by the preexisting legitimacy of the environmental groups, tribal governments, state governments, and other significant stakeholders. At a more local level, the groups protesting Hanford seldom experienced direct opposition from local and state government officials. Often a non-DOE federal agency or state/local agency provided valuable streams of technical and scientific evidence to challenge Hanford's own claims.

In summary, the citizen groups protesting Mayak faced a double disadvantage. In terms of their opportunities for building alliances in Chelyabinsk, there was no network of established interest groups and political allies that could assist the Nuclear Safety Movement, the Democratic Green Party, and other environmental groups. Thus Russian environmental groups had to build their organizations totally themselves, with only limited assistance or experience from others to draw on. Furthermore, the environmental groups in Chelyabinsk also encountered more outright official opposition to their activities and their goals. They were small organizations facing the clear resistance of Russia's conservative political forces that continue to hold considerable power in the Chelyabinsk region.

Political Action

The mobilization of resources is a means to an end, and the end is structuring the public debate about the nuclear weapons facilities and influencing government policy dealing with the facilities. To an extent, these involve potentially conflicting goals. For example, what best motivates citizens to support an environmental group—a protest or a spectacular action—might not be the best method for influencing the formation of government policy. Similarly, the tactics available to a group may vary as a function of the political context; a large part of the democratic political repertoire was not available to Russian environmental groups before Gorbachev's glasnost and perestroika reforms.

While other aspects of the mobilization process differed sharply between environmental groups in the two nations, their tactics and activities displayed many similarities. Both movements had similar goals: to focus public attention on the plutonium production facilities and force government action to clean up the sites and ameliorate the environmental damage. Thus environmental groups in the nuclear weapons arena were not simply protest groups; both movements used a mix of political tactics, ranging from public meetings to protest activities to lobbying government agencies and officials. The Russian groups organized the successful 1991 referendum against the Southern Urals Nuclear Power Station, as well as street protests. HEAL mobilized public pressure for the closure of the N-reactor at Hanford, and acted as an adviser to the DOE on land use issues through its position on the Hanford Advisory Board. Government authorities in Moscow and Chelyabinsk, however, would never consider putting a group like the Nuclear Safety Movement on an advisory board for Mayak or even establishing any meaningful advisory group that included citizen representation.

The broad similarity between the activities of American and Russian groups results from their similar objectives. For instance, both needed to attract public attention to their cause. They used different methods that reflected the specific situations encountered in each nation, but with functionally equivalent goals and effects in mind. For instance, HEAL used mass mailings and the media to spread its message to the public; the Nuclear Safety movement distributed flyers at a Tatar cultural festival. Hanford groups lobbied the government through participation in public hearings and some of the more institutionalized aspects of a democratic policy process; Mayak groups met informally with government officials at both the local and national levels. The forms of political lobbying were necessarily different in Washington State and the Chelyabinsk Oblast; however, the differences often involved functionally equivalent behavior rather than intrinsically different in political strategies.[4]

The similarity in action repertoires may have initially reflected independent judgments about how to achieve the common goals of the Hanford and Mayak groups. However, as knowledge about Western environmental groups and their tactics gradually were absorbed through the Russian environmental movement, there was a conscious effort to adapt aspects of the Western model of protest to the Russian context. NSM officials had come to the United States to learn what worked for the groups protesting against Hanford, then they selected tactics that they thought could be most useful in Russia. The international connections of the Socio-Ecological Union were additional conduits for this exchange of tactics and methods.

The nuclear environmental movements in both countries thus were quite pragmatic in working toward their goals. These were neither entirely "in-groups" nor entirely "out-groups" confined to a narrow range of political activities. Both movements displayed a full action repertoire, and both tried to use the tactics they thought would work. This may reflect their focus on remediation of specific environmental problems, a stance that perhaps forced them toward pragmatism rather than ideologically based action. It is quite striking to uncover such parallels when the institutional, political, and legal context for action was so widely different between the two movements. Context does matter in structuring political action, but perhaps less than one would think given the dramatic social and political contrasts between these two areas in the early 1990s.

In summary, the mobilization of environmental protest in these two cases showed many similarities in the characteristics and activities of these two environmental movements. Both movements began with a similar goal and a similar set of political concerns, and this may explain the convergence we found in many aspects of their behavior. For instance, both movements were dealing with complex scientific issues that necessitated access to expert information and the formulation of a strategy to translate complex science into examples the average citizen could understand. In addition, as groups that challenged the existing national-security establishment, both movements faced many similar organizational questions. A successful confrontational campaign against public authorities requires substantial financial and informational resources to counter the government's efforts. The organizations within both the American and Russian movements struggled to meet their resource needs, albeit by sometimes different methods. Both movements also faced similar challenges in informing and mobilizing the public to support their respective causes. Indeed, these similarities in tactics were encouraged by direct and indirect contact between two movements that shared common concerns.

At the same time, the respective institutional contexts for political mobilization, or what social movement scholars call the "political opportunity structure" (Kitschelt 1986; Jenkins and Klandermans 1995), also influenced how these two environmental movements functioned. Both sets of groups needed resources; but the sources of funding, the methods of recruiting members, and the pattern of alliances differed substantially between American and Russian groups. The groups protesting Hanford were larger, better funded, and better organized than their counterparts protesting Mayak. In addition, groups concerned about Hanford's impact on the surrounding environment could locate willing and supportive allies to aid their cause; the alliance network in Chelyabinsk was much thinner.

These findings lead us to the important conclusion that the ability to mobilize citizen action is likely to be less dependent on the seriousness of the grievance than on the political context for citizen action. The environmental damage at Mayak was much more severe than it was at Hanford, but the ability of environmental groups to mobilize citizen protest depended on such pragmatic factors as funding for the groups, knowledge on how to mobilize citizen protest, and access to scientific expertise. Perhaps even more important, the climate of pluralist politics that existed in the United States provided legitimacy and ample opportunities for citizen protest. These traits were in short supply for the Mayak groups, and this has not improved with the December 1993 Russian Constitution that established a republican form of government with a comparatively strong president, a weak legislature, and a still underdeveloped judicial system.

This finding has ironic implications for the process of citizen mobilization. The activities of the American environmental groups contributed to changing public policy toward the facilities, including the advent of a massive cleanup program. As these efforts have moved ahead, the Hanford environmental groups have remained active, monitoring these efforts and continuing to challenge the government to keep its word. Although the major peak of political mobilization has passed, these groups are still politically active and remain highly influential.

The groups protesting Mayak, in contrast, have seen their situation worsen over this same time period. The democratic euphoria that fueled their efforts in the late 1980s and early 1990s has waned, decreasing the number of activists involved in the movement and lessening public support for environmental action, especially where progress on environmental quality is seen as conflicting with economic progress. This pattern appears common for a variety of citizen groups in Russia today (Fish 1995). It reflects both the organizational weakness of these new citizen groups and the absence of a political culture to sustain voluntary citizen activism. Simultaneously, the political influence of Mayak and its officials has increased, partially as a function of the revival of conservative political forces in Russia and partially due to the nation's economic difficulties. The management of the nuclear facilities has regained much of its former status and political influence, even though government funding for the maintenance of the Mayak facilities and for environmental cleanup activities is sharply restricted. The environmental problems of Mayak remain severe and the governmental response has been limited, yet citizen action has waned. The experience of the Chelyabinsk environmental movement thus carries mixed messages about the prospects for democracy in Russia today.

Citizens and the Environment

The environmental movement has been a major force for social and political change in most Western democracies, offering a new philosophy on how economic activity should be organized, lifestyles altered, and social relations changed (Milbrath 1989; Dalton 1994). One of our research questions was whether environmentalism entailed similar values and political orientations in formerly Communist Russia.

By concentrating on two geographic areas where a common environmental problem has captured public attention, we could compare how selected groups of Americans and Russians think about environmental issues. These two regional samples may yield more meaningful comparisons than even national comparisons between Americans and Russians. Both regions had experienced recent political campaigns by environmental groups, focusing on equivalent ecological problems. The radiation and pollution problems of the facilities were roughly comparable between the two regions.[5] In addition, Russia was in the midst of economic and political crisis in the early 1990s, and this diminished public interest in environmental matters in many other parts of the nation. In Chelyabinsk, however, because of the debates about Mayak and the other environmental problems of the region, we can assume that environmental issues were more salient in the area and, therefore, we are measuring opinions for a relatively interested public. This type of comparative analysis allows us to consider whether there are common elements of environmentalism and "green thought" that transcend the wide ideological and political divide separating the United States and Russia in the early 1990s.

Western scholarship has described environmental values in terms of a "New Environmental Paradigm" (NEP) (Dunlap and Van Liere 1978; Milbrath 1984; Dobson 1995). The NEP combines several distinct elements of environmental thinking: a positive valuation of nature, a biocentric view of nature, concern about the limits to growth and the role of technology, and a more collectivist orientation toward societal needs. These separate elements tend to cluster together for Western publics, creating a new framework for thinking about politics, society, and the natural environment. These orientations also motivate and guide environmental action among their adherents, explaining both the underlying nature of the world's environmental problems and suggesting the solutions in changing social and economic systems.

On first impression, the environmental values of respondents in our American and Russian surveys appeared broadly similar (chapter 7). Both publics were concerned about the risks and limits of economic development, doubted the claims of technology, and supported a more biocentric view of nature. When we probed more deeply, however, basic differences in the structure of beliefs appeared. For our American respondents, the various elements of the New Environmental Paradigm were interconnected; they saw these items as part of a single value system that they either accepted or rejected. In contrast, environmental thinking was considerably less structured for residents of the Chelyabinsk region. Instead of a single value system, there was evidence of separate subclusters of environmentally related values. Russians did not tend to see the connection between nature conservation and a more biocentric view of nature, between the limits of technology and collectivist values. Similarly, while the attitudes and personal characteristics associated with these environmental values displayed predictable correlations in the American survey, the separate subdimensions of environmentalism in Russia displayed sharply varying patterns of social distribution.[6]

Our interviews with Russian environmental activists uncovered evidence of a similarly diffuse pattern of environmental thinking (chapter 8). At the outset, many activists in the Chelyabinsk region saw the environmental problems of Mayak only in terms of local health and safety concerns. The ability or inclination to link this issue to a larger framework of environmentalism was largely missing. When our discussions explicitly turned to the values embedded in the New Environmental Paradigm, environmental activists displayed some of the same intermixing of values and unformed opinions that we observed in our general survey of public opinion. Moreover, other opinion surveys conducted in Russia have found that attitudes are loosely structured on other political dimensions; this is partially due to the turbulence of contemporary Russian politics and partially to the unformed nature of public opinion in a new democratic setting (Miller, Reisinger, and Hesli 1993; Inglehart and Klingemann 1996). In short, in the early 1990s, many Russians were certainly concerned about the environment, but these concerns had not developed into the type of structured and coherent value system that the NEP signifies among Western publics.

We attribute the diffuse nature of Russian environmental thinking primarily to the early stage of the Russian environmental movement. The leaders of Russia's environmental movement are in the process of articulating a new, green philosophy for themselves and for the nascent movement. During the course of our study, we observed the initial development of these orientations. Through discussions among Russian activists and with Western environmentalists, we observed a growing Russian awareness of environmental thinking and writing in the West.[7] In addition, by meeting with environmentalists from other regions within Russia, they gradually developed a realization that environmental issues are indeed interconnected and have common, fundamental causes. In many ways this is probably similar to what would have been found for the American environmental movement in the 1960s: a growing environmental awareness, but a still ill-formed value framework for thinking about environmental issues. For instance, concepts such as sustainable development and biodiversity are fairly recent political constructs that have given focus and a common language to the environmental movement in the West.

We thus expect that a more integrated environmental thinking will continue to develop within the Russian environmental movement. The way in which Russians and Americans conceptualize environmental problems will probably become more similar over time as a result of this process. It is unclear, however, whether this will lead to complete convergence. There may be aspects of environmental thinking that differ between East and West because of other cultural and historic factors. For instance, the legacy of communism and socialism may have implications for how Russian environmentalists think about private enterprise and collective values (see chapter 8). We have taken a snapshot at the beginning of the modern Russian environmental movement; the question is, now how will this movement evolve?

For the present, these differences in environmental thinking have practical consequences for the environmental movement in both nations. American environmentalists and the public are more likely to view Hanford's problems within a broader environmental framework. This means that the presumed causes and preferred remedies of these problems are seen in terms of issues of an irresponsible use of corporate (government) power, excessive secrecy in the context of national security, a lack of governmental responsiveness, and a critique of present sociopolitical structures. The solutions involve more than simple remediation of immediate environmental dangers, but also require changes at more fundamental levels of individual behavior and these sociopolitical structures. The dictum to "think globally, act locally" presumes comprehension of concepts and chains of reasoning that link local conditions to universal concerns of humankind. Such broad environmental thinking is important in mobilizing wide support for the groups protesting Hanford. The critique of Hanford is not limited to those who feel personally threatened (the downwinders), but is shared by those whose environmental values lead them to sympathize with many environmental causes in the United States and even worldwide.

Because environmental thinking was in an early stage in Chelyabinsk, the environmental movement did not (and cannot) use ideological appeals as a basis for mobilizing public support. Citizen protest against the Mayak facility was predominately a movement of victims; that is, individuals who felt harmed or threatened by the health and safety risks of the facilities (chapter 9). This included individuals with known damage to themselves and their immediate families from the past radiation releases from Mayak, as well as those who worried that they, too, were exposed to these health-threatening radiation levels. The potential, additional, NEP-based mobilization appeal that benefited the Hanford groups was largely lacking in the Chelyabinsk case.[8] There was a striking absence of a relationship between adherence to general environmental principles and support for the groups protesting the activities of the Mayak facilities. This pattern limited both the mobilization appeal of the Russian environmental groups and limited their political scope. More so than in the United States, the Mayak movement seemed motivated by specific and narrow grievances directed against the facilities rather than by broader views of environmental change.

Our comparisons may be limited to the environmental movements in these two locales, but there is evidence that these patterns may be typical of other environmental action in both nations. In the West, environmentalism itself has become a broad-based and inclusive vehicle for social change. In Russia it is still predominately a self-interest movement in reaction to the horrific environmental legacies of the Soviet era.[9] Environmental groups in Chelyabinsk are narrowly concerned about Mayak, and this is paralleled in the actions of similar antinuclear movements in Tomsk, Krasnoyarsk, and other Russian cities. When citizens in St. Petersburg protest against the Krishi Biochemical Plant or others demonstrate against nuclear power stations, these protests have a specific and narrow focus. Only in some Moscow-based groups, such as the Socio-Ecological Union, did we find the seeds of a broader environmental agenda and a broader frame of reference for thinking about the environment (chapter 8).[10]

With the passage of time we should expect a convergence between the environmental movements in the United States and in Russia. Russians will necessarily be preoccupied with the terrible environmental problems facing their nation that dwarf America's unresolved environmental problems. As the Russian public and environmental groups begin to deal with these multiple problems on a more sustained basis, we would expect that they also will begin to identify the inevitable links between socioeconomic structures and their environmental consequences. The diffusion of Western environmental thinking to Russian activists should spur this convergence. One of the prime goals for environmental research in Russia should be to monitor this process, and to determine whether the cultural and historical pasts of post-Soviet Russia and the West will lead to different conceptions of environmental thinking and prevent these two paths from converging fully.

Governmental Response

The story of the governmental response to public concerns about the adverse environmental and health consequences of nuclear weapons production has much to teach us. In both the United States and the former Soviet Union, postwar governmental authorities undertook the demanding task of creating the scientific, technological, military, and political infrastructure required to exercise the influence of a global superpower in a primarily bipolar world. Central to this role was the creation of two powerful nuclear bureaucracies—one American and one Soviet—tied to the military establishment and empowered with the mantle of official secrecy. Key to the success of these two public agencies was the production of large supplies of plutonium, created in locations where official secrecy could be maintained and local environmental conditions—particularly access to ample water supplies and reliable, secure transportation corridors—permitted the development of large-scale technological complexes where scientists and engineers could apply their talents to the task of developing weapons of mass destruction.

The ultimate question for our study is how governments dealt with the policy issues raised by the Hanford and Mayak facilities. More specifically, once the human and environmental costs of the facilities entered the public agenda in the 1980s, how did governments respond? Especially for an established democracy like the United States, Hanford raises fundamental questions of openness and democratic accountability and the potential limits of the democratic process. To answer these questions it is useful to divide our results into two periods. We briefly review the first period: the production phase of the facilities. Then we focus our attention on the second period that begins with the end of weapons production in the 1980s. This was the period when the Cold War competition was ending, public concerns about the facilities were mounting, and the facilities shifted their attention from weapons production to cleanup and environmental amelioration.

Production Period

Hanford and Mayak posed major, and to some extent unprecedented, policy challenges for their respective governments. The production of plutonium was first a military program to ensure national security. Viewed from the perspective of many patriotic American and Soviet citizens at the time, their countrymen were engaged in developing their nation's nuclear weapons and therefore deserved virtually unqualified praise and largely unquestioned public and private support. The national security status of Hanford and Mayak created an aura of importance for the production of materials for the nuclear weapons program, and thus subjugated other policy goals, such as environmental quality, to the nuclear arms race. Viewed from the standpoint of the governmental authorities involved, the agencies managing the two facilities deserved both full support from other public agencies and exemption from the normal forms of political and bureaucratic oversight. Viewed from the perspective of the managers of these nuclear weapons facilities—that is, the governmental authorities managing Hanford and Mayak—the task facing them was at once noble and daunting.

Nuclear weapons production occurred behind a veil of strict secrecy. Few in government knew about the details of plutonium production at Hanford, and the facility was exempt from existing environmental legislation and other routine regulatory oversight outside of the AEC/DOE. Mayak's secrecy was even more absolute. The Mayak facilities and the affiliated research city were not included on Soviet maps, and residents of the Chelyabinsk area were generally unaware of the facilities and their products, or even aware of Mayak's past nuclear disasters. In addition, the production of plutonium and nuclear research was a matter of the highest technology. These facilities were pushing forward the frontier of science; they traveled where no one had gone before. Hanford and Mayak raised questions that lacked scientific answers. And as might be expected on the boundaries of science, the initial answers of research were often contradictory or ambiguous. In any case, frontier questions of science and technology were dealt with in strictest secrecy in both countries. As far as environmental protection was concerned, the insulation from normal peer review precluded the operation of the self-correction processes that come from open scientific scrutiny.

From one perspective, the noble task of protecting the nation's security is indeed work of great honor and pride. Hanford and Mayak were largely successful in achieving the primary policy goals set for them by their respective governments. Despite incredible challenges, these facilities produced plutonium and nuclear research to support their primary mission during the Cold War. Even as the accumulated problems of the past mounted, production continued and expanded. The costs were high, but the supposed gain—national security—was even greater.

From another perspective, however, the task of competing with other world-class scientists and engineers in a nuclear arms race—dictated by political motives more than scientific and technological realities—often produced working conditions at these facilities that were hazardous to one's health. The facilities functioned at the very borders of human understanding of fundamental physical and biological processes. As one side produced evidence of a breakthrough in either scientific understanding or engineering solutions to technical problems, the pressure on the competition to match this performance was extreme. Nuclear weapons systems, moreover, required plutonium and related radioactive materials; and each new system required its own reliable sources of the world's most deadly substances.

The benefits of hindsight and an end to the Cold War now instruct us that on too many occasions the pressure to "keep up" was great enough to cause both intentional radioactive releases for test information, and accidents borne of either questionable risk taking or simply venturing into uncharted terrain. For the greater part of the four decades of nuclear warhead material production in the United States and the former USSR, the operative standards for environmental protection were considerably more permissive than they are today.

Unfortunately, in both the American and Soviet cases the pertinent governmental authorities responsible for causing damage to public health and the local environment chose to hide behind the sacred mantle of official secrecy. Both facilities, when confronted with evidence of damage to the environment or public health, issued categorical denials to these claims. Secrecy shielded the facilities from sharing information on their activities with their respective local populations or even with other scientists. Even if the environmental discharges of Hanford were within range of what scientists thought were safe, these discharges put local populations at risk without their full knowledge or consent. The early estimates of limited damage from the plutonium production process vastly underestimated the environmental damage (and the enduring risk from past wastes). Moreover, absent stiff environmental standards and effective public oversight, both the American and the Soviet plant managers (and their political overseers) took dangerously reckless risks with the treatment of nuclear wastes and the testing of nuclear weapons-related materials. The evidence is clear that on occasion both facilities knowingly exposed citizens living nearby to health hazards. When accidents did occur, such as the waste storage tank leaks at Hanford or airborne contaminated dust at Mayak, the veil of secrecy dissuaded plant officials from sharing this information with local governmental officials and the local populace.[11]

The scale of societal risk and the immediate threat to human health admittedly was inestimably greater at Mayak. From the construction of the facilities through their decades of operation, human safety suffered to benefit the Soviet Union's nuclear weapons program and save construction and maintenance costs. Construction of the initial facilities used Gulag labor, and the first production workers were exposed to tremendous doses of radiation. Chapter 3 and other historical accounts (Cochran, Norris, and Bukharin 1995; Donnay et al. 1995) document other health and safety dangers that Mayak created over the years for its downwind neighbors. Under the Soviet system, moreover, secrecy could be transformed into intentional or duplicitous manipulation of the evidence; health officials, for example, were prohibited from recording radiation-related health problems. Numerous researchers in the Chelyabinsk region have encountered reports of a deliberate policy of falsification of death certificates in order to hide evidence of radiation-related illnesses. From the 1957 waste tank explosion to the wards of the Chelyabinsk Children's Hospital that are today filled with children suffering from leukemia, Mayak's legacy is indeed horrific.

The historical record now makes it clear that both governments failed to provide sufficient consideration of the costs of the weapon's facilities to the health and safety of local populations. To some extent, this is not surprising for Mayak. The Soviet Union was an authoritarian state where individual interests carried less weight, and individual needs were more easily subjugated to the state's interests. Therefore, what is more disturbing is the similarity, even if partial, between the Soviet and American experiences. National security and secrecy prevented the citizens of eastern Washington State from exercising their democratic rights. Policy was made without local knowledge or consent. Indeed, information released under the Freedom of Information Act, the testimony of whistleblowers, and other historical evidence now shows that plant officials and government representatives intentionally deceived the public. The tale is similar for other nuclear-related controversies. For example, analysis of the contemporaneous official response to the Three Mile Island accident painted a similar picture of secrecy and official obfuscation distorting the public debate (Walsh 1988; Goldsteen and Schorr 1991).[12] Thus one lesson from our findings is clear—democracy cannot function in the shadows of excessive secrecy and limited public access to information.

How did these distortions of policy making occur? Only to a slight degree was it from a conscious disregard by the Hanford policy makers. Most officials believed that they were not harming local residents in pursuit of national security interests. Instead, secrecy prevented the full disclosure of information, even to employees. Limited knowledge can easily distort the policy process. In addition, scientific research was constrained by the Cold War environment. Nuclear research was initially dealing with many unknowns, and later experiences often proved early assumptions false. For instance, initial estimates that nuclear materials would take centuries to migrate through the Hanford water table were disproven when waste streams spread through groundwater and wildlife food chains in a matter of decades. But it is difficult for science to advance when the sharing of such results is limited by national security concerns. In a context of excessive secrecy, the natural corrective of the scientific process cannot function effectively; other scientists cannot note the weakness of initial assumptions or contradictory evidence if research is not shared. Furthermore, when science is conducted by a sponsor—whether the Hanford facilities or a commercial firm—there is a potential that the research might artificially conform to the sponsor's expectations. As others have observed, it is difficult for science to function in the shadow of secrecy and without sufficient information (Meehan 1988).

Another source of excess at Hanford was the development of a "corporate culture" of the weapons production industry (Gusterson 1996). Managers that met production quotas were rewarded. An apparently excessive concern with safety and the avoidance of risk taking was implicitly or explicitly discouraged. Furthermore, the Hanford community developed its own internal identity and sense of loyalty. Relatively isolated in eastern Washington, Richland became a family for Hanford employees. The local high school adopted a well-known atomic image, a nuclear mushroom cloud, as the school's emblem. Local streets on which many Hanford employees lived had names with a nuclear theme. Nuclear weapons were part of life. Furthermore, whistleblowing was both illegal, because of the national security aspects of weapons production, and a mark of disloyalty to the Hanford family.[13] Such cultural norms for Hanford managers and experts made it extremely difficult for public criticism from nonexperts to be accepted, or for the public's environmental and health concerns to be assigned much credence.

Perestroika and the Freedom of Information Act

In both nations the advent of more open politics at the end of the Cold War began to change this situation. The Gorbachev-era reforms of perestroika and glasnost gave residents of the Chelyabinsk region their first opportunities to discuss publicly the Mayak facilities and their consequences for the environment and public health. The mobilization of HEAL and other citizen groups forced a similar public debate of Hanford and its legacy.

A climate of greater openness associated with the ending of the Cold War pressed the authorities at both Hanford and Mayak to admit to some of the facilities' past and present shortcomings. After years of unquestioned license to use great portions of the public treasury in an all-out effort to assure their nation's security, the authorities at Hanford and Mayak came under unprecedented public scrutiny. In both locations the first reaction of the authorities was to deny culpability, to label the critics as nuts and extremists, and to hide (again) behind the protective cover of national security.

In time, however, both sets of authorities were moved to admit responsibility for more and more policy and operational transgressions. Both Hanford and Mayak authorities ultimately were forced to conclude that they were now controlling some of the planet's most environmentally devastated areas. This admission began at Hanford with the 1986 release of documents under the Freedom of Information Act. Further information was released during the late 1980s and early 1990s. In Russia, the published transcript of statements made to the Soviet Parliament in July 1989 by L.D. Ryabov, deputy chair of the USSR Council of Ministries, was the first official statement on the environmental and health consequences of the Mayak accidents. These revelations sparked the Chelyabinsk protest movement, and started public action against the facilities. By the early 1990s, the spokespersons for Hanford and Mayak seldom advocated the need for maintaining the capacity to produce nuclear weapons materials (though this remains an important role at Mayak); instead, they argued strongly for vast new sums of public resources to manage nuclear wastes and promote environmental remediation.

Despite a very similar governmental response at both Hanford and Mayak during the plutonium production and immediate postproduction phases, the pattern and degree of response varied greatly between the two nations as the facilities came under sustained public scrutiny. As Laurence Lynn (1996) has argued with insight and eloquence, American exceptionalism—the combination of strong democratic norms among the populace and the relative openness of the American political system—made it more possible for Americans to mount effective citizen-based challenges to political authorities than in most other democracies. The governmental response to Hanford, in many respects, supports Lynn's interpretation of American exceptionalism (chapter 10).

As the horrible truths of the true costs of the nuclear arms race began coming to light, the critics of Hanford found many sources of societal and governmental support to promote their cause. American federalism, which arms the states (and their local governments) with considerable powers of governance beyond the control of federal authorities, could be used to call into question Hanford's official version of key historical events and the environmental conditions at the facilities. The state of Washington became an active player in the effort to gain environmental compliance oversight powers for the Department of Energy's Hanford facility. As noted in chapters 3 and 10, the state of Washington, the U.S. Department of Energy, and the U.S. Environmental Protection Agency negotiated the *Tri-Party Agreement* (TPA) in 1989. The TPA sets very demanding targets for environmental remediation, and establishes public oversight processes for research studies, remediation plans, and progress reports.[14]

Another access point arises from the legal status afforded to the tribal governments of the Yakama, Umatilla, and Nez Perce reservations in original treaties and under the "trust relationship" developed by federal law. Tribal claims afforded another effective avenue to legitimately challenge Hanford authorities and served as a channel for access to critical information. The American judiciary, one of the world's most powerful independent judicial systems, assured the Hanford critics of a fair day in court if they could substantiate their claims of environmental and/or public health dangers arising from Hanford's activities. The American common law tradition and the celebration of constitutionally sanctified rights, which places such great store in the judicial protection of individual rights and property claims, allowed environmental groups to use the courts as an effective means to challenge Hanford.

Gradually the authorities at Hanford went from denial of any serious problems to funding major new studies of health effects, environmental damage, and the safety of waste containment facilities. Public opposition groups often served as active parties and beneficiaries of these studies. For instance, a class action lawsuit in behalf of downwinders relied on the results of dosimetry and epidemiological studies that Hanford financed![15] In summary, the direction of public policy did change, albeit slowly, and public pressure played a major role in stimulating this redirection and monitoring its progress.

Though change is evident at Hanford, stakeholder groups continue to see substantial room for improvement. From the perspective of these groups, DOE continues to hide too much information behind national security barriers, is too slow to share what it knows can be shared, and is making too slow progress on many aspects of environmental remediation. The operation of a strong civil service system with its highly protective seniority rights means that many members of the "Hanford family" continue to hold key positions in the agency. These individuals actively resist pressures from the DOE secretary to change the corporate culture of the agency. Furthermore, the federal government's commitment to balance the budget has conflicted sharply with the government's pledge to meet the milestones for progress under the Tri-Party Agreement. This erosion of federal financial support (see chapter 10) has added further fire to the advocacy efforts of environmentalists, local governments, the state governments, and tribal authorities.

The record of governmental response at Mayak is quite different (chapter 11). There is not the legacy of democratic institutions, an independent judiciary, and citizen rights that the critics can call into action. There is a new set of environmental protection legislation that gives environmentalists a legal basis of action. But even as the Yeltsin reform government started to enact nuclear waste legislation, Mayak received executive dispensation from key provisions by Yeltsin himself. The concentration of authority in the Ministry of Atomic Power ensured that the most conservative elements of the Russian national bureaucracy would maintain oversight of the facilities.

During the period between 1993 and 1995, the successor organs of the KGB were used to intimidate foreigners and jail activists within Russia on charges of violating national security laws. Under the Russian system, the accused were incarcerated while under investigation. These actions by authorities engendered caution in activists and their Western collaborators.

Local government did not provide a check on the power of Moscow; in fact, the local authorities in Chelyabinsk are generally more conservative than national political leaders. The Mayak facilities continued to exercise significant autonomy from even their official superiors.[16] In addition, the Tatar and Bashkir minorities in the region enjoy no special treaty protection or legal standing that would assist them in challenging the facilities. In short, virtually all of the official and unofficial paths of access and influence that were available to the groups protesting Hanford were unavailable to the groups protesting Mayak.

Because of this limited influence, it is not surprising that the governmental response to Mayak has been limited. There are insufficient funds for environmental remediation, so progress has been essentially nonexistent. The financial crises of 1998 will only worsen these problems. The government has attempted to seal Lake Karachy, though this has been a slow and incomplete process. There are repeated claims that Mayak is attempting to reprocess spent radioactive fuel from the West to earn hard currency, thus increasing the radioactive materials at the facilities and the danger to the environment. In 1996, a Duma committee warned that the storage ponds for radioactive waste were in disrepair and could collapse and disperse their contents on a massive scale. Even worse, the government has renewed proposals for construction of a new breeder reactor at Mayak. Ostensibly the reactor would use water in such amounts that it would lower the level of wastewater in the storage ponds, but many experts doubt the feasibility of this plan and question the real motivation for the new reactor. In short, much is continuing unchanged at Mayak. The restoration of conservative forces in the Chelyabinsk region, the lack of strong environmental protection policies from Moscow, and a flawed economy will be major brakes on cleanup projects for Mayak.

Conclusion

The infancy of Russian democracy and the exceptionalism of American democracy have shaped the nature of citizen protest against the environmental consequences of Hanford and Mayak, and have shaped the governments' response to the environmental legacy of these facilities. In recent years the environmental groups protesting about Hanford have been fairly successful in using the democratic process to press their claims and bring about change. However, this progress comes after decades of governmental deception and denial. Perhaps the most fundamental lesson of our study is that democracy is fragile—it cannot function, even in the United States, when national security shields the government from citizen oversight, and when secrecy prohibits the public from gaining accurate information about the actions of its government. When government officials deceived the public on Hanford's environmental consequences, when they dissuaded and punished whistleblowers, or when they unnecessarily restricted access to information, these actions had harmful consequences for the environment and for American democracy. Thus Hanford's production period carries ominous lessons for the democratic process.

In comparison, the situation at Mayak was much more severe. The environmental damage wrought by the facility far exceeds Hanford's impact. Many times government officials and plant administrators showed callous disregard for human health and safety. To some extent this was expected; it reflected the flaws in the Soviet political system that dismissed the public interests except as defined by the government. Perestroika, glasnost, and emerging Russian democracy supposedly heralded a new political era. Yet we found that democratic action is still severely restricted in the case of nuclear pollution. Environmental groups protesting Mayak have struggled to survive, and their policy impact has been much more uncertain. The shortfalls of the Mayak groups reflect both the organizational weakness of these new citizen groups and the absence of a political culture to support voluntary citizen activism in Russia (Eckstein et al. 1998). The experience of the Chelyabinsk environmental movement thus carries ambiguous lessons for the prospects of democracy in Russia today.

Chapter 10 introduced a framework for understanding these patterns. The United States is an example of a political system with strong democratic norms, but with a weak-state structure. The former point empowered environmental groups and enabled them to force their issue onto the political agenda. However, the later point means that the fragmentation of political power in the United States limits the ability of government to make authoritative decisions and implement them when a broad political consensus does not exist. Thus as people in the state of Washington lobby for environmental cleanup, other political groups are pressing for budget cuts that include DOE's environmental programs. Interagency and intergovernmental rivalries are another complication; as one agency advocates one strategy for Hanford's cleanup, another favors an alternative. This combination of strong democracy and a weak state thus facilitates the expression of public demands, but hampers the translation of these demands into public policy.

Russia presents a contrasting pattern. We might see Russia as a case of weak democracy, and (now) a weak state. The Russians' limited understanding of the democratic process and the limited infrastructure for a civil and democratic society impede the development of the environmental movement protesting Mayak. As we have repeatedly seen, the green movement is struggling to survive in Russia today. At the same time, there are several key weaknesses in the Russian governing structure that have generated repeated constitutional crises under the Yeltsin administration. The degree of democratic representation in the Russian government is still unclear; does Parliament govern or merely observe the issuance of executive decrees? Moreover, the fundamental economic problems facing Russia today limit the ability of any government to act. Even with strict nuclear safety legislation on the books, the implications of this for Mayak's operation are unclear.

In short, neither pattern—strong democracy/weak state or weak democracy/weak state—present the best conditions for the representation of citizen interests, especially when these interests require a change in the status quo. Thus in both nations the past response by government has been short of what is needed to protect the environment and health of Hanford's and Mayak's neighbors.

Past differences in response need not continue, however. American environmental groups are pressing Hanford and DOE officials to change their corporate culture and respond to public demands. It is also clear that many of the lessons learned by Hanford environmental groups and by Hanford authorities can be shared with Mayak protestors. As Russian democracy explores its future, Western environmental activists can play a constructive role in the development of Russia's democratic institutions. Russian environmentalists also can rely on their American counterparts for a listing of the most critical questions, the most telling evidence to be sought on health and environmental effects, and the best methods to rebuild public support. Our research paralleled the development of these exchanges between Western and Russian environmentalists, funded by Western foundations and government agencies. The further development of Russian democracy and the resolution of Mayak's problems can benefit greatly from a continuation of these exchanges.

Similarly, the managers of both facilities can learn from their counterparts. The end of the Cold War has seen the first technical collaboration between engineers and scientists from Hanford and Mayak. At a 1991 UC Irvine research conference on the environmental consequence of nuclear development, the U.S. Department of Energy prohibited department employees or Hanford officials from participating because Russian or Chinese participants might gain classified knowledge. Within a few years a series of regular exchanges were occurring between Western and Russian nuclear experts, and technology sharing on environmental remediation measures has become more common. As a very modest sign of these changes, the formerly secret city of Chelyabinsk-70 now maintains its own World Wide Web site, initially established with Western aid.[17]

The loss of land and the loss of life were major consequences of nuclear weapons production at Hanford and Mayak. In addition, however, there was a loss of public innocence. Democracy, and civil society more generally, are based on social trust; there is a social contract between governments and the governed that is essential for a political system to function (Goldsteen and Schoor 1991; Putnam 1993). This social trust, and hence the social contract, were violated by the plant managers and government agencies responsible for the Hanford and Mayak facilities. Our surveys in both nations showed widespread doubts about the credibility of the facilities' representatives, despite the attempts by Hanford and Mayak to create a positive and reformed image for themselves. Public trust, once lost by governmental authorities, is very difficult to reestablish. Having been lied to once about matters of personal health and welfare, the public has difficulty believing new claims of honesty and concern. Exchanges between the facilities may also help Mayak and Hanford managers learn that there are constructive ways to work with their critics in order to restore public credibility in the facilities—and not just focusing on technology issues. It would be ironic, indeed, if Hanford officials could provide their counterparts with guidance on how to work in concert with citizen groups to address the facilities' problems, restore public trust, and ultimately aid in the democratization of Russia.

Both the Hanford and Mayak authorities, and their respective governments, face a difficult challenge in recreating the high levels of public confidence and citizen trust that they once enjoyed. The road to reestablishing public trust is likely to be a long one. The progress along this course can serve both as a measure of the vitality of American democracy and as an index of the ability of contemporary Russia to develop a new democratic process.

Notes

1. The Russian economy lacked the simple bank check that was common in capitalist economies.

2. During these visits Mironova and others met with representatives of several Hanford groups as well as other environmental organizations throughout the

United States. Besides the scientific information she gained on this trip, she also received advice on how American groups develop community support for their activities.

3. For a period in 1991 and 1992 the regional government was controlled by liberal proreform forces. The Movement for Nuclear Safety and Democratic Green Party received direct support from the regional government during this time, but this came to an end in 1993.

4. For instance, American groups make frequent use of the judicial and administrative review procedures available in U.S. law; these opportunities are not available to Russian groups. But Russian groups pursued these same objectives by organizing public hearings on environmental issues and informally participating in the drafting of new environmental legislation at the local and national levels.

5. There were, however, marked differences in the other environmental problems between regions, which is discussed in chapter 4.

6. For instance, social class and age are strongly related to the NEP in the American sample, but these characteristics display varying patterns for Russian attitudes toward the environment (see chapter 7).

7. Our project may have contributed to this process by providing the writings of Western environmental scholars, such as Lester Milbrath and Riley Dunlap, to Russian environmental groups. In one specific meeting where environmental values were discussed, it was clear that one Russian activist had read Milbrath (1984) and thus more easily could discuss environmentalism within this framework.

8. Indeed, one might argue that the number of "conscious supporters" of environmental action is regularly larger than the number of those who are directly threatened by a specific environmental problem, and thus are an essential element in the Western environmental movement.

9. During the early Gorbachev reform period the Russian environmental movement was a vehicle for broad social change by those who used it as a shelter for antisystem criticism (White 1991). However, once it became possible to criticize the system or advocate democratic reforms without using the guise of environmentalism, many of these activists left the movement and took up these other causes directly.

10. In their earlier work on environmental groups in Japan, Pierce and his colleagues found evidence of the same pattern in that country (Pierce et al. 1989). The first environmental protests were focused on specific grievances. In relatively short order, however, political action spread to other sites and elements of the Japanese intelligentsia. These elites took up the challenge of providing the broader ecological perspective on Japanese public policy.

11. We should also restate the point that American intelligence services and high-level officials in the American weapons production system knew of the environmental damage wrought by Mayak. The 1957 storage tank explosion was widely chronicled in Western sources, though U.S. government withheld comment. In

addition, the Mayak complex was a prime surveillance target for the American intelligence agencies.

12. The Soviet experience with other nuclear sites is even more negative (Marples 1995; Cochran, Norris, and Bukharin 1995).

13. For instance, it has been said that some workers who were exposed to health-threatening doses of radiation had to sign nondisclosure statements as a condition of their disability claims.

14. Six of the most prominent U.S. Department of Energy's major nuclear weapons facilities are operating under similar tri-party agreements supervised jointly by the EPA and the respective states: Rocky Flats in Colorado, Fernald in Ohio, Savannah River in South Carolina, INEL in Idaho, Oak Ridge in Tennessee, and Hanford.

15. A district court judge dismissed the case in August 1998, but plantiffs say they will file an appeal. And further litigation by various plantiffs seems an inevitable part of Hanford's future.

16. One example is illustrative. Our research team received official approval from the Ministry of Atomic Power to interview residents of Chelyabinsk-65 and meet with plant officials. The deputy manager of the facility would not recognize this approval when our research team visited the facility in 1992.

17. The site is: http://www.ch70.chel.su

Appendix A: Survey Methodology

A portion of the findings in this project are based upon public opinion samples of selected communities surrounding Hanford and Mayak. This appendix provides information on the methodology of these two surveys.

Hanford Survey

The Hanford Survey was conducted by the Division of Governmental Studies and Services (DGSS) at Washington State University, under the direction of Professors Nicholas Lovrich and John Pierce. To the maximum extent possible, the U.S. study was designed in parallel to the Russian study, such as in the content of the survey and the selection of primary sampling sites. DGSS has extensive experience in doing in-person and mail surveys, and has done numerous environmental opinion projects in the past. The division adapted the in-person Russian questionnaire for use in a mail survey of residents in the Hanford region. A portion of the survey also replicated questions from the U.S. *Health of the Planet Survey,* and we appreciate Gallup International's assistance in supplying us with the questionnaire from the American survey.

DGSS used a mail survey primarily to collect opinions in four locations. Richland is directly adjacent to Hanford and is the "host city" for the facility. Many of Richland's 32,000 residents work at Hanford and are familiar with the facility. Moreover, Richland has a distinctive demographic composition because of its relationship to Hanford; the scientific, technical, and administrative requirements of professional employment at the facility result in disproportionately high overall levels of education and economic status among its residents. A sample of Richland residents was drawn from digitized address lists held by Survey Sampling, Inc., and three waves of mailings were done between October 1993 and March 1994. We received 549 usable questionnaires, with an overall response rate in Richland of 54 percent.

The city of Spokane is our "downwinder" site. Spokane has a metropolitan area of more than 220,000 residents and lies approximately 120 miles to the northeast of Hanford, directly downwind along the path of the prevailing weather patterns that customarily pass over Hanford. Spokane also is the largest city and media center for eastern Washington, northern Idaho, and western Montana. The major regional newspaper and the television outlets in Spokane were the primary vehicles through which much of the negative information about Hanford, its history, and its excessive secrecy was disseminated to the public. In Spokane, a mail survey produced 527 completed questionnaires, and a response rate of 51 percent after three waves of mailings.

A third component of the study was a survey of the minority populations living near Hanford. We included three communities on the Yakama Nation Tribal Reservation in our study: White Swan, Toppenish, and Wapato. These small, largely minority communities are located approximately 50 miles downriver from Hanford and have a combined total population of about 12,000. They contain a significant proportion of Native Americans, as well as Hispanics, who work primarily as agricultural workers in the area. Through a combination of mail surveys, hand-delivered questionnaires, and numerous telephone follow-ups gathered by contracted members of the Yakama Reservation community (using a random sample of households provided by the researchers at Washington State University), we collected a sample of 149 residents of these three communities (with a 2 percent refusal rate).

Our fourth location was chosen as a "control site" for the study. We selected a city in the same general geographic region that was similar in demographic composition to the downwinder site, but which was not directly affected by the Hanford facility—either economically or in terms of exposure to radiation-laden pollution. The inclusion of such a control site provides the opportunity to compare the experiences of the downwinder populations relative to residents in a neutral location. The city of Wenatchee is the control site. Wenatchee has 22,000 residents and is about 100 miles north of Hanford, out of the immediate vicinity of the paths of contamination. It serves as a regional economic center for the surrounding agricultural area. The survey efforts conducted in Wenatchee resulted in a final sample size of 178, and a response rate of 48 percent after three waves of mailings.

The use of mail surveys in the American survey and in-person surveys in Russia does raise a serious question about the comparability of results from the two studies. In order to estimate whether cross-national differences were due to variations in response rates attributable to the different survey formats, the DGSS conducted a small-scale methodological test for instrumentation administration effects. The division conducted a conventional telephone follow-up minisurvey with 75 nonrespondents, using an abbreviated form of the questionnaire. Twenty-five telephone follow-ups were conducted in Richland, Spokane, and Wenatchee, respectively, during the process of mail-survey cross-validation. The results of the telephone interviews were most encouraging; the telephone interview results corresponded very closely with the mail-survey results, indicating that there was little evidence of any particular pattern of bias in the mail-survey responses.

Mayak Survey

The Mayak survey was conducted by the Kaluga Sociological Institute under the direction of Andrei Zaitsev, the institute director. The Kaluga Institute was selected because of its prior experience with environmental research of the kind planned for Mayak. The Kaluga Institute first conducted population surveys related to the Chernobyl disaster, and later did research on the social situation at the East Urals Radioactive Trace in Chelyabinsk.

The survey questionnaire was developed in collaboration between the American team at UC Irvine and their Russian coinvestigators. A portion of the survey replicated the 1992 Gallup *Health of the Planet Survey* in Russia, and we are indebted to the Gallup office in Moscow for providing us with the Russian and English copies of their questionnaire. Other questions had been pretested with Russian businesspeople visiting UC Irvine, and with a number of Russian environmentalists in Moscow. The Kaluga Institute conducted a pretest survey in September 1992, and these findings were used to develop the final questionnaire.

The Kaluga Institute organized the fieldwork for the survey, and Paula Garb supervised the administration of the survey. Kaluga hired and trained approximately 120 interviewers in the Chelyabinsk area. Interviewers were limited to a maximum of fifteen respondents, and a sample of completed questionnaires (N = 205) were verified for authenticity by fieldwork supervisors.

The four sampling sites for the survey were selected by the UCI researchers as pairs for the American survey sites. Chelyabinsk-70 (now called Snezhinsk) is the closed city that housed the research lab affiliated with Mayak, called the All-Union Scientific Research Institute of Technical Physics.[1] In 1992, Chelyabinsk-70s population was approximately 45,000. Even more so than people in Richland, the residents of this closed city had direct economic connection to the nuclear weapons production facilities. A total of 203 interviews were completed in Chelyabinsk-70.

The city of Chelyabinsk, with a population of more than 1 million, is our downwinder site. It is the major urban area in the region, the major media center, and home for most environmental activism. It is an industrial center for the Southern Urals, with large metalurgical and chemical industries. Chelyabinsk is located approximately forty miles south of Mayak. The research team from the Kaluga Institute conducted 503 interviews there.

Our surveying of the indigenous populations was based on two communities. Muslyumovo is a small village of approximately 4,000 on the banks of the Techa River; it is the first inhabited community, roughly fifty miles downriver from Mayak. The residents are primarily Tatar and Bashkir, and it is a largely agrarian community. The main occupations in Muslyumovo are related to the operation of the local State Farm, which focuses on stockbreeding and grain growing, and servicing the railway station. Kyshtym is a town of approximately 40,000 residents that lies adjacent to Mayak. Its population is disproportionately Russian, and many of its residents have direct or family ties to individuals employed at Chelyabinsk-65, Chelyabinsk-70, or support activities for the facilities. Most of the population is employed in the city's three major industries—radio production, copper production, and machine-tool building. Kyshtym's proximity to Mayak led to the usage of the town's name in labeling the 1957 nuclear accident. We have 137 completed interviews from Muslyumovo, and another 163 from Kyshtym.

Chebarkul was selected as the control site for the study. It is a city of approximately 250,000 that is located fifty miles southwest of Mayak in the Chelyabinsk Oblast. Chebarkul's economy is based on manufacturing, and it has not been directly affected by past radiation releases from Mayak. We have 163 interviews from residents of Chebarkul.

The Kaluga Institute's staff used a multistage sampling procedure to select survey subjects. Because of its size, the Chelyabinsk City survey began by selecting seven local areas within the city. Each of these local areas was further identified by its electoral districts, of which there were between nine and twelve. An electoral district usually included about 10,000 people, and one district was randomly sampled from each local area. At the final stage of the sampling designation, respondents were selected from the electoral lists maintained by city officials. In the other communities, the initial sampling of local areas was not necessary, and the sampling began with the electoral districts and electoral lists. The fieldwork for the survey was done between October 1992 and January 1993.

The Kaluga Institute reported only fifty outright refusals out of 1,169 completed interviews.[2] At the same time, the last election had occurred fifteen months before fieldwork, and some of the addresses on the electoral rolls were inaccurate and the youngest adults had not yet been added to the rolls. The Kaluga Institute reported that respondents were well disposed toward the survey, interested in the content, and response rates were greater than in their usual surveys.

Notes

1. Initially we had permission from the Ministry of Atomic Power to survey in Chelyabinsk-65. However, Mayak officials refused to allow surveying in this closed city, essentially overruling the decision made by government officials in Moscow.

2. At the same time, surveying in Russia at the time was not without its challenges. At one point the chairman of the Muslyumovo local Soviet stopped an interviewer in the village and "arrested" the questionnaires. On the whole, however, response rates in Muslyumovo were higher than in the other locales.

Appendix B: Survey Index Construction

Biocentric/Anthropocentric Index This index in the Chelyabinsk survey is based on the factor analysis of table 7.2. Three items were combined into the index: plants and animals exist primarily for human use, the good effects of technology outweigh the bad, and humankind was created to rule the rest of the world. The index is a count of the number of items with which the respondent disagreed. Respondents who did not answer one or more items are excluded from the index.

Collective Values This index in the Chelyabinsk survey is based on the factor analysis of table 7.2. Two items were combined into the index: society should use its resources to benefit future generations even at the detriment of today's living standards, and the interests of society should take precedence over individual needs. The index is a count of the number of items with which the respondent agreed. Respondents who did not answer one or more items are excluded from the index.

Democratic Values

The democracy values index in the Russian survey is based on six items that measure acceptance of democratic norms such as freedom of the press, free speech, and freedom of expression (Q115 and Q117 to Q121). The index is a count of the number of items with which the respondent disagreed. Respondents who did not answer two or more items are excluded from the index.

Liberal/Conservative Ideology

The American survey used a ten-point liberal/conservative scale, with each term anchoring an endpoint of the scale. The question wording was: "In political matters, people talk of 'Liberals' and 'Conservatives.' How would you place yourself in terms of these categories?"

The Russian survey used the terms Reformer and Conservative to anchor the scale. In the Russian survey, about 40 percent of the public did not place themselves on this scale, compared with 6 percent in the American survey.

Local Environmental Problems

This index is based on the following question: "Here is a list of environmental problems facing many communities. Please tell me how serious you take each one to be here in your community: very serious, somewhat serious, not very serious, or not serious at all?" For the list of items see figure 4.2. The index is a count of the number of items the respondent felt were "very serious" problems. Respondents who did not answer three or more items are excluded from the index.

Military Support Index

The military support index combines responses to two questions: nuclear weapons are an important part of our national defense, and (America/Russia) must maintain a strong military defense. The index is a count of the number of items with which the respondent agreed. Respondents who did not answer one or more items are excluded from the index.

New Environmental Paradigm (NEP)

This index combined the ten environmental values in table 7.2 into a single index. The index is a count of the number of items with which the respondent gave a proenvironmental response. Respondents who did not answer three or more items are excluded from the index.

Postmaterial Values

This index is based on a variant of Inglehart's (1990) four-item battery. Each respondent was asked to pick his or her two priorities from a list of four items: a) a stable economy; b) progress toward a less impersonal and more humane society; c) progress toward a society in which ideas count more than money; d) the fight against crime. Respondents choosing options a) and d) were designated "materialists"; those choosing b) and c) are labeled as "postmaterialists"; and other combinations were labeled as holding mixed values.

Private Economy Index

The private economy index in the Russian survey is based on several items on the government's economic role: The government should play a larger role in planning economic development, only the widespread development or private property will take the economy out of its crisis, all industrial enterprises in Russia should be under the control of a government ministry, and the five-year plans were an efficient way to run the economy. The index is a count of the responses favoring a private economy on these items. Respondents who did not answer two or more items are excluded from the index.

Valuation of Nature

This index in the Chelyabinsk survey is based on the factor analysis of table 7.2. Three items were combined into the index: mankind is using up the world's natural resources too quickly, humans must live in harmony with nature in order to survive, and wildlife, plants, and humans have equal rights to live and develop. The index is a count of the number of items with which the respondent agreed. Respondents who did not answer one or more items are excluded from the index.

References

Agence France Presse. 1993. Russia abandons second operation to dump nuclear waste. *Agence France Presse* (October 21).

Akleyev, A. V., and E. R. Lyubchansky. 1994. Environmental and medical effects of nuclear weapons production in the Southern Urals. *The Science of the Total Environment* 142:3.

Alm, Leslie. 1993. The policy making process in the American West. In Zachary A. Smith, ed., *Environmental Politics and Policy in the West.* Dubuque, IA: Kendall/Hunt Publications.

Armbruster, Tom. 1998. Personal communication, IAEA Radwaste Priorities (June 17, 1998).

Atomic Energy Commission. 1947. *Second Semiannual Report of the Atomic Energy Commission.* Washington, D.C.: U.S. Government Printing Office.

Atomic Energy Commission. 1949. *The Handling of Radioactive Waste Materials in the U.S. Atomic Energy Program.* AEC 180/2. Washington, D.C.: U.S. Atomic Energy Commission.

Atomic Energy Commission. 1957. *Twenty-first Semiannual Report of the Atomic Energy Commission.* Washington, D.C.: U.S. Government Printing Office.

Barry, Donald D., and Howard R. Whitcomb. 1987. *The Legal Foundations of Public Administration,* 2d ed. St. Paul, MN: West Publishing Company.

Barwich, Heinz. 1977. *Das Rote Atom.* Munich and Berne: Scherz Verlag.

Belyanivov, K. 1992. Russian government is very short of energy. *Komsomolskaya Pravda* (June 2):2.

Benson, Allen B., and Larry Shook. 1985. *Blowing in the Wind: Radioactive Contamination of the Soil around the Hanford Reservation.* Spokane: Hanford Education Action League.

Berry, Jeffrey. 1984. *The Interest Group Society.* Boston: Little, Brown.

Bierschenk, William H. 1959. *Aquifer Characteristics and Ground Water Movement at Hanford.* Hanford: Hanford Works, HW-60601 (RL: HAPO, June 9).

Blush, Steven M., and Thomas H. Heitman. 1995. *Train Wreck along the River of Money: An Evaluation of the Hanford Cleanup: A Report for the U.S. Senate Committee on Energy and Natural Resources.* Washington, D.C. (March).

Boudreau, Robert. 1998. Building blocks to lasting peace. *Air Force Times* (1 June).

Bullard, Robert. *Dumping Dixie: Race, Class, and Environmental Quality.* Boulder: Westview.

Braden, Kathleen. 1992. US-Soviet cooperation for environmental protection. In John Massey Stewart, ed., *The Soviet Environment: Problems, Policies, and Politics.* Cambridge: Cambridge University Press.

Bradley, Don J., Clyde Frank, and Yevgeny Mikerin. 1996. Nuclear contamination from weapons complexes in the former Soviet Union and the United States. *Physics Today:* 40–45.

Bramwell, Anna. 1989. *Ecology in the 20th Century: A History.* New Haven: Yale University Press.

Broad, William J. 1994. Nuclear roulette for Russia: Burying uncontained waste. *New York Times* (November 21): sec. A.

Buhkarin, Oleg. 1997. The future of Russia's plutonium cities. *International Security* 21 (Spring): 126–158.

Butler, William E. 1983. Natural resources law. *Soviet Law.* London: Butterworths.

Butler, William E. 1987. *Collected Legislation of the Union of Soviet Socialist Republics and the Constituent Union Republics.* Dobbs Ferry, NY: Oceana Publications.

Byrne, John, and Steven Hoffman, eds. 1996. *Governing the Atom: The Politics of Risk.* New Brunswick, N.J.: Transaction Books.

Caldwell, J. J. 1971. Letter to T. A. Nemzek. WADCO radioactive waste disposal summary (January 27).

Caldwell, Lynton. 1976. Novus ordo seclorum: The heritage of American public administration. *Public Administration Review* 36:476–488.

Catton, William, and Riley Dunlap. 1980. A new ecological paradigm for post-exuberant sociology. *American Behavioral Scientist* 24:15–47.

Cawley, R. McGregor. 1993. *Federal Land, Western Anger: The Sagebrush Rebellion and Environmental Politics.* Lawrence, KS: University Press of Kansas.

Chernomyrdin, V. 1995. Passport of the federal program: Management of radioactive waste and spent nuclear materials, their utilization and disposal for 1996–2005. (Decree No. 1030, Moscow).

Christensen, G. C., et al. 1997. Radioactive contamination in the environment of the nuclear enterprise "Mayak": Results from the joint Russian-Norwegian fieldwork in 1994. *Science of the Total Environment* 202 (N1–3): 237–248.

Chukanov, V. N., Y. G. Drozhko, A. P. Kuligin, G. A. Mesyats, A. N. Penyagin, A. V. Trapeznikov, and P. V. Volobuev. 1991. Ecological conditions for the creation of atomic weapons at the atomic industrial complex near the city of Kyshtym. Paper presented at the Conference on the Environmental Consequences of Nuclear Weapons Development, University of California, Irvine, CA, April. Translated for Lawrence Livermore Laboratory by the Ralph McElroy Translation Co.

Clarke, Jeanne Nienaber, and Daniel C. McCool. 1996. *Staking Out the Terrain: Power and Performance Among Natural Resource Agencies,* 2nd ed. Albany: State University of New York Press.

Cochran, Thomas B., William Arkin, Robert Norris, and Milton Hoenig. 1987. *Nuclear Weapons Databook: Volume III, U.S. Nuclear Warhead Facility Profiles.* Cambridge, MA: Ballinger.

Cochran, Thomas B., and Robert S. Norris. 1991. *Soviet Nuclear Warhead Production, Nuclear Weapons Databook Working Papers.* NWD 90–3 (3rd revision), Natural Resources Defense Council, Washington, D.C.

Cochran, Thomas B., and Robert S. Norris. 1992. *Russian/Soviet Nuclear Warhead Production.* Washington, DC: Natural Resources Defense Council.

Cochran, Thomas B., and Robert S. Norris. 1993a. *Nuclear Weapons Databook Working Papers.* NWD 93–1 (5th revision), September 8, Washington, DC: Natural Resources Defense Council.

Cochran, Thomas, and Robert S. Norris. 1993b. *Soviet Nuclear Weapons Databook.* Washington, DC: Natural Resources Defense Council.

Cochran, Thomas B., Robert S. Norris, and Oleg A. Bukharin. 1995. *Making The Russian Bomb, From Stalin To Yeltsin.* San Francisco: Westview Press.

Commission for Investigation of the Ecological Situation in the Chelyabinsk Region. 1991. Proceedings of The Commission for Investigation of the Ecological Situation in the Chelyabinsk Region, Vol. I-II. By decree #P1283 of the President of the USSR (3 January 1991).

Conover, Pamela, and Stanley Feldman. 1989. Candidate perceptions in an ambiguous world. *American Journal of Political Science* 33:912–940.

Cotgrove, Steven. 1982. *Catastrophe or Cornucopia: The Environment, Politics and the Future.* New York: Wiley and Sons.

Cotgrove, Steven, and Andrew Duff. 1980. Environmentalism, middle class radicalism, and politics. *Sociological Review* 28:333–351.

Cragg, Chris. 1996. Editorial: Russia. *Financial Times* 180 (October): 1.

Dalton, Russell. 1994. *The Green Rainbow: Environmental Interest Groups in Western Europe.* New Haven: Yale University Press.

Dalton, Russell, and Manfred Kuechler. 1990. *Challenging the Political Order: New Social and Political Movements in Western Democracies.* New York: Oxford University Press/Cambridge, UK: Polity Press.

Dalton, Russell, and Robert Rohrschneider. 1999. The greening of Europe: Environmental values and environmental behavior. In Roger Jowell et al., eds., *British and European Social Attitudes—the 15th Report.* Brookfield, VT: Dartmouth.

D'Antonio, Michael. 1993. *Atomic Harvest: Hanford and the Lethal Toll of America's Nuclear Arsenal.* New York: Crown Publishers.

Davis, Kenneth Culp. 1969. *Discretionary Justice.* Baton Rouge: Louisiana State University Press.

DeBardeleben, Joan. 1985. *The Environment and Marxism-Leninism.* Boulder, CO: Westview Press.

DeBardeleben, Joan. 1992. The new politics in the USSR: The case of the environment. In John Massey Stewart, ed., *The Soviet Environment: Problems, Policies and Politics.* Cambridge: Cambridge University Press.

DeBardeleben, Joan, and John Hannigan, eds. 1995. *Environmental Security and Quality after Communism.* Boulder: Westview.

Degteva, M. O., et al. 1996a. Population exposure dose reconstruction for the Urals region. In S. L. Kellogg and E. J. Kirk. Proceedings from the 1996 AAAS Annual Meeting Symposium. February 12. Washington, D.C.: American Association for the Advancement of Science, 21–34.

Degteva, M. O., et al. 1996b. *Dose Reconstruction for the Urals Population.* Livermore, CA: Lawrence Livermore National Laboratory, UCRL-ID-123713.

DeLeon, Peter. 1997. *Democracy and the Policy Sciences.* Albany: State University of New York Press.

Department of Defense. 1997. Cooperative Threat Reduction Program. Washington, D.C.: Department of Defense.

Department of Ecology, United States Environmental Protection Agency and the United States Department of Energy. 1989. *Hanford Federal Facility Agreement and Consent Order,* 89–10V. Richland: U.S. Department of Energy.

Department of Energy. 1989. *Environmental Restoration and Waste Management: Five Year Plan,* DOE/S-0070. Washington, D.C.: Department of Energy (Available from National Technical Information Service, U.S. Department of Commerce, Springfield, Virginia, 22161).

Department of Energy. 1994. *U.S.-Former Soviet Union Environmental Restoration and Waste Management Activities* (DOE/EM0153P). Washington, D.C.: U. S. Government Printing Office.

Department of Energy. 1995. *The 1995 Baseline Environmental Management Report: Executive Summary.* Washington, D.C.: U.S. Department of Energy.

Department of Energy. 1998a. *The FY 1998 Budget Request in Perspective.* Washington, D.C.: U.S. Department of Energy.

Department of Energy. 1998c. Enhanced work planning at Hanford: An overview of successes. (http://tis.nt.eh.doe.gov/wpphm/ewp/sites/rl/RI_succ.htm).

Department of Energy. 1998d. Lights Out at B plant: A major step toward successful deactivation. (http://www.hanford.gov/press/1998/98-087.html).

Department of Energy. 1998e. DOE announces plan to address groundwater and soil contamination at Hanford to help protect the Columbia River. (http://www.hanford.gov/press/1998/98-006.htm).

Department of Energy. 1998f. MPC&A Program Strategic Plan. Washington, D.C.: Department of Energy.

Department of Energy. n.d. *Hanford Facts.* Richland, WA: U.S. Department of Energy.

Diani, Mario. 1995. *Green Networks: A Structural Analysis of the Italian Environmental Movement.* Edinburgh: University of Edinburgh Press.

Dietrich, William. 1995. *Northwest Passage: The Great Columbia River.* New York: Simon & Schuster.

Dionne, E. J. 1991. *Why Americans Hate Politics.* New York: Simon & Schuster, Publishers.

Dmitrieva, Irina. 1995. Activists decry fallout from decree on waste. *Moscow Times* (September 13).

Dobson, Andrew. 1995. *Green Political Thought,* 2nd ed. London: Routledge.

Doktorov, Boris Z., Boris M. Firsov, and Viatcheslav V. Safronov. 1993. Ecological consciousness in the USSR: Entering the 1990s. In Anna Vari and Pal Tamas, eds., *Environment and Democratic Transition: Policy and Politics in Central and Eastern Europe.* Dordrecht: Kluwer Academic Publishers.

Donnay, Albert et al. 1995. Russia and the territories of the former Soviet Union. In Arjun Makhijani, Howard Hu, and Katherine Yih, eds., *Nuclear Wastelands: A Global Guide to Nuclear Weapons Production and its Health and Environmental Effects.* Cambridge: MIT Press.

Downey, Gary L. 1986. Ideology and the Clamshell identity: Organizational dilemmas in the anti-nuclear power movement. *Social Problems* 33:357–373.

Drozhko, Evgonii et al. 1995. Technical Report, Project on Radiation Safety of the Biosphere. Laxenburg, Austria: International Institute for Applied Systems Analysis.

Dunlap, Riley. 1994. International attitudes toward environmental development. In Helge Ole Bergesen and Georg Parmann, eds., *Green Globe Yearbook of International Cooperation on Environment and Development 1994.* Oxford: Oxford University Press.

Dunlap, Riley, and Willian Catton. 1979. Environmental sociology. *Annual Review of Sociology* 5:243–273.

Dunlap, Riley, George H. Gallup, Jr., and Alec M. Gallup. 1993. *Health of the Planet Survey.* Princeton, NJ: Gallup International Institute.

Dunlap, Riley, and Angela Mertig. 1992. *American Environmentalism.* Philadelphia: Taylor and Francis.

Dunlap, Riley E., and Marvin E. Olsen. 1984. Hard-path versus soft-path advocates: A study of energy activists. *Policy Studies Journal* 13:413–428.

Dunlap, Riley, and Kent Van Liere. 1978. The new environmental paradigm. *Journal of Environmental Education* 9:10–19.

Dunlap, Riley et al. 1993. Local attitudes toward siting a high level nuclear waste repository at Hanford, Washington. In Riley E. Dunlap, Michael Kraft, and Eugene A. Rosa, eds., *Public Reactions to Nuclear Waste*. Durham, NC: Duke University Press.

Eaton, Jason H. 1995. Kicking the habit: Russia's addiction to nuclear waste dumping at sea. *Denver Journal of International Law and Policy* 23:287–312.

Ebbin, Steven, and Raphael Kasper. 1974. *Citizen Groups and the Nuclear Power Controversy: Users of Scientific and Technological Information*. Cambridge, MA: The MIT Press.

Eckersley, Robyn. 1992. *Environmentalism and Political Theory: Toward an Ecocentric Approach*. Albany: State University of New York Press.

Eckstein, Harry et al. 1998. *Can Democracy Take Root in Post-Soviet Russia?: Explorations in State-Society Relations*. Lanham, MD: Rowman & Littlefield.

Energy Economist. 1996. Kola: It's the real thing. *Energy Economist* (May): 12–15.

Environmental Law Institute. 1987. Environmental law and policy in the USSR. *Environmental Law Report* (March).

Essig, T. H. 1971. *Hanford Waste Disposal Summary-1970*. BNWL-1618. Richland: Battelle Pacific Northwest Laboratories.

Faden, Ruth. 1995. *Advisory Committee on Human Radiation Experiments—Final Report*. U.S. Department of Energy. Washington, D.C.: U.S. Government Printing Office.

Fallows, James. 1996. *Breaking the News*. New York: Pantheon.

Feshbach, Murray, and Alfred Friendly, Jr. 1992. *Ecocide in the USSR*. New York: Basic Books.

Fish, Steven. 1995. *Democracy from Scratch: Opposition and Regime in the New Russian Revolution*. Princeton: Princeton University Press.

Fleischer, Christian Calmeyer. 1974. *The Tri-City Nuclear Industrial Council and the Economic Diversification of the Tri-Cities, Washington, 1963–1974*. Unpublished Master's Thesis, Washington State University.

Fleming, A. G. 1973. *Report on the Investigation of 106-T Tank Leak at the Hanford Reservation*. TID-26431. Richland: U.S. Atomic Energy Commission.

Fleras, Angie, and Jean Leonard Elliot. 1992. *The Nations Within: Aboriginal-State Relations in Canada, the United States, and New Zealand*. Toronto: Oxford University Press.

Fomichev, Sergey. 1997. *Raznotsvetnyie Zelyonyie* (Multicolored Greens). Moscow–Nizhnyi Novgorod: Tretii Put.

Forbes, Ian A., Daniel F. Ford, Henry W. Kendall, and James J. MacKenzie. 1972. Cooling water. *Environment* 14:40–47.

Forester, John. 1989. *Planning in the Face of Power*. Berkeley: University of California Press.

Foster, Heath. 1997. Hanford bomb work plan protested, money for tritium would be taken from already reduced cleanup budget. *Seattle Post-Intelligencer* (April 4): C2.

Foster, Heath. 1998a. Hanford start-up plan gets loud 'no,' crowd of 400 packs hearing in Seattle. *Seattle Post-Intelligencer* (January 21): A1.

Foster, Heath. 1998b. Hanford gearing up for new life, a new fight, restart proposal draws fire. *Seattle Post-Intelligencer* (January 12): A1.

French, Hilary F. 1991. Restoring the East European and Soviet environments. In *The State of the World.* New York: W.W. Norton.

Garb, Paula. 1997. Complex problems and no clear solutions: Difficulties of defining and assigning culpability for radiation victimization in the Chelyabinsk region of Russia. In Barbara Rose Johnston, ed., *Life and Death Matters: Human Rights at the End of the Millennium.* Walnut Creek: Altamira Press.

General Accounting Office. 1995. *The Galvin Report and National Laboratories Need Clearer Missions and Better Management: A GAO Report to the Secretary of Energy.* Washington, D.C.: U.S. Government Printing Office.

General Accounting Office. 1998. Nuclear Waste: Understanding of Waste Migration at Hanford Is Inadequate for Key Decisions. GAO/RCED-98-80, Washington, D.C.: Government Printing Office.

Geranois, Nicholas K. 1996. Judge throws attorney off Hanford case. *Columbian* (November 15): 1.

Geranois, Nicholas K. 1997a. Hanford blast released plutonium traces. *Columbian* (July 10): 1.

Geranois, Nicholas K. 1997b. Hanford blast wrecks more than storage tanks. *Columbian* (August 4): 1.

Gerber, Michele. 1992a. *On the Home Front: The Cold War Legacy of the Hanford Nuclear Site.* Lincoln: University of Nebraska Press.

Gerber, Michele S. 1992b. *Legend and Legacy: Fifty Years of Defense Production at the Hanford Site,* WHC-MR-0293. Richland: Westinghouse Hanford Company.

Gerber, Michele S. 1992c. Why Hanford's history is important. *Perspective* 10/11 (Summer/Fall): 19.

Gerber, Michele. 1993. Lessons learned at the Hanford site. *Universe* (Fall).

Gerrard, Michael. 1994. *Whose Backyard, Whose Risk? Fear and Fairness in Toxic and Nuclear Waste Siting.* Cambridge: The MIT Press.

Gibson, James, and Ray Duch. 1995. Postmaterialism and the emerging Soviet democracy. *Political Research Quarterly* 47:5–39.

Gluck, Ira. 1993. The mobilization of citizen protest in Chelyabinsk. Paper presented at the Conference on Critical Masses. Irvine: University of California, Irvine.

Goldsteen, Raymond, and John Schorr. 1991. *Demanding Democracy after Three Mile Island.* Gainesville: University of Florida Press.

Gordon, Joshua, and Mark Knapp. 1989. *Consequences of a Nuclear Accident.* Washington, D.C.: Public Citizen Press.

Government Accountability Project. 1994. Challenging whistleblower retaliation in the forest service. *Bridging the GAP* (Summer).

Government Accountability Project. 1995. Incinerating without a safety net. *Bridging the GAP* (Spring).

Government Accounting Office. 1993. *Nuclear Health and Safety: Examples of Post World War II Radiation Releases at U.S. Nuclear Sites.* Washington, D.C.: U.S. Government Printing Office. GAO/RCED-94-51-FS.

Government Accounting Office. 1996. *Nuclear Waste, Management, and Technical Problems Continue to Delay Characterization of Hanford Tank Waste* (GAO/RCED-96-56). Washington, D.C.: U.S. Government Printing Office.

Grainey, Michael W., and Dirk Dunning. 1995/1996. Federal sovereign immunity: How self-regulation became no regulation at Hanford and other nuclear weapons facilities. *Gonzaga Law Review* 31:137–146.

Green, Eric. 1991. *Ecology and Perestroika: Environmental Protection in the Soviet Union.* A report for the American Committee on US-Soviet Relations.

Greider, William. 1992. *Who Will Tell the People: The Betrayal of American Democracy.* New York: Simon & Schuster.

Groves, Leslie. 1962. *Now it Can be Told: The Story of the Manhattan Project.* New York: Harper.

Gundersen, Adolf. 1995. *The Environmental Promise of Democratic Deliberation.* Madison: University of Wisconsin Press.

Gusterson, Hugh. 1996. *Nuclear Rites: A Weapons Laboratory at the End of the Cold War.* Berkeley: University of California Press.

Hamilton, Lawrence C. 1985. Concern about toxic wastes: Three demographic predictors. *Sociological Perspectives* 28:463–486.

Handler, Joshua. 1993. No sleep in the deep for Russian subs. *The Bulletin of Atomic Scientists* 49 (3):7–9.

Hanford Environmental Dose Reconstruction Project, Phase I Summary Report. 1991. Richland, WA: Battelle Memorial Institute.

Hanford Health Information Network. 1997. The Release of Radioactive Materials from Hanford: 1944–1972 (http://www.doh.wa.gov/hanford/publications/history/release.html).

Hanford Update. 1994. Protection of the Columbia River emphasized in new agreement. 6 (1):1–2.

Hanford future site uses working group. 1992. *The Future for Hanford: Uses and Cleanup: The Final Report of the Hanford Future Site Uses Working Group.* Richland, WA: U.S. Department of Energy.

Hanford Thyroid Disease Study. 1999. Summary of the Hanford Thyroid Disease Study Results Newsletter (January 1999) (http://www.fhcrc.org/science/phs/htds/news/99.htm).

Harlan, W. E., D. E. Jenne, and Jack Healy. 1950. Dissolving of twenty day metal at Hanford. Hanford Works, Atomic Energy Commission. HW-17381 (ACHRE no. DOE-050394-A-4).

Healy, Jack W. 1953. *Release of Radioactive Wastes to Ground.* HW-28121 (RL: HAPO, May 20).

Healy, Tim. 1989. Tug of war. *Seattle Times* (February 13):1.

Heeb, C. M. 1994. *Radionuclide Releases to the Atmosphere from Hanford Operations, 1944-1972.* HEDR (January): PNWD-2222.

Heeb, C. M., and D. J. Bates. 1994. *Radionuclide Releases to the Columbia River from Hanford Operations, 1944–1971.* HEDR (January): PNWD-2223.

Helmer, John. 1998. Rouble crisis pays dividend. *Moscow Times* (June 17):1.

Hibbs, Mark. 1996. RT-1 Operation faces cost crisis, uncertain future demand schedule. *Nuclear Fuel* 21 (January 1): 10.

Hirschman, Alberto. 1981. *Shifting Involvements.* Princeton: Princeton University Press.

Hohryakov, V. F., C. G. Syslova, and A. M. Skryabin. 1994. Plutonium and the risk of cancer: A comparative analysis of PU body burdens due to releases from nuclear plants. *Science of the Total Environment* 142 (N1-2): 101–104.

Holdren, John P. 1996. Reducing the threat of nuclear theft in the former Soviet Union. *Arms Control Today* (March): 14–20.

Holloway, David. 1994. *Stalin and the Bomb.* New Haven: Yale University Press.

Hosking, Geoffrey. 1991. *The Awakening of the Soviet Union.* Cambridge: Harvard University Press.

Houff, William H. 1992. Nuclear science as a black art. *Perspective* 10/11 (Summer/Fall): 3.

Hunter, Robert. 1979. *Warriors of the Rainbow.* New York: Holt, Rinehart and Winston.

Impact Assessment. 1987. *Socioeconomic Impact Assessment of the Proposed High-level Nuclear Waste Repository at Hanford Site, Washington.* La Jolla, CA: Impact Assessment Inc.

Inglehart, Ronald. 1990. *Culture Shift in Advanced Industrial Society.* Princeton, NJ: Princeton University Press.

Inglehart, Ronald. 1997. *Modernization and Postmodernization.* Princeton, NJ: Princeton University Press.

Inglehart, Ronald, and Hans-Dieter Klingemann. 1996. Dimensionen des Wertewandels: Theoretische und methodische Reflexioned anlässlich einer neuuerlichen Kritik. *Politische Vierteljahresschrift* 37: 319–340.

Ingram, Helen, and Dean Mann. 1989. Interest groups and environmental policy. In James Lester, ed., *Environmental Politics and Policy.* Durham, NC: Duke University Press.

Interfax News Agency. 1997. Russian nuclear waste disposal programme underfunded. May 14, BBC Monitoring International Reports. Record Number: 00805*19970514*00684.

International Physicians for the Prevention of Nuclear War (IPPNW). 1992. Plutonium: Deadly gold of the nuclear age, Massachusetts: International Physicians for the Prevention of Nuclear War and Institute for Energy and Environmental Research.

Inter Press Services. 1998. Environmental activists face former KGB's unwelcome attention. *The Indian Express* (January 27).

Irvine, Sandy, and Alec Ponton. 1988. *A Green Manifesto: Politics for a Green Future.* London: Macdonald Optima.

Ivanov, Andrei. 1997. Nuclear poisoned village must endure a while longer, *Inter Press Service* (August 22), Record Number: 01084*19970822*00008.

Ivanov, A., and J. Perera. 1997. Secret service scores victory against eco-investigators. *Interpress Service* (October 17).

Jamieson, Robert L. 1997. Radioactive waste found in aquifer at Hanford, U.S. report called too little, too late. *Spokesman-Review* (October 26):A1.

Jancar, Barbara. 1990. Democracy and the environment in Eastern Europe and the Soviet Union. *Harvard International Review* 12:13.

Jancar-Webster, Barbara, ed. 1992. *Environmental Action in Eastern Europe: Responses to Crisis.* Armonk, NY: M.E. Sharpe.

Jancar-Webster, Barbara. 1993. Eastern Europe and the Soviet Union. In Sheldon Kamienecki, ed., *International Environmental Policy.* Albany, NY: State University of New York Press.

Jenkins, J. Craig. 1983. Resource mobilization theory and the study of social movements. *Annual Review of Sociology* 9:527–553.

Jenkins, J. Craig, and Bert Klandermans, eds. 1995. *The Politics of Social Protest.* Minneapolis: University of Minnesota Press.

Johnson, T. 1993. Russian nuclear dumping deepens Japanese mistrust. *Japan Economic Newswire* (October 23) LEXUS-NEXUS Library, JEN file.

Jones, Gary L. 1998. Letter of transmittal B279142 (May 13, 1998): Washington, D.C.: General Accounting Office.

Jones, Robert Emmett, and Riley E. Dunlap. 1992. The social bases of environmental concern: Have they changed over time? *Rural Sociology* 54:28–47.

Jones, Vincent C. 1985. *Manhattan: The Army and the Atomic Bomb.* Washington, D.C.: U.S. Center for Military History.

Jurji, Judith. 1998. Some lessons learned from ten years of Hanford activism. *HHIN Connections: Hanford Health Information Network Newsletter* 4 (Winter/Spring): 4.

Kachenko, Y. 1993. Russian factory gets rid of nuclear waste products. *TASS* (January 5).

Kaplan, Louise. 1992. The History of HEAL, *Hanford Education Action League Perspective* (Summer/Fall).

Kassiola, Joel J. 1990. *The Death of Industrial Civilization: The Limits to Economic Growth and the Repoliticization of Advanced Industrial Society.* Albany: State University of New York Press.

Kempton, Willett, James Boster, and Jennifer Hartley. 1995. *Environmental Values in American Culture.* Cambridge, MA: The MIT Press.

Kitschelt, Herbert. 1986. Political opportunity structures and political protest. *British Journal of Political Science* 16:58–95.

Kitschelt, Herbert. 1989. *The Logics of Party Formation: Ecological Politics in Belgium and West Germany.* Ithaca: Cornell University Press.

Klandermans, Bert. 1990. Linking the 'old' and 'new' movement networks in the Netherlands. In Russell Dalton and Manfred Kuechler, eds., *Challenging the Political Order.* New York: Oxford University Press.

Knoke, David. 1988. Incentives in collective action organizations. *American Sociological Review* 88:311–329.

Knoke, David. 1990. *Organizing for Collective Action: The Political Economies of Associations.* New York: de Gruyter.

Koshurnikova, N. A. et al. 1996. Mortality among personnel who worked at the Mayak complex in the first years of its operation. *Health Physics* 71 (N1): 90–93.

Kossenko, M. M. 1996. Cancer mortality among Techa River residents and their offspring. *Health Physics* 71 (N1):77–82.

Kossenko, M. M., M. O. Degteva, and M. A. Petrushova. 1992. Estimate of leukemia risk to those exposed as a result of nuclear incidents in the Southern Urals. *PSR Quarterly* 2(4):189.

Kossenko, M. M., M. O. Degteva, O. V. Vyushkova, D. L. Preston, K. Mabuchi, and V. P. Kozheurov. 1997. Issues in the comparison of risk estimates for the population in the Techa River region and atomic bomb survivors. *Radiation Research* 148 (N1):54–63.

Kraft, Michael E., Eugene A. Rosa, and Riley E. Dunlap. 1993. Public opinion and nuclear waste policymaking. In Riley E. Dunlap, Michael E. Kraft, and Eugene A. Rosa, eds., *Public Reactions to Nuclear Waste.* Durham, NC: Duke University Press.

Kriesi, Hanspeter, and Philip van Praag. 1987. Old and new politics. *European Journal of Political Research* 15:319–346.

Kudrik, Igor. 1997. The Current Status, September 1997: The Nikitin Case. (http://www.bellona.no/e/russia/status/9709.htm).

Kuklinski, James H., Daniel S. Metlay, and W. D. Kay. 1982. Citizen knowledge and choices on the complex issue of nuclear energy. *American Journal of Political Science* 26:615–642.

Laraña, Enrique, Hank Johnston, and Joseph Gusfield, eds. 1994. *New Social Movements: From Ideology to Identity*. Philadelphia: Temple University Press.

Laverov, N. 1993. Vice President, Russian Academy of Sciences, quoted in *Nuclear Waste News* (February 25).

Laverov, N. P., B. I. Omelianenko, and V. I. Velichkin. 1994. Geological aspects of the nuclear waste disposal problem. Berkeley: Lawrence Berkeley Laboratory.

Lavine, Steven D. 1996. Russian dumping in the Sea of Japan. *Denver Journal of International Law and Policy,* 24 (Spring):417–450.

League of Women Voters of Spokane. 1985. *Radiation in Eastern Washington: Following the Trail of Nuclear Waste.* Spokane: League of Women Voters of Spokane.

Lens, Sydney. 1982. *The Bomb.* New York: Lodestar Books.

Leslie, Stuart W. 1993. *The Cold War and American Science: The Military-Industrial-Academic Complex at MIT and Stanford.* New York: Columbia University Press.

Lewis, Jack. 1980. Civil protest in Mishima. In Kurt Stuher, Ellis Krauss, and Scott Flanagan, eds., *Political Opposition and Local Politics in Japan.* Princeton: Princeton University Press.

Liebow, Edward B., Catherine Younger, and Julie A. Broyles. 1987. *A Synthesis of Ethnohistorical Materials Concerning the Administration of Federal Indian Policy Among the Yakama, Umatilla and Nez Perce Indian People.* Seattle: Battelle Human Affairs Research Center, Pacific Northwest Laboratory, Battelle Memorial Institute. [Report prepared for the Basalt Waste Isolation Project, under U.S. Department of Energy contract OE-AC06-76 RLO 1830].

Lipset, Seymour Martin. 1979. *The First New Nation.* New York: W.W. Norton.

Lipset, Seymour Martin. 1990. *Continental Divide: The Values and Institutions of the United States and Canada.* New York: Routledge.

Lipset, Seymour Martin. 1996. *American Exceptionalism: A Double-Edged Sword.* New York: W.W. Norton & Co.

Loeb, Paul. 1986. *Nuclear Culture.* Philadelphia: New Society Publishers.

Lowe, Philip D., and Wolfgang Rudig. 1986. Political ecology and the social sciences. *British Journal of Political Science* 16:513–550.

Lowi, Theodore J. 1979. *The End of Liberalism: The Second Republic of the United States,* 2nd ed. New York: W.W. Norton.

Lowi, Theodore J. 1995. *The End of the Republic Era.* Norman, OK: University of Oklahoma Press.

Lugar, Richard. 1997. The Cooperative Threat Reduction Program, MPC&A Program Strategic Plan. Washington, D.C.: Department of Defense.

Lynn, Laurence E. 1996. *Public Management as Art, Science, and Profession.* Chatham, NJ: Chatham House Publishers.

Makhijani, Arjun, Howard Hu, and Katherine Yih, eds. 1995. *Nuclear Wastelands: A Global Guide to Nuclear Weapons Production and its Health and Environmental Effects.* Cambridge: The MIT Press.

Makhijani, Arjun, Howard Hu, and Katherine Yih. 1995. The United States. In Arjun Makhijani, Howard Hu, and Katherine Yih, eds., *Nuclear Wastelands: A Global Guide to Nuclear Weapons Production and its Health and Environmental Effects.* Cambridge: The MIT Press.

Maloney-Dunn, E. 1993. Russia's nuclear waste law: A response to the legacy of environmental abuse in the former Soviet Union. *Arizona Journal of International and Comparative Law* (November):379–381.

Marone, Joseph G., and Edward J. Woodhouse. 1989. *The Demise of Nuclear Energy, Lessons for Democratic Control of Technology.* New Haven, CT: Yale University Press.

Marples, David. 1995. The post-Soviet nuclear power program. In Joan DeBardeleben and John Hannigan, eds. *Environmental Security and Quality after Communism: Eastern Europe and the Soviet Successor States.* Boulder: Westview Press.

Martin, Todd. 1993. Milestone review: Progress toward what end? *Perspective* 12 (Winter): 16.

Martinez, Cecilia, and John Byrne. 1996. Science, society and the state: The nuclear project and the transformation of the American political economy. In John Byrne and Steven Hoffman, ed., *Governing the Atom.* New Brunswick, NJ: Transaction Books.

Mazur, Allan. 1981. *The Dynamics of Technical Controversy.* Washington, D.C.: Communications Press, Inc.

Mazur, Allan. 1990. Nuclear power, chemical hazards, and the quantity of reporting. *Minerva* 28:294–323.

McAdam, Douglas. 1994. Culture and social movements. In Enrique Laraña, Hank Johnston, and Joseph Gusfield, eds., *New Social Movements: From Ideology to Identity.* Philadelphia: Temple University Press.

McCarthy, John, and Mayer Zald. 1977. Resource mobilization and social movements. *American Journal of Sociology* 82:1212–1241.

McCormick, John. 1993. International nongovernmental organizations: Prospects for a global environmental movement. In Sheldon Kamieniecki, ed., *Environmental Politics in the International Arena.* Albany: State University of New York.

McFarland, Andrew S. 1984. *Common Cause: Lobbying in the Public Interest.* Chatham, NJ: Chatham House Publishers.

McKean, Margaret. 1981. *Environmental Protest and Citizen Politics in Japan.* Berkeley: University of California Press.

Medvedev, Z. A. 1979. *Nuclear Disaster in the Urals.* New York: W.W. Norton.

Meehan, Eugene. 1988. *The Thinking Game: A Guide to Effective Study.* Chatham, NJ: Chatham House Publishers.

Melucci, Alberto. 1980. The new social movements: A theoretical approach. *Social Science Information* 19:199–226.

Merton, Robert. 1973. *The Sociology of Science.* Chicago: University of Chicago Press.

Milbrath, Lester W. 1984. *Environmentalists: Vanguard for a New Society.* Albany: SUNY Press.

Milbrath, Lester. 1989. *Envisioning a Sustainable Society.* Albany: SUNY Press.

Miller, Arthur, William Reisinger, and Vicki Hesli. 1993. *Public Opinion and Regime Change: The New Politics of Post-Soviet Societies.* Boulder: Westview Press.

Mironenko, M. V., et al. 1995. The cascade reservoirs of the Mayak Plant: Case history and the first version of a computer simulator. Berkeley: Lawrence Berkeley Laboratory.

Mitchell, Robert Cameron. 1979. National environmental lobbies and the apparent illogic of collective action. In Charles S. Russell, ed., *Collective Decision Making: Applications from Public Choice Theory.* Baltimore: Johns Hopkins University Press.

Mitchell, Robert Cameron. 1984. Rationality and irrationality in the public's perception of nuclear power. In William R. Freudenberg and Eugene A. Rosa, eds., *Public Reactions to Nuclear Power.* Boulder, CO: Westview Press.

Monroe, S. D. 1992. Chelyabinsk radioecological conference sets precedent for cooperation. *CIS Environmental Watch* 2 (Spring):13.

Mueller, Carol McClure. 1992. Building social movement theory. In Enrique Laraña, Hank Johnston, and Joseph Gusfield, eds., *New Social Movements: From Ideology to Identity.* Philadelphia: Temple University Press.

Muller, Edward, and Karl-Dieter Opp. 1986. Rational choice and rebellious collective action. *American Political Science Review* 80:471–488.

Nash, Roderick. 1967. *Wilderness and the American Mind.* New Haven: Yale University Press.

National Research Council. 1995. *A Review of Two Hanford Environmental Dose Reconstruction Project (HEDR) Dosimetry Reports.* Washington, D.C.: National Academy Press.

National Research Council. 1996. *Nuclear Wastes: Technologies for Separations and Transmutation.* Washington, D.C.: National Academy Press.

Nealey, Stanley M., Barbara D. Melber, and William L. Rankin. 1983. *Public Opinion and Nuclear Energy.* Lexington, MA: D.C. Heath & Co.

Nichols, K. D. 1987. *The Road to Trinity.* New York: William Morrow.

Nikipelov, B. V., et al. 1989. Accident in the Southern Urals of 29 September 1957. *International Atomic Energy Information Circular* (May 28).

Nikipelov, B. V., and Y. G. Drozhko. 1990. Explosion in the Southern Urals. *Priroda* (May):48–49.

Nuclear Energy Committee 1996. Report of the Nuclear Energy Committee U.S.-Russian Commission on Economic and Technological Cooperation. Washington, D.C., January, 5.

Nuclear News. 1996. Construction of the Bn-800 fast reactor for the South Urals Power Station is to resume this year. *American Nuclear Society Nuclear News* (June):59.

Nuclear News. 1997. Last of Finnish spent fuel sent to Russia. *American Nuclear Society Nuclear News* (February):45.

Nunn, Sam. 1996. Senator Nunn's valedictory. *Air Force Magazine* (December):41.

Nye, Joseph, Philip Zelikow, and David King. 1997. *Why Americans Mistrust Government.* Cambridge: Harvard University Press.

O'Connor, Sally, Nicholas Lovrich, William Serbain, and Chequita Webb. 1995. *Inventory of Public Concerns at the Hanford Site.* New Orleans: Xavier University of Louisiana.

Offe, Claus. 1990. Reflections on the institutional self-transformation of movement politics. In Russell Dalton and Manfred Kuechler, eds., *Challenging the Political Order: New Social and Political Movements in Western Democracies.* New York: Oxford University Press/Cambridge, UK: Polity Press.

Office of Environmental Management, DOE. 1995. *Closing the Circle on the Splitting of the Atom.* Washington, D.C.: Department of Energy.

Office of Environmental Management, DOE. 1995. *Estimating the Cold War Mortgage.* Springfield, VA: U.S. Department of Energy, U.S. Department of Commerce (Technology & Administration), and the National Technical Information Service.

Office of Technology Assessment, U.S. Congress. 1991. *Complex Cleanup: The Environmental Legacy of Nuclear Weapons Production,* OTA-O-484. Washington, D.C.: U.S. Government Printing Office.

Oge, Margo T. 1994. Testimony before the House Subcommittee on Energy and Mineral Resources by Margo T. Oge, Director of the Office of Radiation and Indoor Air, U.S. Environmental Protection Agency, 3/8/94.

Olson, Mancur. 1965. *The Logic of Collective Action: Public Goods and the Theory of Groups.* Cambridge: Harvard University Press.

Olson, Mancur, and Hans Landsberg, eds. 1973. *The No Growth Society.* New York: W.W. Norton.

Omelianenko, B. I., B. S. Niconov, B. I. Ryzhov, and N. S. Shikina. 1994. Weathering products of basic rocks as sorptive materials of natural radionuclides. Berkeley, CA: Lawrence Berkeley Laboratory.

Owen, F. E. 1971. Private communication (unpublished data). Richland, WA: Douglas-United Nuclear.

Pacific Northwest Laboratory. 1990a. *Draft Summary Report: Phase I of the Hanford Environmental Dose Reconstruction Project* PNL-7423 HEDR UC 707. Richland: Pacific Northwest Laboratory.

Pacific Northwest Laboratory. 1990b. *Draft Columbia River Pathway Report: Phase I of the Hanford Environmental Dose Reconstruction Project* PNL-7411 HEDR UC 707. Richland: Pacific Northwest Laboratory.

Pacific Northwest Laboratory. 1991. *Summary Report: Phase I of the Hanford Environmental Dose Reconstruction Project.* PNL-7410 HEDR Rev.

Parker, Herbert M. 1945. *Status of Problem of Measurement of the Activity of Waste Water Returned to the Columbia.* HW-7-2346 (RL:HEW, September 11).

Parker, Herbert M. 1948. *Speculations on Long-Range Waste Disposal Hazards.* HW 8674 (RL: HEW, January 26).

Paulson, Michael. 1998a. Locke, Gregoire threaten suit over Hanford cleanup. *Seattle Post-Intelligencer* (February 28):A1.

Paulson, Michael. 1998b. Hanford violations will bring hefty fine: Contractor blamed in plutonium handling. *Seattle Post-Intelligencer.* A1.

Pearce, Fred. 1991. *Green Warriors: The People and Politics Behind the Environmental Revolution.* London: Bodley Head.

Perera, Judith. 1992. Russian government may press parliament to accept waste. *World Environment Report* (October 13) 18(21):176.

Peterson, D. J. 1993. *Troubled Lands: The Legacy of Soviet Environmental Destruction.* Boulder, CO: Westview Press.

Petrov, A. V., et al. 1994. Numerical modeling of the groundwater contaminant transport for the Lake Karachai area: The methodological approach and the basic two-dimensional regional model. Berkeley: Lawrence Berkeley Laboratory.

Pevar, Stephen L. 1992. *The Rights of Tribes: The Basic ACLU Guide to Indian and Tribal Rights,* 2nd ed. Carbondale, IL: Southern Illinois University Press.

Pierce, John C., and Nicholas P. Lovrich. 1986. *Water Resources, Democracy and the Technical Information Quandary.* New York: Associated Faculty Press.

Pierce, John C., et al. 1989. *Public Knowledge and Environmental Politics in Japan and the United States.* Boulder, CO: Westview Press.

Pierce, John C., et al. 1986. Vanguards and rearguards in environmental politics: A comparison of activists in Japan and the United States. *Comparative Political Studies* 18:419–448.

Pierce, John C., et al. 1988. Public information on acid rain in Canada and the United States. *Social Science Quarterly* 69:193-202.

Pierce, John C., et al. 1992. *Political Communication and Environmental Interest Groups: The United States and Canada.* Westport, CT: Praeger.

Polikarpov, Dimitry. 1998. Scientists join miners at White House. *Moscow Tribune* (June 17).

Porritt, Jonathan. 1984. *Seeing Green: The Politics of Ecology Explained.* London: Basil Blackwell.

Pryde, Philip. 1972. *Conservation in the Soviet Union.* New York: Cambridge University Press.

Pryde, Philip. 1991. *Environmental Management in the Soviet Union*. New York: Cambridge University Press.

Pryde, Philip, ed. 1995. *Environmental Resources and Constraints in the Former Soviet Union*. Boulder, CO: Westview Press.

Putnam, Robert. 1993. *Making Democracy Work: Civic Traditions in Italy*. Princeton: Princeton University Press.

Rankin, William L., Stanley M. Nealey, and Barbara Desow Melber. 1984. Overview of national attitudes toward nuclear energy: A longitudinal analysis. In Eugene Rosa and William Freudenburg, eds., *Public Reactions to Nuclear Power*. Boulder, CO: Westview Press.

Ratliff, Jeanne Nelson. 1994. *Challenging the Establishment: A Rhetorical Case Study of the Hanford Education Action League*. Unpublished Master's Thesis, Washington State University.

Reed, Richard, David J. Lemak, and W. Andrew Hesser. 1997. Cleaning up after the Cold War: Management and social issues. *Academy of Management Review* 22:614–642.

Rees, Joseph V. 1994. *Hostages of Each Other: The Transformation of Nuclear Safety Since Three Mile Island*. Chicago: University of Chicago Press.

Reeves, Merilyn. 1998. A progress report from the Hanford Advisory Board: Tank waste treatment. *Hanford Update: A Bulletin on Hanford Cleanup and Compliance of the Tri-Party Agreement* 10 (April/May):3–4.

Rhodes, Richard. 1986. *The Making of the Atomic Bomb*. New York: Simon & Schuster.

Richland Office, Department of Energy. 1996. *Hanford Cost Savings*. Richland: Department of Energy, DOE/RL-96/113.

Richland Office, Department of Energy. 1997. *Hanford Works! Fiscal Year 1997 Annual Report*. Richland: Department of Energy.

Riggs, Fred W. 1998. Public administration in America: Why our uniqueness is exceptional and important. *Public Administration Review* 58:22–31.

Rimmerman, Craig A. 1997. *The New Citizenship: Unconventional Politics, Activism, and Service*. Boulder, CO: Westview.

Robinson, Nicholas. 1992. International law in the ex-USSR: Pollution controls developing. *National Law Journal* (April 13).

Rogers, Joel. 1996. Hanford: A surprising natural legacy. *Washington Wildlands: The Magazine of the Nature Conservancy of Washington* 2 (Spring-Summer): 6–12.

Rogovin, Mitchell. 1980. *Three Mile Island: A Report to the Commissioners and the Public*. [Nuclear Regulatory Commission Special Inquiry]. Washington, D.C.: U.S. Government Printing Office.

Rohr, John A. 1986. *To Run a Constitution*. Lawrence: University Press of Kansas.

Rohrschneider, Robert. 1988. Citizen attitudes toward environmental issues: Selfish or selfless? *Comparative Political Studies* 21:347–367.

Rohrschneider, Robert. 1990. The roots of public opinion toward new social movements. *American Journal of Political Science* 34:1–30.

Rohrschneider, Robert. 1991. Public opinion toward environmental groups in Western Europe. *Social Science Quarterly* 72:251–266.

Romanov, G. 1992. Radioecological conditions accounted for the 1957 and 1967 accidents and production activities of the industrial complex. Paper presented at the International Radiological Conference, Chelyabinsk, Russia, May.

Rosa, Eugene A., and William R. Freudenburg. 1984. Nuclear power at the crossroads. In Eugene Rosa and William Freudenburg, eds., *Public Reactions to Nuclear Power*. Boulder, CO: Westview Press.

Rosa, Eugene A., and William R. Freudenburg. 1993. The historical development of public reactions to nuclear power: Implications for nuclear waste policy. In Riley E. Dunlap, Michael E. Kraft, Eugene A. Rosa, eds., *Public Reactions to Nuclear Waste*. Durham, NC: Duke University Press.

Rose, Richard. 1993. *Lesson-Drawing in Public Policy: A Guide to Learning Across Time and Space*. Chatham, NJ: Chatham House.

Rosenbaum, Walter A. 1993. Energy policy in the West. In Zachary A. Smith, ed., *Environmental Politics and Policy in the West*. Dubuque, IA: Kendall/Hunt.

Rosenbaum, Walter. 1995. *Environmental Politics and Policy,* 3rd ed. Washington: Congressional Quarterly Press.

Rothman, Stanley, and S. Robert Lichter. 1982. The nuclear energy debate: Scientists, the media and the public. *Public Opinion* 5:47–52.

Rudig, Wolfgang. 1991. *Anti-nuclear Movements: A World Survey of Opposition to Nuclear Energy*. London: Longmans.

Salisbury, Robert. 1969. An exchange theory of interest groups. *Midwest Journal of Political Science* 13:1–32.

Sandel, Michael J. 1996. *Democracy's Discontent: America in Search of a Public Philosophy*. Cambridge: Harvard University Press.

Sanger, S. L. 1989. *Hanford and the Bomb: An Oral History of World War II*. Seattle: Living History Press.

Sanger, David E. 1993. Nuclear materials dumped off Japan. *New York Times* (October 19):A1.

Schattschneider, E. E. 1960. *The Semisovereign People*. New York: Holt, Rinehart and Winston.

Schneider, Keith. 1991. Nuclear industry plans ads to counter critics. *New York Times* (November 13).

Schneider, William. 1986. Public ambivalent on nuclear power. *National Journal* 18:1562–1563.

Schwartz, E. A. 1991. Student Druzhinas movement. In *All OurLife*. Moscow: Master Publisher.

Science and Global Security. 1989. Kyshtym and Soviet nuclear materials production, vol. 1, nos. 1–2, p. 174 (a fact sheet containing technical information collected during a visit to Chelyabinsk-40 by an NRDC/Soviet Academy of Sciences delegation July 7–8, 1989).

Sergerstahl, Boris, Alexander Akleyev, and Vladimir Novikov. 1997. The long shadow of Soviet plutonium production. *Environment* (January/February):12–20.

Sheldon, Charles H. 1988. *A Century of Judging: A Political History of the Washington State Supreme Court*. Seattle: University of Washington Press.

Shulyakovskaya, Natalie. 1998. Luzhkov dreams of nuclear-free city. *Moscow Times* (April 30).

Simon, Jim. 1997. Downwinders' test hitting a roadblock, Feds argue over Hanford-related costs. *Seattle Times* (December 10):A1.

Smith, Rogers M. 1993. Beyond Tocqueville, Myrdal, and Hartz: The multiple traditions in America. *American Political Science Review* 87:549–565.

Sneider, Daniel. 1992. Baker focuses on atomic waste in visit to Urals. *Christian Science Monitor* (February 6).

Soden, Dennis L., Nicholas P. Lovrich, and John C. Pierce. 1985. City/Suburb perceptions of groundwater issues. *Proceedings of the American Water Resources Association Symposium on Groundwater Contamination and Reclamation*. Tucson, AZ: American Water Resources Association.

Soran, D., and D. B. Stillman. 1982. *An Analysis of The Alleged Kyshtym Disaster*. Los Alamos, NM: Los Alamos National Laboratory.

Spicer, Michael W. 1995. *The Founders, the Constitution, and Public Administration: A Conflict in World Views*. Washington, D.C.: Georgetown University Press.

Steger, M. A., et al. 1989. Political culture, postmaterial values, and the New Environmental Paradigm. *Political Behavior* 11:233–254.

Steele, Karen Dorn. 1984. Group voices radiation safety concern. *Spokesman-Review* (October 24):A14.

Steele, Karen Dorn. 1985a. Critics of nuclear power, weapons to speak at symposium in Spokane. *Spokesman-Review* (October 20):B2.

Steele, Karen Dorn. 1985b. Spokane anti-nuclear group criticized. *Spokesman-Review* (November 10):C6.

Steele, Karen Dorn. 1989. Hanford: America's nuclear graveyard. *The Bulletin of Atomic Scientists* (October).

Steele, Karen Dorn. 1990. Perseverance finally revealed Hanford secret. *Spokane Spokesman-Review* (July 25).

Steele, Karen Dorn. 1995. Judge rules Hanford study should stay secret. *The Spokesman-Review* (April 13):B3.

Steele, Karen Dorn. 1996. Judge urges dismissals of attorney, U.S. magistrate calls for investigation in lawyers handling of downwinders' lawsuit. *Spokesman-Review* (August 14):B1.

Steele, Karen Dorn. 1997. Radioactive waste from Hanford is seeping toward the Columbia. *High Country News* (September 29):1, 10–11.

Steinhardt, Bernice. 1995. *Nuclear Weapons Complex, Establishing a National Risk-Based Strategy for Cleanup.* Testimony before the Committee on Energy and Natural Resources, U.S. Senate, March 6, 1995. GAO/T-RCED-95-120.

Stillman, Richard J. II, ed. 1987. Symposium: The American Constitution and the administrative state. *Public Administration Review* [Special Issue], vol. 47.

Sun, M. 1993. Japan to help Russia with nuclear waste problem. *National Public Radio* (October 28), Transcript #1284-5 in LEXUS-NEXUS script file.

Tamplin, Arthur R. 1971. Issues in the radiation controversy. *Bulletin of the Atomic Scientists* 27:25–27.

Tanner 1997. Russia struggling with pollution blight. *Reuters* (January 31), Reuters America, Inc., Record Number: 0849*19970131*00608.

Thomas, Craig W. 1998. Maintaining and restoring public trust in government agencies and their employees. *Administration and Society* 30:166–193.

Thomas, James. 1990. Shadows of Hanford's past. *Perspectives* 1 (Fall).

Thomas, James. 1998. Prescription for healing Hanford's past—Tell the whole truth. *HHIN Connections: Hanford Health Information Network Newsletter* 4 (Winter/Spring):4.

Tilly, Charles. 1978. *From Mobilization to Revolution.* Reading, MA: Addison-Wesley.

Timoshenko, Alexandre. 1988–89. Developments in environmental law in the Soviet Union. *Journal of Environmental Law and Litigation* 3–4:131–134.

Tolchin, Susan J. 1996. *The Angry American: How Voter Rage Is Changing the Nation.* Boulder, CO: Westview Press.

United Press International. 1993. Officials and environmentalists criticize radiowaste waste dumping. In LEXUS, NEXUS Library, UPI File.

Usui, N. 1993a. Russia won't stop sea dumping but supplies details to Tokyo. *Nucleonics Week* (May 20):17.

Usui, N. 1993b. Mikhailov in Tokyo, Pledges no more waste dumping in Japan Sea. *Nucleonics Week* (October 28):10.

van Buren, Lisa. 1995. Citizen participation and the environment. In Joan DeBardeleben and John Hannigan, eds., *Environmental Security and Quality after Communism: Eastern Europe and the Soviet Successor States.* Boulder: Westview Press.

Verba, Sidney, Norman Nie, and Jae-on Kim. 1978. *Participation and Political Equality.* New York: Cambridge University Press.

Vraalsen, Tom. 1998a. Letter to Deputy Secretary of State Strobe Talbot (28 January 1998).

Vraalsen, Tom. 1998b. Letter to Deputy Secretary of State Strobe Talbott (5 June 1998).

Wadham, N. 1997. Alleged spy Pasko still in jail. *Vladivostok News* (December 11).

Wagoner, John D. 1992. *Cleanup More Than Paperwork. In A Look at Hanford: 1992 Progress Report.* Richland: Office of Communications, DOE.

Walsh, Edward. 1988. *Democracy in the Shadows: Citizen Mobilization in the Wake of the Accident at Three Mile Island.* Greenwich, CT: Greenwood Press.

Walsh, Edward, and Rex Warland. 1983. Social movement involvement in the wake of a nuclear accident. *American Sociological Review* 48:764–780.

Watanabe, Teresa, and Richard Boudreaux. 1993. Russian nuclear waste sparks feud: Japan is angry after Moscow dumps toxic liquid in nearby waters. *Los Angeles Times* (October 19):A7.

Weslowsky, Tony. 1994. Duma to vote on nuclear waste. *Moscow Tribune* (November 11):2.

Weiner, Douglas. 1988. *Models of Nature: Ecology, Conservation and Cultural Revolution in Soviet Russia.* Bloomington: Indiana University Press.

Wheeler, Michael. 1989. *Nuclear Weapons and the National Interest: The Early Years.* Washington, D.C.: National Defense University Press.

White, Stephen. 1991. *Gorbachev and After.* New York: Cambridge University Press.

Whiteley, John. 1991. *Environmental Consequences of Nuclear Weapons Production.* Irvine: School of Social Ecology.

Whiteley, John, and Yevgeny Gontmacher. 1998. Tilting against the environment: The Soviet governmental response to environmental problems and to the nuclear environment in the Chelyabinsk region. Irvine, CA: Center for the Study of Democracy.

Wildavsky, Aaron. 1995. *But Is It True? A Citizens Guide to Environmental Health and Safety Issues.* Cambridge: Harvard University Press.

Wilhelm, Steve. 1993. Uncommon interest, common ground: Hanford's diverse political factions are learning to cooperate. *Puget Sound Business Journal,* November 26.

Wynne, Brian. 1982. *Rationality and Ritual.* Chalfont St. Giles, U.K.: British Society for the History of Science.

Yablokov, Alexei. 1993. Dangerous trends in the Ministry of Ecology. *Zelyony Mir* 14.

Yanitsky, Oleg. 1993. *Russian Environmentalism: Leading Figures, Facts, and Opinions.* Moscow: Mezhdunarodnyje Otnoshenija Publishing House.

Yanitsky, Oleg. 1995. Industrialism and environmentalism: Russia at the watershed between two cultures. *Sociological Research* 33:48–66.

Yanitsky, Oleg. 1996. The ecological movement in post-totalitarian Russia: Some conceptual issues. *Society and Natural Resources* 9:65–76.

Yosie, Terry. 1998. Environmental perestroika. *Environmental Forum* 5 (May–June):9–11.

Zabelin, Sviatoslav. 1991. Socio-Ecological Union. In *All Our Life.* Moscow: Master Publisher.

Zaitsev, Andrei. 1993. Chelyabinsk-Central Russia Survey. Kaluga: Kaluga Institute of Sociology.

Zald, Mayer. 1992. Looking backward to look forward. In Aldon Morris and Carol McClure Mueller, eds., *Frontiers in Social Movement Theory.* New Haven: Yale University Press.

Zald, Mayer, and John McCarthy. 1987. *Social Movements in an Organizational Society.* New Brunswick, N.J.: Transaction Books.

Ziegler, Charles. 1987. *Environmental Policy in the USSR.* Amherst: University of Massachusetts Press.

Zimmerman, William. 1994. Markets, democracy and Russian foreign policy. *Post-Soviet Affairs* 10:103–127.

About the Authors

Russell J. Dalton is Professor of Political Science and Director of the Center for the Study of Democracy at the University of California, Irvine. He is author of *The Green Rainbow: Environmental Groups in Western Europe, Citizen Politics,* and *Politics in Germany;* coauthor of *Germany Transformed;* and editor of *European Politics Today, Germans Divided, The New Germany Votes, Challenging the Political Order,* and *Electoral Change.*

Paula Garb is Adjunct Associate Professor in the School of Social Ecology and Associate Director of the Global Peace and Conflict Studies at the University of California, Irvine. She is author of *Where the Old Are Young: Long Life in the Soviet Caucasus, They Came to Stay: North Americans in the USSR,* and numerous articles on ethnic issues and the Caucasus.

Ira Gluck is a Compliance Examiner for NASD Regulation, Inc. He received his Masters of Social Science from the University of California, Irvine, after completing his studies in Russian politics at the University of Pennsylvania.

Yevgeny Gontmacher is a section director in the Russian Prime Minister's Office. He is the author of several articles and book chapters, including "Social Problems of Northern Cities," "Social Policy as a Policy of Broad Dialogue," and "Russian Pension Reform."

Galina A. Komarova is a senior researcher in the Department of Interethnic Relations Studies at the Institute of Ethnology and Anthropology in Moscow, Russia. She is the author of several books, including *A Chronicle of Interethnic Conflicts in Russia. Vols. 1–3,* the co-author of *Ethnic-Cultural Processes in the Urals-Volga Region,* and the author of numerous articles on ethnic and gender issues in Russia and other countries of the former Soviet Union.

Nicholas P. Lovrich is the Claudius O. and Mary W. Johnson Distinguished Professor of Political Science at Washington State University. He has served as the Director of the Division of Governmental Studies and Services in the Department of Political Science since 1977, and holds an appointment as "Local Government Specialist" in WSU Cooperative Extension. Lovrich has served as Editor-in-Chief of the *Review of Public Personnel Administration* since 1990.

John Pierce is Professor of Political Science and Vice Chancellor for Academic Affairs at the University of Colorado at Colorado Springs. He has written scholarly books on public opinion and environmental politics in the United States, Japan and Canada, and his work has appeared in *The American Political Science Review, The Journal of Politics, The American Journal of Political Science, Social Science Quarterly, The American Behavioral Scientist,* and other journals.

William D. Schreckhise is a Visiting Assistant Professor of Political Science at the University of Arkansas. He has authored articles on jury selection and environmental policy formulation and is currently conducting a study on administrative law judge decision making.

John Whiteley is Professor of Social Ecology at the University of California, Irvine. Professor Whiteley has been involved since 1988 in increasing cooperation and dialogue among community members and scholars in Irvine and their counterparts in the former Soviet Union. In addition to this project, he is the principal investigator of a longitudinal study of the development of character in college students, known as the Sierra Project. Two volumes have appeared in this series: *Character Development in College Students,* Vol. I and II. A third volume, *Moral Action in Young Adulthood,* has also been published.

Andrei K. Zaitsev is Director and founder of the Kaluga Institute of Sociology in Russia, which studies social problems related to nuclear environmental disasters in Chernobyl, Chelyabinsk, and Semipalatinsk. He is also Editor-in-Chief of the Russian journal *Social Conflict,* Vice President of the Russian Sociological Society, and Vice President of the Russian Consultants' Association. He has published over 100 books and articles.

Index